Technology and economic development

Technology and economic development:
the dynamics of local, regional, and national change

Edward J. Malecki

Longman
Scientific &
Technical

Copublished in the United States with
John Wiley & Sons, Inc., New York

Longman Scientific & Technical,
Longman Group UK Ltd,
Longman House, Burnt Mill, Harlow,
Essex CM20 2JE, England
and Associated Companies throughout the world.

Copublished in the United States with
John Wiley & Sons, Inc., 605 Third Avenue, New York, NY 10158

First published 1991

British Library Cataloguing in Publication Data
Malecki, Edward J.
  Technology and economic development : the dynamics of
local, regional and national change.
  1. Economic development. Role of technology
  I. Title
  330.9

  ISBN 0-582-01758-0

Library of Congress Cataloging-in-Publication Data
Malecki, Edward J., 1949-
  Technology and economic development : the dynamics of local,
regional, and national change / Edward J. Malecki.
    p. cm.
  Includes bibliographical references and index.
  ISBN 0-470-21723-5
  1. Technological innovations--Economic aspects. 2. Economic
development. 3. Regional economic disparities. 4. Technology and
state. I. Title.
  HC79. T4M346 1991
  338.9--dc20   90-23690
                                                                      CIP

Set in Sabon

Printed in Malaysia
by Vinlin Press Sdn. Bhd.,
Sri Petaling, Kuala Lumpur

Contents

To Cindy and Mike

Preface

This book is primarily a synthesis of the large literature which has grown around the topics of economic development and technology. It is intended to be a resource for further inquiry into the topics covered in its pages. Consequently, there are many references to the large and diverse literature which has contributed to development studies, broadly defined. My core field, economic geography, provided the starting point for a wider survey of theory and empirical findings in other disciplines, beginning in conventional fields such as economics and planning, but directed primarily at fertile interdisciplinary areas, such as management and entrepreneurship studies, regional science, and science and technology policy.

The expansive sweep of research which informs investigations of economic development adds to the growing realization, far beyond that recognized a decade or two ago, that technology and the management decisions of global business enterprises affect localities as well as national economies. This book brings together existing knowledge about economic development, especially at the scale of regions within nations. It is research at the regional scale which is most useful for addressing the realities of economic development in the 1990s, for the openness and vulnerability of regions and nations to outside influences, so evident from the early research into regional development, are now evident as well at national scales.

Technology is a central ingredient in economic development. It is behind the proliferation of products and of weapons. It permits entrepreneurs to get an edge on competitors, and it allows one region to be more prosperous than another. Much of what we call technological change is the process of learning by people and, through them, by organizations and nations. The skills embodied in people result in some companies – and some regions and nations – being more prosperous and successful than others.

This book treats the process of technological change as a central part of economic development. Its relevance to the national scale, recognized since the postwar era began, has spread to the regional and local levels. There, and at the national level as well, local or indigenous entrepreneurship, networks, and policies affect the level of economic well-being. The behaviour of large corporations and of new, small firms contribute to what we see as development. Organizational and locational changes accompany corporate responses to technological change. As firms have had to deal with global competition, so communities, regions, and nations must confront rivals for development in distant locations.

Over the years, my thinking on technology and development issues has benefited greatly from interaction with a number of people. These include Morgan Thomas, John Rees, Geoffrey Hewings, John Britton, Andy Isserman, Peter Nijkamp, Richard Walker, Ann Markusen, Guy Steed, Nancy Ettlinger, Lay Gibson, and many others. My debt to their ideas will be evident in these pages. I am grateful to Professor Peter Dicken for his constructive comments on a draft of the manuscript. I also wish to thank Jan Coyne and Steve Rogers for their assistance with the figures.

Finally, this book is dedicated to my wife, Cindy, and our son, Mike, who have seen this book grow and develop from an idea to reality.

List of abbreviations

ASEAN	Association of South East Asian Nations
BRITE	Basic Research in Industrial Technologies for Europe
CAD	computer-aided design
CAM	computer-aided manufacturing
CERN	Conseil Européaen pour la Recherche Nucléaire
CGIAR	Consultative Group on International Agricultural Research
CIM	computer-integrated manufacturing
CIT	convergent information technology
CRITT	Centre régional pour l'innovation et le transfert technologique
DARPA	Defense Advanced Research Projects Agency (USA)
DREE	Department of Regional Economic Expansion (Canada)
EC	European Community
EUREKA	European Research Coordinating Agency
ESPRIT	European Strategic Programme for Research and Development in Information Technologies
FMS	flexible manufacturing system
GM	General Motors Corporation
GRE	government research establishment
HDTV	high-definition television
IBJ	Industrial Bank of Japan
ISNAR	International Service for National Agricultural Research
IT	information technology
JESSI	Joint European Submicron Silicon
JIT	just-in-time
MCC	Microelectronics and Computer Technology Corporation
MESBIC	Minority Enterprise Small Business Investment Company
MIT	Massachusetts Institute of Technology
MITI	Ministry of International Trade and Industry (Japan)
MNC	multinational corporation

NASA	National Aeronautics and Space Administration (USA)
NIC	newly industrializing country
NTT	Nippon Telephone and Telegraph
OECD	Organisation for Economic Co-operation and Development
OPEC	Organization of Petroleum Exporting Countries
RACE	R&D in Advanced Communication Technologies for Europe
R&D	research and development
RAs	Research Associations
ROI	return on investment
S&T	science and technology
SBIC	Small Business Investment Company
SMEs	small and medium-size enterprises
SDI	Strategic Defense Initiative
SHPEC	super high performance electronic computer
UN	United Nations
VCR	video cassette recorder

Acknowledgements

We are grateful to the following for permission to reproduce copyright material:

Academy of Management Journal and the author, Prof. F. Hull for fig. 4.6 (Hull & Collins, 1987); American Enterprise Institute for Public Policy Research for fig. 5.1 (Pisano, Russo & Teece, 1988); The American Geographical Society for fig. 1.3 (Taaffe, Morrill & Gould, 1963); the author, Prof. B. Ashcroft for table 3.4 (Ashcroft, 1978); *Business Week* for table 6.1 (Lee, Engardio & Dunkin, 1989) copyright © 1989 by McGraw-Hill, Inc.; Paul Chapman Publishing Ltd. for fig. 1.1 (Dicken, 1986); Mouton De Gruyter (A Division of Walter de Gruyter & Co.) for table 1.1 (Bivand, 1981); Minister for Development Co-operation (Ministry of Foreign Affairs, The Netherlands) for fig. 4.11 (Reddy, 1979); The Economist Newspaper Ltd. for fig. 9.3 (*The Economist*, 1989) © 1989 The Economist Newspaper Ltd.; the author, Prof. C. Edquist for table 6.5 (Edquist & Jacobsson, 1988); Elsevier Science Publishers (Physical Sciences & Engineering Div.) and the respective authors for fig. 4.12 (Pavitt, 1984) & table 4.5 (Kim, 1980); Elsevier Science Publishing Co. Inc. for table 6.4 (Tani, 1989) copyright 1989 by Elsevier Science Publishing Co. Inc.; the author, Prof. C. Freeman for table 7.3 (Freeman, 1987); The Free Press (A Div. of Macmillan, Inc.) for fig. 6.1 (Shapero, 1985) copyright © 1985 by The Free Press; Gower Publishing Group-Avebury and the author, Prof. M. Fischer for fig. 9.2 (Fischer, 1990); the author, Prof. P. Hall for table 5.1 (Hall & Preston, 1988); Industrial Research Institute for table 6.3 (Goldhar, 1986); Japanese Ministry of International Trade & Industry (Industrial Location Policy Div.) for fig. 7.3 (MITI, 1990); *Journal of Regional Science* and the author, Prof. R. Bolton for fig. 2.3 (Bolton, 1985); Longman Group UK Ltd. for fig. 6.2 (Kaplinsky, 1984a) © 1984 Raphael Kaplinsky; the author, Prof. A. Markusen for fig. 4.4

(Markusen, 1985); National Academy Press for fig. 4.2 (Kline & Rosenburg, 1986); National Council for Geographic Education (Indiana University of Pennsylvania) for fig. 4.10 (Souza, 1985); Organisation for Economic Co-operation and Development (OECD) for fig. 4.7 (Vickery, 1989); the author, A. Pred for fig. 2.2 (Pred, 1977); Sage Publications for table 1.2 (Friedmann, 1986a); Sage Publications, Inc. for table 3.1 (Hansen, 1988) copyright 1988 by Sage Publications, Inc.; the author, R. Samuels for fig. 7.1 (Samuels & Whipple, 1989); the author, Dr. C. Stevens for table 7.5 (Stevens, 1990); the author, Dr. E. Swyngedouw for table 6.2 (Moulaert & Swyngedouw, 1989); *Technology Review* for fig. 4.5 (Abernathy & Utterback, 1978) copyright 1978 *Technology Review*; United Nations Centre of Regional Development (UNCRD) for fig. 1.2 (Lo, Salih & Douglass, 1981); University of Texas Press for table 2.3 (Singlemann, 1981); Wadsworth Publishing Co., Inc. for fig. 2.5 (Abler, 1975) © 1975 by Wadsworth Publishing Co., Inc.; John Wiley & Sons, Inc. for table 7.7 (Linge & Hamilton, 1981) copyright © 1981 John Wiley & Sons, Inc.

Whilst every effort has been made to trace the owners of copyright material, in a few cases this has proved impossible and we take this opportunity to offer our apologies to any copyright holders whose rights we may have unwittingly infringed.

Chapter 1

Economic growth and decline: theories and facts

Economic growth and development have always been geographically uneven. Early observations were of relatively conspicuous contrasts, such as thriving cities and backward rural areas, advanced European countries and lagging colonies and former colonies. These and other, more subtle, disparities are the outcomes of a capitalist economic system which appears more influential in the 1990s than it has been for several decades. Corporations and entrepreneurs carry out their tasks within this system, and the results of their decisions affect people and places. The capitalist system is a global one, but it both shapes and responds to local circumstances. The growth of part-time work, the rise and decline of industrial areas, and the highly selective growth within the Third World – to mention only a few familiar examples – are impacts of economic change, viewed at a regional (less than global, more than localized) scale.

The 1970s and 1980s presented a time of dramatic change in the structure of economic change at all scales. Growth in international competition has thrust firms and regions which were previously isolated from competition into a new economic environment where interdependence and interaction are the norm. Trade became an immensely prominent consideration for nations and localities alike. Far-flung corporate networks engendered 'global factories' that shifted goods and components from country to country but under the control of corporate management. 'Strategic alliances' and other coalitions of giant firms strengthened both the technological and geographical scope of large enterprises. The theoretical basis for studies of economic growth through the 1960s and 1970s – there was little study of decline – tended to assume that growth and development would inevitably occur, if only the correct policies were chosen, if the proper economic factors were available, and if the population would accept the appropriate social and political changes (Landes 1989; Reynolds 1983).

1

Some of these changes were under way earlier, but their full impact was not felt until the 1980s. A series of unanticipated events seemed to unfold at the same time, affecting both advanced post-industrial economies and underdeveloped nations simultaneously, if in distinct ways. For the Western countries, the ascent of Japan to economic forerunner, the stagnation of Europe in jobless growth during the 1980s, and the meteoric rise of the newly industrializing countries (NICs) all set existing theories and policies on their heads. What had worked in the 1950s, when the economic pie was growing, did not appear to work in the 1970s and 1980s, when the pie was constant or even shrinking.

The measurement of seemingly simple concepts such as employment became more difficult as full-time jobs and fringe benefits became less assured. Firms in several countries also shifted production from urban locations with unionized workforce to peripheral regions, where industrialization and unionization had little, if any, history and where, consequently, wages, benefits, and work rules were more congenial for employers. More alarmingly, jobs by the thousands disappeared as new plants opened in Asia and elsewhere. Within advanced economies, sharper contrasts emerged between regions. The 'North–South divide' in the UK, the 'bi-coastal economy' and a persistently stagnant rural region in the USA, the booming Tokyo metropolis and lagging peripheral regions in Japan, the Third Italy between the industrial North and the perpetually backward Mezzogiorno all became symptomatic of a general set of problems pervading economic change world-wide (Dicken 1986). At the same time, high technology industry, robotics, and entrepreneurship were hailed as solutions to such problems, although in reality they often seemed merely to reinforce inequality (Bessant and Cole 1985; Castells 1985a; Keeble and Wever 1986).

The plight of underdeveloped countries, with a few notable exceptions, also worsened. Growth became more difficult, and imports rose dramatically while production and living standards fell. Oil, on which machinery and the other trappings of essential industrialization depend, rose in price, making the cost of economic transformation even more difficult. Markets for products became more global – with standards, distribution, and service often being more important than price – again placing peripheral countries without close ties to multinational corporations (MNCs) out of the system by which economic growth occurs. International transfers of new technology – a primary means of technological change, but one carried out largely by private organizations – thus often perpetuate patterns of dependence (Ernst 1980). By any measure, stark disparities in human welfare are present at a global scale (Tata and Schultz 1988).

New forms of production organization, such as flexible specialization, economies of scope, strategic alliances among firms, and widespread subcontracting, have transformed economic activities into forms that are directly challenging theories and policies of all types. Flexible production systems are radically different from the mass-production assembly line, relying mainly on unskilled workers and routine tasks, on which Marxist critiques have focused (Braverman 1974; Sayer 1985; Scott 1988a). Most importantly, technology and product innovation are central elements in these complex new patterns of corporate activity and make any simple explanation less tenable (Hyman and Streeck 1988).

Global production

An especially important set of contributions to the understanding of economic activity has been research on multinational corporations, stemming in large part from the work of Hymer (1975). Multidivisional firms operating in many countries at once have a great deal of flexibility with respect to products and markets. However, Hymer's central point was that the division of labour within corporations corresponds to a geographical pattern – what is now called a spatial division of labour – that creates and reinforces uneven development. The hierarchical nature of large firms, reinforced by information technology (Ch. 5), led many researchers to conclude that widespread economic development cannot occur (Fröbel, Heinrichs, and Kreye 1980; Müller 1979). Much of this argument focuses on the technological dependence associated with the concentration of corporate research and development (R&D) in home countries and production in peripheral countries and regions. In addition to the division of labour, restrictive transfers of technology seem to perpetuate technological dependence (Germidis 1977). In truth, the situation of multinationals is very complex, and has eluded simple explanations (Biersteker 1978; Dicken 1986; Dunning 1979).

Multinational firms have evolved through a sequence of three distinct stages. The earliest global firms were those organized to extract from other places natural resources and primary products not available at home. Spices, silk, furs, gold, and silver, among the earliest such items, are still in demand today. However, the list has expanded to include tropical fruits and other agricultural crops, petroleum, and other minerals.

The second stage was the expansion into other markets primarily for the sale of manufactured products, although this also took the form of production as well, beginning in the second half of the nineteenth century (Dicken 1986: 57–9). This market stage occurs among the wealthy countries (as well as the élites of Third World countries) where markets – consumers, industries, and governments – can afford the products. Thus, trade between Europe, Japan, and North America takes place largely among multinational firms in these countries seeking markets for their manufactured products and services. Both the resource and the market objectives of multinationals maintain the traditional international division of labour embodied in Vernon's (1966) influential product cycle thesis. Only advanced economies could produce manufactured goods while they were profitable; underdeveloped countries had little role in the world economy other than as suppliers of primary inputs.

The third stage of multinational activity is very different. It relies on the 'new international division of labour', in which production tasks can be broken down into activities which can be done by unskilled workers in countries where labour costs are very low relative to the advanced countries (Fig. 1.1). Asian workers, outside Japan, earned less than one-fifth the level enjoyed by workers in the USA or Europe. The emergence of manufacturing in Third World countries contrasts greatly with the previous role of those countries in the world economy. Indeed, direct investment in the form of factories and, probably even more important, the surge of subcontracting arrangements with Third World firms has altered the way in which economic development is viewed.

The rise of the NICs to economic powerhouses has been especially remarkable (OECD 1988a). The NICs include a first tier, all located in Asia, also known as the 'four tigers' – Hong Kong, South Korea, Singapore, and Taiwan – and a larger, more diverse group, which includes India, Malaysia, the Philippines, and Thailand in Asia, Brazil and Mexico in Latin America, and, in some lists, Israel and South Africa. (For some issues on defining NICs, see Ingalls and Martin 1988.) Although NIC manufacturing began in unquestionably low-tech products such as textiles, clothing, and shoes, more technology-intensive products, such as electrical and electronic products and transport equipment, are now common (J Henderson 1989; OECD 1988a).

The new spatial division of labour which has grown out of global corporate activity has produced regional economies of three broad types (Hamilton and Linge 1983: 24):

1. Those with a highly technological environment;
2. Those with a significant proportion of skilled personnel but lacking in a diversified and modern industrial structure; and

Hourly Labour Costs, 1981 (US = 100)

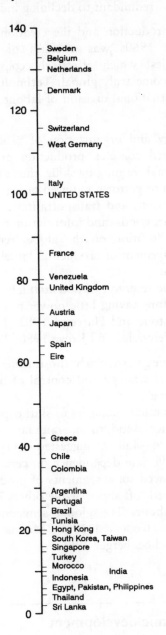

Figure 1.1 Hourly labour costs in selected counties, 1981. *Source*: Dicken (1986: 240 (fig. 7.8))

3. Those with reserves of unskilled labour which is either surplus to farming and rural occupations or redundant to declining industries.

The great surge in international production and the creation of the 'global factory' in the 1950s and 1960s was a result of several 'permissive' or 'enabling technologies' which effectively compressed space and allowed large firms to become truly global. Technology has been an enabling element for the international division of labour, taking place from three main sets of changes:

1. Developments in the technology and organization of production that subdivided and standardized complex production processes into simple units requiring minimal training or skills, thus allowing workers in virtually any location to perform required tasks;
2. Improvements in telecommunications and transportation since the Second World War permit people, goods, and information to travel efficiently and cheaply to any location on the globe, rendering industrial location and the management of production largely independent of geographical location;
3. The development of a world-wide reservoir of potential labour, not penetrated by unions, and therefore having little power to demand wages and fringe benefits (Bluestone and Harrison 1982: 115–18; Dicken 1986: 106–16; Fröbel, Heinrichs, and Kreye 1980: 37–44).

These developments have allowed firms, particularly those firms which were already large, to develop a global strategy and control of product development, production, and marketing.

These 'permissive technologies' permitted managers to shift capital (as well as products) across long distances, and to operate far-reaching networks of production facilities. The 'de-skilling' suggested in (1) above is a scientific and deliberate way to alleviate dependence on certain sets of skilled workers, but it has also allowed some segments of production requiring low levels of skill to be moved 'off shore' while others remain near needed pools of specialized labour. The global semiconductor industry exemplifies this tendency (J Henderson 1989; Molina 1989; Scott 1987; Scott and Angel 1988; E Schoenberger 1986).

Technology, regions, and economic development

The dynamics of technological change provide the focus of this book. Technology is central to regional change, positive and negative, and to economic change, job-creating and job-destroying. It is the most obvious cause and effect of the cumulative wealth of rich nations. Technology also promises, more than any other phenomenon, to bring poor nations out of poverty. As a 'chronic disturber' of comparative advantage, it has provided the principal source of change for firms, regions, and nations alike (Chesnais 1986). The concept of technology behind the book is a broad one that encompasses knowledge in all its forms, from simple and routine procedures of everyday life, to the methods of organization and management in enterprises large and small, from the machines that produce in enormous quantities what formerly required many workers, to the complex scientific investigations that create ever newer inventions and products. Capitalist firms profit from new products as much, if not more than, from new, labour-saving processes, although far more attention has been paid to the latter (Sayer 1985).

This chapter introduces the concept of *regional development* as a combination of qualitative and quantitative features of a region's economy, of which the qualitative or structural are the most meaningful. The perspective of regional development will not be familiar to all readers, but it has played an important bridging role for social scientists attempting to explain disparities in economic structure and well-being. Economists, geographers, planners, political scientists, and sociologists all contribute to regional development as a field of study (Malecki 1983).

Growth and decline, as quantitative changes in economic activity, gauge the impact on a region, in jobs and incomes, of decisions both from within and from external sources. The qualitative attributes include the *types* of jobs – not only their number – and long-term and structural characteristics, such as the ability to bring about new economic activity and the capacity to maximize the benefit which remains within the region. The standard theory of economic growth and development has concentrated on quantitative changes, despite an increasing awareness that regional growth depends, often critically, on aspects that are understood only in comparison with other regions or nations. The facts of regional development suggest that it is not enough to rely on concepts of growth without an equivalent concern for the forces which permit growth to take place or prevent it from occurring. These are the concerns of regional development, whether examined at the national, subnational, or local scale.

Conventional theories and policies regarding regional development focus in one way or another on the capital-labour 'production function' and responses by the state via various policies. Early policies derived almost directly from the neoclassical model, which 'is an ideology, a religion' on which virtually all post-war economic policies have been grounded (Vining 1987). Labour is especially problematic, since it is seen by firms as an input to production and by governments as an outcome of regional growth. Labour skills and wages vary tremendously and these greatly influence location decisions of firms and the ultimate spatial division of labour. The interplay between capital, labour, and the state (or public sector) is fundamental to the dynamics of economic development (Storper and Walker 1989). For large enterprises, many other factors in addition to labour enter into their decision-making, and support the view of complexity in economic change (Clarke 1986).

Competition is keen for technological competence and superiority, both on a local level and internationally (Botkin, Dimancescu, and Stata 1982; Miller and Coté 1987; Office of Technology Assessment 1984). This is the latest arena where state intervention has interacted with private interests to generate policies that perceive and address the competitiveness of firms and of regions. The supply-side approach, which emphasizes the creation of firms, jobs, and wealth based on internal resources, is a part – albeit an important part – of the development process (Sweeney 1987). However, new firms, and high technology, involve critical steps, barriers, and synergies which determine the pattern of growth and development (Stöhr 1982, 1986a).

Regions: the arena of economic change

What do we mean by 'regional'? Regions can be subnational spaces, especially of large countries, or aggregates of several nations. For example, Europe after 1992 will be more a single region than it has ever been. The flexibility of definition is not imprecision; rather, it reflects the various geographic scales at which economic change occurs.

Traditionally in regional economic analysis, regions were defined in three ways (Boudeville 1966; Meyer 1963). First, regions may be homogeneous with respect to some physical, social, or economic characteristic. In the USA, the corn belt and the manufacturing belt represent two such regional designations that fail to hold up to long-term scrutiny (Smith and Dennis 1987). Second, nodality or polarization around a central urban place has the advantage of focusing attention on economic attributes of cities and their dominance over surrounding space, especially as labour markets (Berry 1973). Finally, policy-oriented regions adhere to administrative boundaries that correlate with political or state institutions and their well-defined spheres of influence.

Policies, whether at a national or a subnational level, generally ignore urban nodes in favour of a smaller set of broad regions, and data collection is typically done only for such administrative areas (Tiebout 1962). At the international scale, data are typically available only for the nation as a whole, and perhaps for a few large cities or urban areas. Subnational data are frequently very difficult to come by. In advanced economies, regional data for administrative units are often better, but problems of detail and frequency still persist when compared to the national scale. In the end, 'we have to make do with the regions we have' (Brown and Burrows 1977: 16).

Regions are decidedly not well defined by political boundaries, and Smith and Dennis (1987) show that the American 'manufacturing belt' evolved from a number of smaller regions, each specializing in the production of different products (Meyer 1983). A corresponding 'new manufacturing belt' has not emerged in the US sunbelt despite high levels of manufacturing employment; its relative industrial diversity distinguishes it from the North as it evolved (Smith and Dennis 1987). In the UK, South Wales has evolved from a coal-mining region into a manufacturing region for new technology that coalesces into the M4 corridor eastward towards London (Hall *et al.* 1987; Sayer and Morgan 1986). The transformations of South Wales in the UK and of the New England region in the USA have attracted a great deal of attention as places where labour has largely defined and moulded the regions and regional change (Cooke 1985a; Harrison 1984; Morgan and Sayer 1985; Sayer and Morgan 1986). Indeed, the local nuances of economic change have led to a surge of 'locality studies' which examine even smaller areas than traditional regions (Cooke 1986; McArthur 1989).

In place of these definitions, Markusen (1987: 16–17) proposes an alternative definition: 'A region is an historically evolved, contiguous territorial society that possesses a physical environment, a socioeconomic, political, and cultural milieu, and a spatial structure distinct from other regions and from the other major territorial units, city and nation.' Regions are differentiated by distinct class structures which are unevenly distributed, by the separation of finance and ownership from production, and the economic and sectoral specialization that often endures over long periods (Markusen 1987; Massey 1984).

Gilbert (1988) summarizes recent thinking on regions by outlining three definitions of region. In the first, the region is the spatial organization of the social processes associated with the mode of production. The social division of labour, the process of capital accumulation, the reproduction of the labour force, and political and ideological processes of domination used to maintain the social relations of production all define the region as 'the concrete articulation of *relations of production* in a given time and place' (Gilbert 1988: 209). A second view of region uses *local culture* as a means by which people and groups are linked through specific communication processes which enhance their collective way of thinking about places and space. The third perspective sees the region as the setting for *social interaction* of all types, but particularly those which create or enforce domination and dependence.

Region formation, which involves elements of the first and third approaches above, occurs through regional social interaction while being both the condition and the outcome of the social relations between individuals, groups, and institutions in regional space (Gilbert 1988: 216–17). The role of the individual in this setting is problematic, but the dominant view at present is to posit individuals as actors (human agents) who are constrained by social structures (Gilbert 1988: 218; Storper 1988). The interactions between the capitalist system and localities and regions define the process of development.

It is clear that regions are not static, but are 'as mutable and malleable as the economic and social relations that comprise them and that they comprise' (Smith and Dennis 1987: 169). In their example of the northern core region of the USA, Smith and Dennis maintain that the decline of that region began in the 1920s with the flight of the textile industry. Similar movements took place in other sectors in response to massive technological changes, such as dramatically lower transportation costs, standardization of production, and increased minimum efficient scale of plants. The result, however, also had an important spatial dimension: regions became not simply subsets of national space or markets, but units of international space. Comparisons of costs, wages, and work conditions shifted from being made merely on a local basis or intraregionally to international in scope. This internationalization, although recognized somewhat in terms of the global arena of finance and production investment, has markedly changed 'the scale at which regions are constituted as coherent and integrated economic units' (Smith and Dennis 1987: 171; see also Graham *et al.* 1988). Douglass (1988) documents a similar transformation of urban areas in Japan since 1970 as a consequence of that country's transnational corporations.

Regional disparity: cores and peripheries

Regions and nations undergo similar processes at work in all spatial scales (Smith 1984). The division of labour within and among large firms, and dominance and dependence relationships among firms and among places are common mechanisms that affect regions of all sizes. In particular, peripheral areas at different spatial scales – national, regional, and local – exhibit similar structural and qualitative impacts and side-effects of traditional economic development mechanisms (Stöhr 1982). Comparisons of differences in economic activity across nations are often dubious, given the vast differences in population, land area, resources, and history (Walsh 1987), but the commonalities within the global economy or world system suggest that much more is similar than different (Drucker 1989; Taylor 1985).

In the world-system view, three basic elements comprise the 'world economy': (1) a single world market, (2) a multiple state system, and (3) a three-tier structure (Wallerstein 1979). Wallerstein's view of a tripartite international division of labour imposes a rather strict *core – semi-periphery – periphery* structure on to the 'capitalist world system' that may well be too strict in practice (Knox and Agnew 1989; Lipietz 1986). However, through the process of capitalist development, including colonialism, regions of the world have become defined within an international division of labour. For Wallerstein, the core is characterized by free labour engaged in skilled tasks. In the periphery, characteristically involved in a single cash crop or mining production, are the countries in which coerced labour, under colonial state power, is found. Between the core and the periphery is what Wallerstein calls the *semi-periphery*, composed of countries that have regressed from core status through undergoing a process of deindustrialization and those heading for core status as they experience rapid industrial development (Cooke 1982: 153; Wallerstein 1979). Southern Europe is a frequent illustration of semi-periphery, resting uneasily between the core economies of northern Europe and the periphery of the Third World (Arrighi 1985; Hudson and Lewis 1985; Seers, Schaffer, and Kiljunen 1979).

Such a core–periphery structure is not static, as the rapid rise of Japan suggests. The world-system view also maintains that 'core' and 'periphery' are not areas, regions, or states, but spaces where core or peripheral processes dominate. 'In simplest terms, core processes consist of relations that incorporate *relatively* high wages, advanced technology and a diversified production mix whereas periphery processes involve low wages, more rudimentary technology and a simple production mix' (Taylor 1985: 17). The semi-periphery is the dynamic category within the world economy. The restructuring of spaces involves regions rising and sinking through the semi-periphery (Taylor 1985: 17–18). Shifts in the relative status of nations in international competition are frequent observations from many perspectives (Cohen and Zysman 1987; Kennedy 1987; Pavitt 1980; Porter 1990; Prestowitz 1988).

Regional units within nations pose a more difficult problem, in part because the boundaries are less precise but also, more importantly, because a region is more 'open' than a nation; that is, a larger proportion of its economy will depend on flows of imports from and exports to other regions. These flows are notoriously more difficult to measure, because the usual restrictions on commerce via customs barriers, immigration control, exchange controls, and trade quotas do not exist, and thus even the most rudimentary information on flows is either unavailable or superficial (Richardson 1973).

Regional economic problems – that apply equally across nations and to regions within nations – include inequality of income, unemployment, and migration losses (Brown and Burrows 1977). Growth rates, of income or of jobs, are customary indicators of regional economic differences. The fact that regions do not grow at equal rates, do not provide equal numbers of jobs or jobs sufficient for those seeking employment, is a complex issue. Consider employment first. Jobs can be provided either from local or indigenous firms, or they can be brought into a region from outside, to produce products or services for markets elsewhere. Some jobs will be concerned with the provision of goods and services to organizations and consumers in the local region, others to distant, even foreign, customers. The variety of possible markets complicates the conventional export base approach to regional development. Even if unemployment is slight, not all jobs are equal, either in pay, in security, or in upward mobility. Likewise, what is produced in a place is perhaps less important than who is producing it, and at what stage in the production process (Clark, Gertler, and Whiteman 1986: 20–38).

Dominance and dependence

Development is fundamentally 'a dominance/dependence relationship that is expressed in a great many ways' (Brookfield 1975: 1). Because of its varied expression, it is not a neutral or scientific concept on which there is any clear agreement (Toye 1987; Arndt 1987). Peattie (1981: 11–12) provides an example:

> The subject of economic development is inherently complicated and difficult to treat from a single perspective. . . . The transformation of political and social institutions, the confrontations both within and between nations of the haves and the have-nots, and the struggle over scarce resources have made it more and more difficult to deal with the subject in any tidy framework, like that provided by economics. Once a subject is seen to involve worldwide and interrelated processes of economic and social transformation, and to be a way of designating the inter-action of the human species with the natural environment, it becomes a topic so broad and complex that it is difficult to generalize.

Within underdeveloped countries, some problems persist that are fundamentally geographic. *Dualism,* the severe contrast between traditional and modern sectors, is nowhere so evident as in the giant cities that characterize underdeveloped nations (Armstrong and McGee 1985; Santos 1979). In fact, it is the highly structured dualism found in the Third World that demonstrates how poorly integrated multinational corporations are with the nations in which they operate. The 'core–periphery' dichotomy identified by Friedmann (1966) stands as one of the '"dominant metaphors" of our time' (Strassoldo 1981). Its further realization in the broad phenomenon of dualism is well captured by Santos (1979) and Brookfield (1975).

Bivand (1981) characterizes the asymmetry between 'pure' periphery and 'pure' centre or core across a number of dimensions (Table 1.1). The distinction between periphery and core geographically concerns levels of linkage and access. Economically, the core is a set of regions where complexity, technology, and control are the norm. Culturally and politically, as well, core regions are dominant and peripheral regions are dependent.

In fact, the situation in Third World settings is considerably more complex than the dualistic or core–periphery model is able to capture. Lo, Salih, and Douglass (1981) have proposed a 'macro-spatial development framework' which consists of five components:

1. A *world market* largely composed of developed countries buying primary products from the Third World countries and exporting manufactured goods, particularly modern technology embodied capital goods, to them;

Table 1.1   Characteristics of 'pure' core and 'pure' periphery

Periphery	Core
Geographical:	
Coupled to few means of transport. High absolute and relative contact costs. Poor position on the transport network. Difficult access to other peripheral areas	Coupled to all means of transport. Low absolute and relative contact cost. Strong position on the transport network
Economic:	
Raw material production. Simple processes, one-sided, vulnerable production. Export of labour. Import of finished wares	Finished wares and services produced. Expansion, agglomeration economies. Complex control processes. Import of labour. Adaptable business community. Control over capital. Contact with other economic agents
Cultural:	
Accepts others' language. Forced to take the consequences of others' models of society. Consumes symbols created elsewhere	Produces and spreads the symbol system. Represents expertise. Control of information media. Rejects symbols from the periphery as irrelevant or unimportant
Political:	
No strategic resources. Absence of élites, or only agents of centre in administration. Poorly represented in the centre. High costs incurred in assembling and putting forward views. Few initiatives	Control of strategic resources. Concentration of élites. Over-represented in formal administrative organs. Low costs incurred in assembling and putting forward views. Many initiatives

Source: Bivand (1981: 221 (table 2)).

2. An *urban formal sector* dominated by enclave foreign and domestically financed modern manufacturing and business of the corporate type;
3. An *urban informal sector* consisting of a wide range of traditional activities, small in scale and characterized by such occupations as hawkers, vendors, daily labourers, and services which are distinct from the enclave sector and its related professional and white-collar occupations (Armstrong and McGee 1985; Portes, Castells, and Benton 1989; Santos 1979);
4. A *rural export sector* generated in many cases from the plantation economy developed during colonial rule together with post-independence natural resource exploitation such as mineral extraction, oil, and timber;
5. A *rural peasant economy* historically isolated from the national and world market and dominated by peasants and landlords engaged mostly in food-crop production.

This framework still simplifies the reality and variety of Third World countries, but it captures effectively the interrelationships among different groups in an economy (Fig. 1.2). In particular, Fig. 1.2 illustrates the contrast between the formal sector, linked to the world economy through trade and commerce, and the informal sector, to which modern technology scarcely belongs, whether in cities or in rural areas.

The actual result in Third World cities has been described by Stretton (1978: 104):

> Governments build grand quarters for themselves in their cities. They encourage private investments in central city offices, hotels, hospitals, affluent residential quarters with western standards of space and service, national and international airlines and airports, and fast motorways from the airports to the city centres. This sort of development is sometimes defended as creating conditions which will attract foreign businessmen and therefore further their investment – though any further investment it attracts is quite likely to be of the same unequalizing kind.

The core–periphery paradigm is manifested in the global urban system as well. One city may become a 'world city' while the rest of the nation – and entire countries – remain largely outside the world economy (Friedmann 1986a; Friedmann and Wolff 1982). Friedmann (1986a) presents seven interrelated theses:

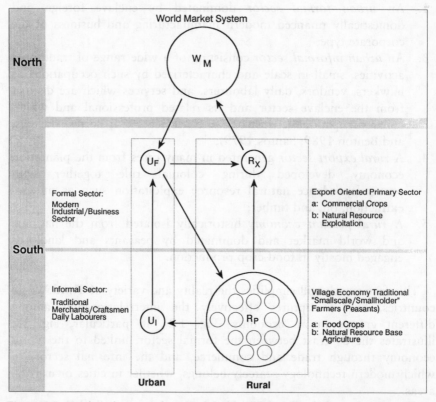

Figure 1.2 Dualism between formal and informal, urban and rural sectors.
Source: Lo, Salih, and Douglass (1981:20 (fig. 2.6))

1. The form and extent of a city's integration with the world
    economy, and the functions assigned to the city in the new spatial
    division of labour, will be decisive for any structural changes
    occurring within it.
2. Key cities throughout the world are used by global capital as
    'basing points' in the spatial organization and articulation of
    production and markets. The resulting linkages make it possible to
    arrange world cities into a complex spatial hierarchy.
3. The global control functions of world cities are directly reflected in
    the structure and dynamics of their production sectors and
    employment.
4. World cities are major sites for the concentration and
    accumulation of international capital.
5. World cities are points of destination for large numbers of both
    domestic and/or international migrants.

6. World city formation brings into focus the major contradictions of industrial capitalism – among them spatial and class polarization.
7. World city growth generates social costs at rates that tend to exceed the fiscal capacity of the state.

The spatial arrangement of Friedmann's tentative hierarchy of world cities is somewhat linear along an East–West axis, with three distinct sub-systems: an Asian subsystem centred on the Tokyo–Singapore axis, with Singapore playing a subsidiary role as regional metropolis in South-east Asia; an American subsystem based on the three primary core cities of New York, Chicago, and Los Angeles, linked to Toronto and to Mexico City and Caracas, thus bringing Canada, Central America, and the Caribbean into the American sphere of influence; and a West European subsystem focused on London, Paris, and the Rhine Valley axis from Randstad to Zurich. The southern hemisphere is linked into the European subsystem via Johannesburg and São Paulo (Table 1.2).

Table 1.2 Friedmann's world city hierarchy

Core countries		Semi-peripheral countries	
Primary	Secondary	Primary	Secondary
London	Brussels		
Paris	Milan		
Rotterdam	Vienna		
Frankfurt	Madrid		
Zurich			Johannesburg
New York	Toronto	São Paulo	Buenos Aires
Chicago	Miami		Rio de Janeiro
Los Angeles	Houston		Caracas
	San Francisco		Mexico City
Tokyo	Sydney	Singapore	Hong Kong
			Taipei
			Manila
			Bangkok
			Seoul

Source: Friedmann (1986a: 72 (table 1)).

Dualism also is present in advanced economies, as illustrated by contrasts in growth, especially in new sectors such as services (Cooke 1987; Green 1988; McKee 1974), and the presence of informal, irregular work (Sharpe 1988). It is not a new situation, but the regional differences found in developed countries can be summed up in the observation: 'Prosperity is tied more to where you live' (Clark 1988). The disparities continue to hit rural areas the hardest, as they have historically (Averitt 1979; McKee 1974). New industries and good jobs gravitate towards cities. However, the mosaic of inequality has become more fine-grained, and the economic prospects vary more from place to place. The decline of large cities, from Newcastle to New Orleans, since the 1960s, and the rise in many countries of sunbelt regions, are evidence of patterns which are not easy to explain.

As an example, the state development strategy in Venezuela has been described as

> a strategy that produced steel but few jobs; that organized the city for the advantage of foreign corporations, but made it difficult for local business people to get established; that concentrated power in the hands of a national elite and left local leaders and community members ignorant of what was being planned; that made some people rich, while appearing likely to leave many . . . to pick up the 'trickle down' of growth as an unskilled, intermittently employed subproletariat (Peattie 1981: 6).

A historical perspective is helpful in shedding light on the concept of development. Economic development as a term denoting a process which societies undergo was hardly used before the Second World War. 'Material progress' or economic growth was the more common policy objective, and development economics as a field of research and policy mirrored the Western preoccupation with employment, income, and other economic indicators (Arndt 1981, 1987: 1–2).

Through the evolution of social as well as economic thought, an intellectual framework emerged for thinking about development as a world-wide process of modernization, involving the spread of rationality and individualism, powered by the transformation of the technical order, and submerging other, 'less aggressive' cultures (Dube 1988; Headrick 1988; Peattie 1981: 35–6). This trend peaked in the early 1940s, when a series of studies on backward or underdeveloped countries presaged the formation of a single international body of thought on economic growth and development (Mandelbaum 1945). International comparisons of national and per capita income figures became the prime objective of development theory and policy (Arndt 1987: 43–54). Even where other goals were acknowledged, such as a shift from agriculture to other sectors or distribution of income, 'these tended to be qualifications. . . . The touchstone, if not the essence, of economic development was taken to be growth in output and income per head in the less developed countries' (Arndt 1987: 53).

The legacy of colonialism

Colonialism presented a different set of problems, which we now know as underdevelopment. Colonies were 'made by and for the mother country', and the economic concern of the colonials was primarily the development of the natural resources of the colonies for Western countries' benefit (Arndt 1987: 24–9). From the beginnings of the 'capitalist world system' that spread first through such colonial ventures, and later through the seemingly less malevolent endeavours of multinational corporations, the underdeveloped nations have been dominated by outside entities and dependent upon an economic system over which they have no control (Amin 1974; Clark 1975; Moudoud 1989; Villamil 1979). The destruction of precolonial economies through the import of products against which local producers could not compete and the construction of a highly concentrated infrastructure in effect created the disparities which later policies sought to reduce (Moudoud 1989: 103–33).

The structure of colonialism was technological as well as political and cultural. Headrick (1988) lists five ways in which technology was involved in the imperialism of Europeans in Asia and Africa between 1850 and 1940. The first was a small set of important inventions which allowed Europeans to conquer inland sections previously impenetrable – steamships, railways, firearms, and telegraph permitted control to be imposed over large territories. The second was indirect. The 'imperialism of free trade' was that 'the industrialization of the Western nations stimulated a growing demand for the products of the tropics' (Headrick 1988: 6). Falling transport costs permitted the shipment of bulky commodities such as cotton and indigo for cloth, palm oil to lubricate machinery, copper and gutta-percha for electric and telegraph lines, tin for canned goods, and rubber for clothing and automobiles. An affluent population also demanded increasing amounts of coffee, tea, sugar, cocoa, and other tropical goods.

Third, Western colonialists transferred technology to Asia and Africa in efforts to increase production and lower production costs by applying Western industrial and scientific methods to commodity production. Fourth, an increase in demand for Western manufactured products was prompted by their introduction into Asia and Africa. 'Motor vehicles, television, and modern weapons have become irresistible but barely affordable temptations for people of poor countries' (Headrick 1988: 7).

Finally, nearly all the technological changes which affected the relations between the West and the tropics originated in the West or from the work of Western scientists and engineers, for the benefit of Western society. The propensity to find substitutes for goods in short supply was remarkable (e.g. aniline dyes for indigo and other natural colourings, petroleum for palm-oil, synthetic rubber for natural rubber, synthetic fibres for silk). 'In these and other ways, Western scientists and engineers have prevented the demand for tropical products from growing in proportion to the growth in industrial production or in tropical population' (Headrick 1988: 8).

Colonialism also brought about changes in the way native societies worked and produced, but the changes did not originate internally. The extraction of mineral resources, the construction of ports, and the creation of urban centres in these nations changed what had formerly been a diffuse constellation of family plots and small-town centres to urban areas with a minimum of planning (Weitz 1986: 32). This sort of 'development' pattern became part of the lore and methodology for economic development. Taaffe, Morrill, and Gould (1963) proposed a sequence of transportation and urban development that mirrored the sequence common to such colonial ventures (Fig. 1.3). This idealized pattern was very optimistic; urban growth concentrated in far fewer centres in most post-colonial settings where primacy was – and still is – the rule.

For much of the Third World, then, technological underdevelopment has been the rule. Colonialism brought education, medical care, and mechanical equipment to the colonies, but did not pass on essential skills and knowledge to the indigenous populations. When former imperial possessions attained political independence, their people and firms lacked the skills and experience to sustain the infrastructure left by the colonial powers or to build on and develop from this base (Fransman and King 1984; Headrick 1988; Weitz 1986).

Despite these conditions, optimism prevailed in the immediate post-colonial era. No better example exists than that of Walt W. Rostow's *The stages of economic growth* which encapsulated the single focus of economic thinking at the time (Rostow 1960; Heilbroner 1963). Growth and development were considered synonymous, and a simple 'iron law' of economic growth was believed to hold: so long as the amount of savings, coupled with the fruitfulness of those savings, results in an output which is faster than the rise of population, cumulative economic growth will take place (Heilbroner 1963: 86). Rostow

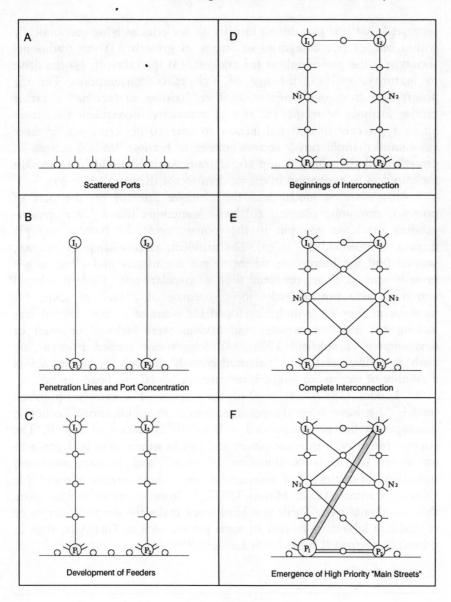

Figure 1.3 Idealized scheme of infrastructure development over time in a developing country. *Source*: Taaffe, Morrill, and Gould (1963: 504)

suggested that it is possible to identify all societies as lying economically within one of five categories or stages of growth: (1) the traditional society, (2) the preconditions for take-off, (3) the take-off, (4) the drive to maturity, and (5) the age of high mass consumption. For the progression from one stage to another, Rostow in fact had a rather precise formula or recipe: the rate of productive investment rises from about 5 per cent of national income to over 10 per cent; one or more substantial manufacturing sectors emerge to become 'leading sectors' in growth; and the political and social framework is modified to exploit the impulses to expansion in the modern sector (Rostow 1960: 39).

Rostow's stages model was the ultimate attempt on the part of post-war economic planners to devise something like a 'development-vending machine: you put in the money, press the button, and get growth' (Brookfield 1975: 29). The problem, as Brookfield points out, was to find the button, or where to put the money and effort to get growth started. The problem was a circular one. Underdeveloped countries were not underdeveloped because of a lack of scope for investment; they were underdeveloped for want of capital; capital was lacking for want of savings, and savings were lacking for want of development (Brookfield 1975: 34). Suggestions ranged from a 'big push' in a single sector, to 'balanced growth' through investment across a number of sectors, to public investment in 'social overhead capital'.

As Landes (1989: 24) says of the impact of Rostow's optimistic model: 'We have been disappointed since' its publication. Economic development has not responded well to the injections of capital. The reasons for growth in some places and not in others may boil down to an as-yet unknown combination of social and human requisites, including 'obstacles of a managerial and administrative nature' (de Oliveira Campos 1982; Mason 1982). Culture seems to matter more than a colonial past. Thrift and hard work make the simple concepts of capital and labour go further in some places, such as East Asia, than in others, like Africa (Crook 1989; Landes 1989).

Growth or development?

Despite the intertwining of growth and development, the two are very dissimilar in the degree to which they are understood. Economic growth in the developed world is 'unambiguous in its meaning', whereas development 'has meant almost all things to all men and women' (Arndt 1987: 6).

Economic development is plainly different from economic growth. Growth – increases in population within a specific area, or increases in the quantity or the value of the goods and services produced in a local economy – does not necessarily lead to qualitative improvements in life. However, it is just such measures which have been conventionally used to gauge regional and national prosperity (Power 1988). In his classic work on urban economies, Thompson (1965: 1) proposed three goals for local economies, but which fit other scales as well: affluence, equity, and stability, for which 'aggregate growth is more a process than a goal . . . more a strategy than a goal – more a means than an end' (Thompson 1965: 1–2).

Economic development refers to increases in the quality of life associated with changes, and not necessarily increases, in the size and composition of the population, the quantity and nature of local jobs, and the quantity and prices of goods and services produced locally (Conroy 1975: 1). This distinction, made by Conroy in an urban context, is equally applicable in any regional setting. 'Economic development is a process of structural change, implying something different if not something more' (Flammang 1979: 50). Structural change focuses attention on issues of *structure*, a factor that affects not only the quantitative level of the economy of a place, but also its stability (Conroy 1975). Thus Thompson's (1965) goals of affluence, equity, and stability address growth, distribution and structure simultaneously. Definitions which focus on total or per capita income tell us about increases, but not about change. They measure growth more than development, although structural change may have accompanied the growth or made it possible (Flammang 1979).

An example derives from Rostow's (1960) stages of economic growth, where growth takes place with some sectors leading, some sectors lagging, and the rate of growth at least partially depending on how fast resources are effectively shifted out of lagging sectors and into the leaders. Flammang (1979: 51) asks:

> Is this not the same thing as saying that development (structural change) is necessary to keep overall production increasing? With no structural change, the economy would eventually smack into diminishing returns in a set of given sectors, and growth would have to slow down. But the reverse also applies: it is growth in the older sectors that accelerates the supplies of savings for investment in newer ones. So development supports growth by phasing out the old and bringing in the new, and growth supports development by supplying many (if not most) of the resources the newer sectors

require. Alternating processes? Yes. But both are occurring at the same time, most of the time, and so mask each other to some degree.

Weitz (1986: 20–1) presses the point further: 'Development is the progressive change in the society's status quo that takes place as a result of new relationships between different socioeconomic forces.' Development dynamism is the product of change involving all infrastructures – economic, spatial, institutional, and social – of society, and is a phenomenon that can restructure a nation's socio-economic base.

The process of economic development is extremely complex, especially in the long run. Development consists of the structural changes which take place in an economy and society – in the technological skills of its population and the technological capability of its firms and institutions that allow them to adapt to competition and change. 'The very essence of long-run growth is, in fact, the transition – sometimes orderly, sometimes chaotic – of the local economy from one export base to another as the area matures in what it can do, and as rising per capita income and technological progress change what the national economy wants done' (Thompson 1965: 3).

One type of structural change, the sectoral shifts from agriculture to industry, has long preoccupied development thinkers (Mandelbaum 1945; Brookfield 1975) and continues to do so in a 'post-industrial' and 'information economy' context (Cohen and Zysman 1987). This issue is pursued further in Chapter 2, where methods and concepts of economic base are discussed in some detail.

Development as qualitative change includes dimensions that go beyond simple number-counting exercises. It involves the changes necessary in social relations and in cultural orientation to stimulate in a cumulative fashion a greater diversity and specialization in the division of labour. This diversity and specialization, in turn, prompts a creative capacity in these regions and urban areas (Perrin 1974: 33; Sweeney 1987). Development also entails 'modifications in the system of values' (Weitz 1986: 15). Creativity, as part of the development process itself, demands values that promote, encourage, and reward creativity rather than those that sustain the status quo (Weitz 1986). However, especially in developing countries, regional planning is 'innately conservative', serving to perpetuate the bureaucratic structures and processes of the status quo (Gore 1984: 221).

A focus on *structure* poses different problems for understanding economic and social change. It is difficult to put structural change into the same terms and analytical frameworks as growth, around which a bevy of theories and models have arisen. The concept of structure generally is 'abhorred by economists who view the economy as a homogeneous species within which molecules move about under the action of prices' (Perroux 1983: 25). Social or cultural factors enter mainly as obstacles to development, or as a sphere of social problems created by the strain of development on a traditional order. 'Cultural factors' remain an oft-cited explanation of last resort to account for success and failure in economic development (Landes 1989). Economists typically acknowledge that some non-economic factors are important, but they are ignored in most analyses in favour of more familiar variables such as capital and labour (Doeringer, Terkla, and Topakian 1987; Weitz 1986). This is less true of development studies in the Marxist tradition, which, in 'thinking about "the problem of under-development"' does not 'locate the relevant conditions, and the obstacles to be overcome, within the backward society itself' (Peattie 1981: 45).

Regional planning

Within advanced and underdeveloped countries, regional planning strategies tended to concentrate on spatial elements, such as infrastructure, population distribution, and spatial interaction. They failed to be successful when the underlying social and economic interactions and conditions changed (Gore 1984: 211–19). 'Spatial policies cannot achieve their objectives unless they work in concert with sectoral and macro-economic policies' (Gore 1984: 223).

Regional planning in advanced economies dates back at least to the 1920s, both as a means of controlling metropolitan growth and as a decentralized set of programmes intended to cure economic backwardness (Friedmann and Weaver 1979). It faded in importance somewhat during and after the Second World War, when the focus shifted to more scientific economic and regional science models of economic growth applicable to developing nations (Isard 1956, 1960). In this context, 'backward regions' in advanced countries were a separate, and relatively minor, concern (Friedmann and Weaver 1979).

The birth of regional development as a field of study is often dated as 1958, corresponding with the independent publication of Gunnar Myrdal's (1957) *Economic theory and underdeveloped regions* and Albert Hirschman's (1958) *The strategy of economic development*.

Although both books had as their context the underdeveloped world, much of the operational influence into regional planning came from Walter Isard's more general view of 'regional science', contained in his *Location and space-economy* (Isard 1956) and *Methods of regional analysis* (Isard 1960), which were used to train legions of regional planners for the Third World. The apolitical and context-free tool kit of analytical theories and methods gave rise to optimistic policies that should solve problems of regional disparities in any setting. Chapter 2 discusses this traditional tool kit and its shortcomings in the light of new sectors and new realities.

Regional planning is based largely on growth as a means of change. The association of machines and technology with modernization and progress after the Second World War became a more widespread rationale for scientific and technical progress than had been the case prior to that (Adas 1989: 402–18). Growth and, with it, technological progress, as a goal of social policy simply overwhelmed competing objectives (Mishan 1976; Wenk 1986). Regional planning has tended to deal with economic structure only little, primarily because regional growth theory has borrowed directly from national growth models (Richardson 1973).

Regional development in most advanced economies waited until the late 1950s for any central government attention, which was greatly influenced by the notion of growth poles (Higgins and Savoie 1988). Chapter 3 outlines both the growth framework for regional change and the ways in which experience with growth poles provided empirical evidence of the technological and organizational complexity of economic change.

Technological capability

Technological change is perhaps the most important source of structural change in an economy, because it alters the mix of products, industries, firms, and jobs which make up an economy. It causes these changes in a subtle manner, creating new jobs and firms, destroying old ones, disturbing the equilibrium (Schumpeter 1934). The post-war economic growth models, in their view of fixed capital, effectively ruled out human capital formation or enhancement of people's creativity, technological skills, technical change, and entrepreneurship (Toye 1987).

Technological capability is not fixed or permanent, since both technology and the abilities of competitors are constantly changing. It relies on firms, on their activities, and how close they are to the state of the art, or *best practice*, at any point in time. Best-practice technology typically refers to process technology (machinery, equipment, management practices) which ranges from that of the most efficient (best-practice) producers to the most inefficient (worst-practice) (Le Heron 1973). The spectrum from best-practice to worst-practice technology is largely a function of the age (or vintage) of machinery and capital equipment employed. Newer equipment will incorporate or embody newer concepts, techniques, and knowledge which tend to give an advantage to firms – and regions – where this technology is employed. Firms and countries which are not near the current 'technology frontier' both in science and in production find it increasingly difficult to keep up with changes in other places (Cohen and Zysman 1987; Katz 1982b; Spence and Hazard 1988). The large quantities of information and the rapid pace of technological change pose serious problems for Third World countries which are late. 'The new technologies are so esoteric and difficult as to be almost unlearnable' (Landes 1989: 27). Yet they must be learned, and the rich must help the poor countries, now with technology as in the past with capital (Colombo 1988).

A more accurate perspective on national and regional economic differences, then, is not a focus on differences in resource endowment or differences in the rate of growth of capital or labour. 'It is the growth and accumulation of useful knowledge, and the transformation of knowledge into final output via technical innovation, upon which the performance of the world capitalist economy ultimately depends' (Griffin 1978: 14; Stewart 1978: 114–40). Capital transfers have not narrowed the gap among nations in technological activities and assets; instead, it has, if anything, widened. The Third World gets almost all its technology from advanced countries, and it is thereby dependent on the advanced countries where most of the search for new technology takes place. The technology available determines the boundaries of what it is possible for a country to do (Stewart 1978). While a small group of countries, the NICs, have attained surprisingly rapid growth, the bulk of the Third World remains desperately poor (Tata and Schultz 1988; Toye 1987). The success of the Asian countries owes more to technological development than to low wages, although the latter certainly were part of their initial growth (Westphal 1987). Technological capability, then, is at the heart of regional change. As outlined in Chapter 4, it applies at the level of the firm, of regions, and of nations, and relates directly to notions of competitiveness and competition.

Much of the process of technical change is an outgrowth of the way in which technology is produced and exploited by capitalist firms, the topic of Chapter 5. Technology does not create itself, of course. Rather, it is a direct outcome of the choices and decisions of people and organizations. Often, these choices are misguided, pouring resources into military expansion and trivial improvements of consumer goods. Awareness of technology as a mixed blessing has accompanied technical progress since the beginning. Both the development of new products in research and development laboratories and the transformation of production processes depend on firms' strategies, production relations, and their interrelationships with their competitors. Some of this topic is encompassed within the various 'cycle' models, ranging from product and profit cycles to the 'long waves' which seem to embody global technological change in a long-term manner.

Technological capability also interacts with industrial structure or industrial mix in affecting income levels, job prospects, and potential for future growth. A concentration of low-income industries, for example, effectively predestines that a region will employ workers with few skills and therefore have lower-than-average incomes. The innovation gap among regions is a primary source of regional development disparities, and it depends on not only the sectors, but also on the firms, their organizational structure, and the extent of their markets (Brugger and Stuckey 1987). Flexible forms of manufacturing appear to be supplanting mass production as a means of organizing production. They employ technology in new ways that combine product technology and process technology in a framework of global competition among firms. Chapter 6 deals with the various influences on firms' locations for production and non-production activities. The division of labour provides a succinct framework for the decisions, locally and globally. It is important not to see this tendency as occurring only at the global scale. Firms, small and large alike, seek out locations where unions and labour power are weaker or less well developed as part of the society.

Competition for technological superiority, locally and internationally, has become a major arena for government intervention. The actions and strategies of individual firms and competition among them affect the competitiveness of regions. Transnational firms diminish the likelihood that such policies can accomplish what is intended, at any scale of region, as Chapter 7 demonstrates. The experiences of high tech regions throughout the world are assessed in this chapter, in the context of desired versus actual development outcomes. The transfer of technology is one of the few means available for nations to obtain technological capability which they lack, yet that is not an easy task, since effort and expertise are needed to absorb technology.

Most models and policies, however, ignore perhaps the key type of economic change: the creation of new goods and services which require new production functions. This view of economic development – often termed Schumpeterian after Schumpeter's (1934) vivid description of the 'creative destruction' of capitalism – perhaps explains best the abrupt shifts in structure, the mobility of labour and capital, and the fact that labour is not simply eliminated from capitalist production entirely. The role of new firms and of small firms is dealt with in Chapter 8, where entrepreneurship as a process in regional economic change is examined.

From a long-term perspective, economic structure connotes the endogenous characteristics of a region's economy which make up the capacity for economic growth. The presence of some characteristics, such as a relatively new stock of capital equipment, allows higher productivity and output for a given level of investment. Similarly, an excellent education system provides a pool of skilled workers who are better able to adapt to new methods of work. Regarding the need to make investments for the long term, Storper (1989: 236–7) notes: 'It is the latter – that is, investments in new products and process development and in worker training and organizational skills – that have long-term payoffs in the form of increasing market shares for firms and generation of high-skill, high-wage employment for the society.' The issue of skills is addressed in Chapter 9.

The introduction of changes in the technological base of society affects every other aspect of society as well (Weitz 1986: 21). Technological change itself encompasses a broad array of dynamic elements of economic life. As a system of information, methods, machinery, and skills, it is among the fundamental or global forces that link regions and nations, and thus make development interdependent (Brookfield 1975; Perroux 1983). Large-scale technological changes have transformed socio-economic structures in profound ways. The first Industrial Revolution allowed the substitution of machinery, usually steam- or water-powered, for human labour. Workers, formerly largely impoverished farmers scattered in rural areas, were drawn to the towns where their labour was needed. Their labour was replaced by machines, as their mental powers have been more recently replaced by semi-conductors in the 'second industrial revolution' (Leontief 1983).

Within short- and long-term transformation, some stability or inertia also holds. As a given level of technology is embodied in machines and in transportation infrastructure, it has a measure of durability (Wegener 1986). Gertler (1988a) notes that the more general problem in understanding regional economic change is the temptation to implant inevitability and irreversibility on observed processes. His list of examples of flawed premises is excellent (Gertler 1988a: 153):

1. Production is increasingly decentralizing away from established urban centres and regions, and will continue to do so.
2. Large firms continue to integrate themselves vertically, if left to their own devices.
3. Oligopolies or monopolies necessarily grow more extensive over time, and certainly never collapse.
4. All products pass through the different phases of a common life cycle.
5. A firm's profits, once high, will remain high.
6. Production processes become increasingly capital intensive over time.

In fact, each of these 'truisms' is questionable, having some merit in particular cases but little generality. However, they continue to provide the impetus for much research on industrial and economic change.

Industrial transformation in localities, regions, and nations thus becomes the pivotal process to explain. The locally specific ways in which such transformations take place make the global view of this book useful by presenting an array of locally based empirical findings. The next chapter examines economic structural change.

Chapter 2

The measurement of regional economic activity

Measurement of economic development has evolved for the most part quite separately from advances in the theory of development sketched in Chapter 1. The temporal emphasis of applied models and measurements has been short term rather than long term, and applications typically have focused on the impacts of specific changes in a regional economy, such as increases in production or in population. Models for measurement have also been much more tied to evaluation of local, regional, or national policies, which are themselves generally short term and often only slightly or tangentially related to either theory or empirical evidence.

This chapter addresses the measurement of regional growth as it has evolved from calibration of the economic base of a regional economy to input–output and econometric models, more complex sets of tools that require large amounts of data and generate large amounts of information in return. Attempts via policies and investments to create industrial growth, such as growth poles and territorial production complexes, largely rely on the structure of production relationships embodied in input–output models. Policies for industrial targeting and regional strategic planning require knowledge about a region's economic base. But they are restricted in their ability to deal with shifts from one sector to another, and especially the increasing dominance of the service sector in many regions. The growing importance of service activities, especially producer services, as a regional economic base is the final topic of the chapter.

The economic base and regional multipliers

At the local and regional levels, economic growth is fundamentally a process of multiplier effects. The production and distribution of goods and services create employment and other income-earning opportunities which attract and keep people in a city or region. But no region is isolated from the rest of the world. Demand for a region's products and services substantially determines whether a place grows economically or declines. This *export* orientation suggests that the economic base of a region lies in what it is able to export to other places. Even if a regional economy is highly specialized, as in the case of a unique tourist attraction, it is dependent on its comparative advantage over other, competing tourist regions, rather than being independent of other places.

The standard means both of estimating local economic impacts of export activity and of forecasting future effects relies on the *multiplier effect*. As workers spend their incomes on local services and firms purchase supplies and other inputs, additional jobs are supported by these expenditures. Rounds of local spending result in additional, but successively lower, spending impacts as workers in the local firms in turn spend their incomes in the local economy. As Table 2.1 illustrates, if people tend to spend 60 per cent of their incomes in shops and businesses locally, the successive rounds of spending result in decreasing amounts of local impact. However, the impacts accumulate to total 2.5, including the initial 1.0.

Table 2.1 Effect of successive rounds of spending on local multiplier

Initial impact	1.0
Round 1	0.6
Round 2	0.36
Round 3	0.22
Round 4	0.13
Round 5	0.08
Round 6	0.05
Round 7	0.03
Round 8	0.02
Round 9	0.01
Total multiplier effect	2.50

The new economic activity and its multiplier effect help to expand the service (or tertiary) sector, especially as local growth increases the size of the economy sufficiently to meet the threshold for new service activities (Fig. 2.1). Increases in population in turn can spark a secondary multiplier effect as new investment enters the economy to serve expanded demand (Keeble 1967; Myrdal 1957; Pred 1977). This *circular and cumulative causation* process captures the dynamics of regional growth. Conceptually, this effect captures the process whereby sectors are linked through flows of money and jobs. The effects therefore are primarily, if not exclusively, concerned with growth rather than development. Income, capital flows and especially employment are the principal variables considered. The size of a region's multiplier, however, varies with economic structure; it will be larger where control, information, and diversity are found, and will be smaller when regional structure is limited or narrow.

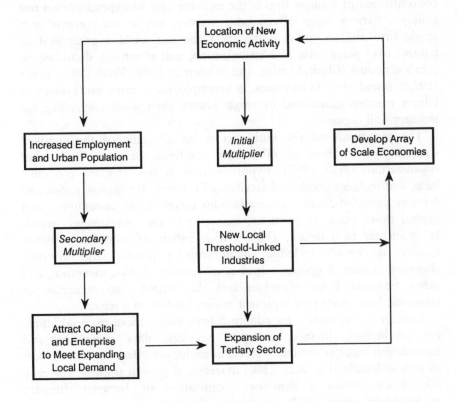

Figure 2.1 The circular and cumulative nature of local economic growth

In many regions, the export sector represents a significant portion of its total economic activity. Smaller regions will tend to export more in order to import and satisfy the demands of inhabitants for items not produced locally. Large, urban regions with diverse economies and many different industries, on the other hand, can trade internally because they produce a larger proportion of local requirements. In addition, large regions are better able to weather economic fluctuations because, even if the demand for some industries' products declines, this is unlikely to affect all industries at the same time. Finally, the larger the region, the more likely it is that a large portion of manufacturing industry will be serving the local market.

The economies of larger regions usually encompass a greater number of firms in a variety of exporting and supplier industries. However, this is not always the case, and concentration in a small number of sectors plagues large as well as small regions. The 'company town' problem of concentration in a single firm is the extreme case of dependence on not simply a narrow range of economic activity, but on the fortunes of a single firm and sector. A dominant plant can wield a great deal of control over wage rates and unionization, and effectively dominate an area's economy (Clark, Gertler, and Whiteman 1986; Sloan 1981). Lever (1979) found that fluctuations in unemployment rates are greater in labour markets dominated by single plants, even when controlling for industry and region.

The magnitude of the multiplier is an outcome of the complex structure of location of sales and purchases of firms and other organizations (Pred 1977). Export activity is the standard economic base, and includes goods and services sold outside the region, goods and services provided locally to people who travel to the community, and capital flows (such as pensions, loans, and other investments) which bring income to a region. Thus, a wide variety of economic activities qualify as 'basic', including manufacturing plants, large regional shopping centres, hospitals, airports, universities, banks, insurance, and other financial firms. Tourism and the retired also comprise an economic base, since they represent money inflows to a region.

Linkages of all sorts – for administrative information, material inputs for production, finance, transport services, data processing and information transfer services – result in multiplier effects in other places as well as locally (Fig. 2.2). This interregional growth transmission may take place within a firm or organization or between different organizations (Pred 1977). In essence, the profitable economic opportunities present in a place initiate a flow of revenues and employment that may disperse widely from their point of origin. At the places where the multiplier effects end up, such as headquarter sites and world cities, local employment and income are increased. This non-local effect has become much more prominent in the context of foreign ownership and production in branch manufacturing plants.

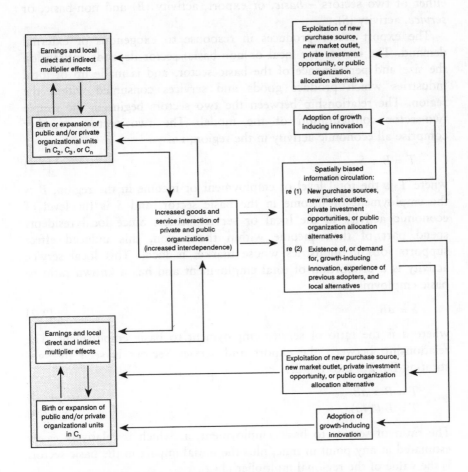

Figure 2.2 Pred's expanded version of circular and cumulative growth, based on information and innovation. *Source*: Pred (1977: 175 (Fig. 3.13))

The economic base model

Most analyses of regional economic activity ignore interregional linkages in favour of examining an isolated region or, at most, limited flows of capital and labour between two regions. (Chapter 3 considers regional production functions and interregional linkages.) The simplest model for measurement of changes in a region's economic activity, the economic base (or export base) model, rests on some profound assumptions about the nature of the linkages and interconnections among sectors in an economy. Economic base analysis classifies all economic activity into either of two sectors – *basic*, or export, activity (B) and non-basic, or *service*, activity (S).

The export sector produces in response to exogenous or external demand. The service (or local or non-basic) sector depends entirely on the size and performance of the basic sector, and is made up of local industries which produce goods and services consumed within the region. The relationship between the two sectors begins in the simple two-sector configuration of the model. The two sectors together comprise all economic activity in the region, i.e.

$$T = B + S \qquad\qquad\qquad\qquad [2.1]$$

where T is the total level of employment or income in the region, B is the employment or income in the basic sector, and S is the level of economic activity in the local or service sector. Since local residents spend part of their income within the region, this induced effect supports jobs in industries whose market is local. This local service activity is a proportion of total employment and has a known ratio to basic employment:

$$S = aB \qquad\qquad\qquad\qquad [2.2]$$

where a is the ratio of service employment to basic employment. The relationship between the export and service sectors is seen by sub-stituting Eq. 2.2 into Eq. 2.1:

$$T = B + aB \qquad\qquad\qquad\qquad [2.3]$$
$$T = B(1+a) \qquad\qquad\qquad\qquad [2.4]$$

The ratio of service to basic employment, a, which is relatively easily estimated at any point in time, plus the initial impact in the basic sector, is the value of the regional multiplier $(1+a)$.

Most explanations of economic growth pay tribute to economic base theory which, as sketched above, stresses the extent to which an area is dependent on demand from outside the region (Andrews 1953a; Lane 1966; Perloff *et al.* 1960; Richardson 1985). It is a demand-side model that emphasizes the dependence of a regional economy on demands for the goods and services produced within it. Such demand-driven models assume that the supply of factors of production – the inputs needed to produce goods and services – is infinitely elastic. If demand for a region's products grows, then an indirect increase in employment in the supplying industries follows. This employment, in turn, generates income within the region in that sector, leading to an induced increase in demand for other goods and services through the multiplier effect.

External forces in addition to demand greatly influence multiplier effects. Oil prices, for example, largely determine the regional level of employment and income in the oil and gas industry in regions like Louisiana. Significant fluctuations in both basic and local service employment result from these linkages of a regional economy to global markets (Andrews and Tate 1988). Because of these linkages, the rapid price increases set by the Organization of Petroleum Exporting Countries (OPEC) for crude petroleum during the 1970s led some regions to attempt to benefit from the increase in local spending. The oil supply industry, like the oil industry itself, is comprised largely of multinational firms based in the USA. This has meant that much of the benefit within any oil-producing region has been less than if local firms were active. In Aberdeen, Scotland, the base for most of the drilling activity in the North Sea, little local oil supply industry has developed (Hallwood 1988).

In this framework, 'regional problems are balance of payments problems' (Thirlwall 1980). Exports propel the regional economy, but the benefit is reduced by the amount of imports. The degree of importing and other leakage thus determines the regional multiplier (Black 1981; Leven 1964). In an international context as well, a nation's ability to export is considered an essential foundation of economic growth (Krueger 1983).

Import substitution is an issue raised about the structure of an economy primarily, but not exclusively, at the national level. If a region begins to produce goods and services which were previously imported, then it is able to keep additional income and jobs within the area. This is most likely to happen as a town or city acquires new stores and services that increase local, rather than non-local, consumption. Import substitution can occur in other ways, especially through technological changes as new production processes or lower transportation costs make local production less costly than previously (Tiebout 1962). Economic growth in some regions, however, must be complemented by a focus on activities which will serve local, rather than only export, markets (Huskey 1985).

The economic base model transfers the demand-driven Keynesian model of a national economy, based on sources of income, to the regional setting (Hewings 1977: 32–4; Hewings and Jensen 1986; Pleeter 1980). The Keynesian income multiplier may be made more complex than the economic base formulation by adding indirect taxation and government spending to the standard model (Archer 1976). More importantly, this sort of model can also adjust for 'leakages' of a regional multiplier because of (1) consumers who spend a portion of their incomes outside the region, and (2) producers of goods and services who purchase their inputs outside the region. Archer (1976: 72) notes that much of the initial round of expenditure, especially in the case of tourism but in other sectors as well, leaks immediately out of a regional economy without generating even one round of income to the region and its residents. This happens most frequently when rental and other retail activities are owned by people or organizations elsewhere.

The attraction of the economic base model is that it provides a simple means, with relatively low information requirements, of analysing a regional economy and of estimating impacts of gains or losses of basic activity. The model's major shortcoming also results from this simplicity and paucity of information. It is extremely difficult to know for certain how much of an activity is sold outside an arbitrary region. Survey research, the best solution, is time-consuming and expensive, and so there is always a demand for short cuts (Gibson and Worden 1981; Hewings 1977). The most naïve solution, but a very common one, has been simply to postulate or assume that primary and secondary economic activities (agriculture, extractive industries, and manufacturing) are basic and that all others are non-basic. This 'assumption method' will, of course, render a value for the regional multiplier, from which conclusions can be drawn and estimates made about the impact of new export industry (Andrews 1953b). However, the error involved 'can be enormous' (Tiebout 1962: 47). Much manufacturing is locally oriented, especially for certain types of firms such as bakeries, brickmaking, and printing and publishing. In addition, many services are non-locally oriented such as the home offices of insurance companies.

The propensity of consumers to spend locally is an influence on the economic base multiplier which has received little attention. Tiebout (1962: 69) listed three principal influences on consumer spending patterns. First, community size is a major factor, since large cities have more items (including cultural and sporting events) on which to spend income locally than do small towns. Second, geographic isolation, even for small communities, tends to increase the amount of income spent locally, since if larger cities are nearby, a larger amount of spending is likely to take place there. Third, higher-income communities are likely to have a smaller proportion of income spent locally, because well-to-do households are inclined to spend more outside the community on such things as vacations and luxury shopping trips. This is a more difficult issue to prove, since it requires detailed data on geographical shopping and spending habits (Boehm and Pond 1976; Erickson 1978). Actual multipliers are reduced because of other leakages, primarily from expenditures that are made outside the region, both by firms on capital equipment and production inputs and by consumers on luxury goods and other imports. When such flows are taken into account, the multipliers that result are largest in regions where a wider range of goods and services are produced (Black 1981; Tiebout 1962).

Gibson and Worden (1981) illustrate the details necessary to do a thorough analysis of economic base. Even regional employment may not be a known entity, since it typically includes full-time, part-time, and seasonal employees. In their study of Arizona, the latter two categories accounted for as much as 40 per cent of all employment. Adjusting for in- and out-commuting, wage-level differences across industries, and (perhaps most importantly) transfer payments (including those for unemployment, old age pensions, and low incomes) resulted in lower multipliers by about 20 per cent. Gibson and Worden's study also demonstrates the value of survey studies for economic base research; they found that a 21 per cent sample of firms resulted in fairly reliable (±10%) estimates of multipliers, compared with a 100 per cent survey, provided small firms (5–10 employees) are also included. Studying large firms alone yielded significantly larger errors. However, Richardson (1985, 1988) suggests that the burden of a survey is great enough that one should eschew an economic base framework in favour of a regional input–output model (to be discussed below).

An alternative to the assumption method for the delineation of the export sector is the use of *location quotients* (*LQ*), or coefficients of specialization (Hildebrand and Mace 1950; Isserman 1977; Mayer and Pleeter 1975). If a region is self-sufficient in a product – neither exporting nor importing it – then its local employment in this industry would be expected to be the same as the national average:

$$LQ_{ir} = (E_{ir}/E_{tr}) \ / \ (E_{in}/E_{tn}) \hspace{2cm} [2.5]$$

where E_{ir} is the employment in industry i in region r, E_{tr} is the total employment in all industries in the region, and E_{in} and E_{tn} are industry and total employment, respectively, in the nation. If a local industry has more than the national proportion of total employment, then *LQ* will be greater than 1.0 and the amount *over* the national proportion is assumed to be for export; the remainder is considered necessary to meet local demand. If a region has less employment than the national percentage, $LQ_{ir} < 0$ and it is likely that these items are imported into the community. As economies have become more interdependent and global in recent years, however, it is not at all certain that any nation is the proper comparison for self-sufficiency in a product.

Of several criticisms lodged against the location quotient method, the most significant concerns product mix (Tiebout 1962). Increasingly, establishments produce highly specialized products rather than a bundle that mirrors the national economy. Most communities will produce a very specific item – probably only one brand – and export most, if not all of it, rather than export only that above the national level. At the other extreme, firms may produce almost solely for other local customers, a situation commonly seen in apparel and electronics complexes (Scott 1983a, 1984). For highly disaggregated industries, product mix is a larger problem, since data on production and export of specific products may be lost within a larger aggregate category. For example, a region may have a number of makers of computer disc drives, but no actual computer manufacturers or even any producers of other non-electrical machinery, as the entire sector including computers is called. If all of the output is sold to firms outside the region, its estimated exports from the location quotient method will be less, perhaps much less, than its actual exports. Tiebout (1962: 49) found underestimates as high as 99 per cent, but others suggest that the location quotient method, coupled with adequate disaggregation and some judicious local survey information, is a viable way to estimate basic activity in a region (Norcliffe 1983; Pfister 1976; Pleeter 1980). Keil and Mack (1986) and Gilmer (1990) have continued the application of location quotients to estimate the role of services as export industries, an approach pursued further by Bloomquist (1988) regarding tourism – normally considered as a group of services – as an export sector.

The *minimum requirements* method is a variation on the location quotient technique. In this case, the minimum expected for local self-sufficiency is based on the actual minimum for a large set (more than 100 in most cases) of cities or regions (Ullman and Dacey 1960). This method takes explicit account of region size, in that the larger the region, the larger is the minimum in any industry. The regression line $Y = a + b \log P$, where Y = minimum employment requirement in industry i, and P = population, yields a different slope, b, for each industry.

There are shortcomings with this technique, including the assumption that all regions are exporting but that none is importing (Pratt 1968), that all regions in a size category are regarded as having identical service components (Tiebout 1962), and that if the industry data used are highly disaggregated, virtually all economic activity is classified as exports (Pfister 1976). Despite these concerns, it is still the method of choice for multiplier estimation in some situations. In developing countries for which more extensive data are not available, minimum requirements provide a useful estimate of exports and plausible estimates of regional multipliers that increase with region size (Brodsky and Sarfaty 1977). The advantage of the minimum requirements approach is that it is 'an extremely convenient method for estimating total exports, particularly when compared to the need for highly disaggregated data with the location quotient approach' (Isserman 1980: 164). It continues to be used as a 'quick and dirty' method of multiplier estimation (Erickson, Gavin, and Cordes 1986; Moore 1975; Moore and Jacobsen 1984). Mulligan and Gibson (1984) have found that it provides quite accurate multipliers for small communities, especially if government transfer payments are also taken into account.

Impacts of economic change

In addition to their use as a description of economic activity, multipliers are used for prediction of population and other impacts arising from new economic activity. Thus, 'impact analysis' typically involves either a simple version of an economic base multiplier or a more disaggregated Keynesian multiplier, along with a population multiplier to yield an estimate of population (Oppenheim 1980: 90–2). This estimate is then used to forecast needs for housing, schools, and urban services, particularly in the context of rapid growth in regions where an identifiable exogenous economic change is occurring (Glasson, van der Wee, and Barrett 1988; Lewis 1986; Devine *et al.* 1981).

The use of economic base methods in impact analysis deals little, if at all, with the questions of measurement or technique, and focuses instead on the value of the multiplier and the time lag over which change takes place (Lewis 1976). Bender (1975) found tremendous variation in multiplier values, which ranged from 0.59 to 3.71, depending on both the region and the basic industry analysed. Mellor and Ironside (1978) and Lewis (1976) also found that multipliers varied widely, depending on the method used in the analysis. Location quotients, for example, yield significantly larger multipliers than those based on any arbitrary categorization, especially for small regional economies (Gibson and Worden 1981). Isolated areas, where tourism or large, advanced technology projects are located, have a rather high degree of leakage because of supplies imported from elsewhere (Archer 1976; Glasson, van der Wee, and Barrett 1988; Devine *et al.* 1981). Multiplier estimates depend critically on the leakages of profit income that often vanish to other regions, especially from small or peripheral regions. The analysis of leakage from regional economies is not well developed, many years after the pioneering work of Steele (1972) and Archer (1976).

As suggested earlier, multiplier effects do not take place all at once. For example, Moody and Puffer (1970) estimated the multiplier for San

Diego, California, at 5.45, a relatively high value realized only after a long time–lag. The issue of the time-lag before a multiplier effect is complete is rarely addressed, although it is a central finding in research that has considered it (Ashcroft and Swales 1982; Gerking and Isserman 1981; McNulty 1977; Martin and Miley 1983). In any event, it is frequently several years before the entire measured multiplier effect is experienced, even in large and developed economies.

Regional decline: the flip side of growth

A little-studied aspect of regional change is contraction or decline – negative impacts which are the opposite of those encompassed by regional growth (Buhr and Friedrich 1981). Schaffer (1981: 16) lists several factors which can lead to decline: (1) a limited, inelastic, or unresponsive market, (2) constant or decreasing returns to scale, either internal or external, (3) a shift in technology, (4) resource limitations, (5) entrepreneurial deficiencies, and (6) lack of capital. Any of these factors limits the potential for endogenous or exogenous stimuli to generate multiplier effects. A downward shift resulting from decline will similarly affect other sectors by reducing demand for their goods and services, lowering employment, and setting into motion a downward spiral of economic repercussions.

The effects of decline are not symmetric with those of growth, since it can be expected that public expenditures will increase (for example, in social work, health, training, and education) and thus reduce the magnitude of the multiplier during contraction compared to its value during an expansion (Brownrigg 1980). The indirect impact on linked industries and firms and the induced effect on local service industries also will contract with a lag and these may well maintain an overstaffed size. The case of a major cut-back or closure by a large multinational manufacturer in Scotland studied by Brownrigg had very low negative multipliers – the largest estimate was 1.14 – because of the limited linkage to local firms by the multinational firm. This compared to values ranging from 1.4 to 1.8 for positive impact multipliers. The long-term effects, such as higher rates of outmigration by young people and long-term unemployment for those losing jobs, are not part of the multiplier effect *per se*, although they are part of the regional reality of stagnation and decline.

Despite its shortcomings, the economic base model has no close substitutes. For the best provision of useful facts about a regional economy, Pfister (1976: 112) makes some useful suggestions:

1. Use the model only for small local economies;
2. Do not assume that the basic/service ratio or multiplier is constant; use time-series data to estimate the multiplier;
3. Do not use it as a long-term growth model;
4. Use highly disaggregated employment data to estimate export employment with the location quotient method;
5. Allocate each industry's employment between local and non-local production; and
6. If used for forecasting, make an attempt to forecast exports.

The best estimates of growth – or decline – multipliers come from a detailed knowledge of the employers in a community, and their degree of intraregional linkages.

Beyond considerations of local linkages and leakages of multiplier effects, knowledge of a region's industries and firms is essential for understanding a regional economy over the long run. Each industry's growth prospects are different. A growing industry is likely to generate more jobs over some future period than a static or declining one, all other things being equal. In addition, local firms vary. There are strong firms and weak firms in any business, and their degree of exporting from the region or the nation may vary greatly (Hayter 1986; Zumeta 1966). Capital-intensive or highly automated investments, generally induce few jobs (Summers *et al.* 1976). Indeed, it is the base of locally owned firms, whose investments concentrate in a region, on which an economy depends most (Coffey and Polèse 1984, 1985; Power 1988).

Watkins (1980: 124) makes the important point that multipliers generally deal only with growth, not with development. If a region changes qualitatively in economic structure, but does not grow (or decline), then the economic base model registers no change. The growth prospects for an industry, and the relative competitiveness of a region's industries and firms, are structural features commonly given short shrift by models of regional growth. Yet multipliers remain important indicators. Most researchers believe that if additional effort is to be spent, it should be put into more complex and accurate models, such as input–output and econometric models. Richardson (1985, 1988) insists that the effort demanded to get accurate multipliers is quite great and he recommends that short-cut input–output models are nearly as costly and are more accurate.

Input–output analysis

Input–output analysis expands economic base and Keynesian multiplier analysis to focus on inter-industry linkages, in which each sector's output is either purchased by another sector or is used in final consumption by households or government. This 'social accounting system' is able to measure indirect effects of change in the output of a given industry (Leven 1964; Hewings 1977, 1985). The appeal of input–output analysis is its greater degree of disaggregation: each sector is identified individually. The 'indirect effects' reflect the logic that if greater demand for industry i's production increases, its purchases of the output of industry j, and j's purchases from its input sectors, will also increase. The magnitude of these rounds of spending declines over several iterations through the matrix, and results in a long-term multiplier over all rounds for the originating sector. (See Hewings (1985) for an introduction to the procedures used in input–output surveys, and Hewings and Jensen (1986) and Richardson (1972, 1985) for more comprehensive reviews.)

Each sector's output is another sector's input, and vice versa (Table 2.2). Since not all transactions take place within a region, purchasers outside the other producing sectors provide closure to the system. These final demand sectors include households, government, and foreign trade. 'Demoeconomic' models combine income growth, migration, and labour supply into the input–output framework (Batey 1985; Isserman 1985). Multiregional input–output models trace not only flows among the sectors within a region, but also between sectors across regions (Miller and Blair 1985: 69–85). Input–output models, however, are based on an assumption of linear production functions for each industry. Changes in technology, the addition of new industries, and changes in the location of purchases generally are difficult to incorporate in the model (Pleeter 1980: 25).

As with economic base analysis, a typical objective of input–output analysis is a set of multipliers that can be used to forecast economic activity (Miller and Blair 1985: 100–48). Their advantage over economic base multipliers is that they are disaggregated and, thus, will vary according to the sector which experiences the initial change in activity and expenditure (Richardson 1972: 31). However, input–output models generate a variety of multipliers, each reflecting a greater degree of precision or an estimate of impacts on income or employment, through impacts on the household or labour sector. One can estimate effects on outputs of the sectors in an economy, on value added, on income earned by households, and on employment, and these provide information that can be used to simulate or forecast the impacts on a region (Beyers 1983; Pullen and Proops 1983). Although this variety might seem advantageous, 'the mass of multipliers and other results produced by most input–output systems may sometimes constitute overkill . . . the typical set of tables from an input–output model run can be confusing or even misleading' (Stevens and Lahr 1988: 95).

Table 2.2 A hypothetical input–output model

(a) A simple input–output table

Sectors	Sectors				Final demand	Total outputs
	A	B	C	D		
A	40	20	20	15	20	115
B	10	30	50	30	5	125
C	35	40	25	5	10	115
D	15	10	5	50	20	100
Total inputs	100	100	100	100	55	455

(b) Inter-industry flows

Sectors	Sectors			
	A	B	C	D
A	0.40	0.20	0.20	0.15
B	0.10	0.30	0.50	0.30
C	0.35	0.40	0.25	0.05
D	0.15	0.10	0.05	0.50

Quite clearly, the potential advantage of input–output models relies entirely on the quality of the data on which they are based. A complete survey of firms in a region, to determine the sectors and regions from which each firm purchases its inputs and to which it sells its outputs, is the standard data base for the input–output approach. Depending on the degree of disaggregation, the sectoral specificity could be as great as that in the US input–output table, which contains 531 sectors (US Bureau of Economic Analysis 1984). Because of computational complexity, this model or matrix is more typically aggregated to a more manageable number of sectors – 39 in the case of the American data.

The existence of input–output data at the national level has permitted the development of non-survey-based regional input–output tables, which contain estimates of linkage based on national data and survey-based tables from other regions (Round 1983). Commonly, some regional data, such as the output of industries in a region, are used to complement the national intersectoral information. A variety of methods, including location quotients (despite the criticisms regarding their use in economic base analysis), are typically used to estimate sectoral inputs and outputs in a region (Hewings 1985: 46–57; Richardson 1985: 618–28). Jensen and Hewings (1985) believe that most 'short-cut' input–output multipliers are fraught with errors. Although researchers have been tempted to use national intersectoral transactions in regional applications, survey-based studies confirm that regional input–output structures vary significantly both from those of other regions and from their national counterparts (Harrigan, McGilvray, and McNicoll 1980; Kipnis 1976; Emerson 1971; Emerson and Ringleb 1977).

The information in any input–output table, even if derived from painstakingly gathered surveys, loses accuracy over time as an economy changes. (The time-lags for data availability are a related issue. The US input–output table based on 1977 transactions among sectors was not generally available to researchers until 1984 (Miller and Blair 1985: 266).) Among the changes that occur are modifications in the techniques of production as a result of several reasons (Miller and Blair 1985: 267):

1. Technological change, or process changes, may replace labour with purchased machinery.
2. Increased economies of scale may distort the measured relationship between (initially high) inputs for a given value of output.
3. New products invented create an entirely new sector (a new row and column in the matrix), which will have its own inputs and may replace an older product. The example of plastics and plastic packaging replacing glass and other products is especially vivid.
4. Relative price changes can alter the inputs used, as occurred when oil prices rose sharply in the 1970s.
5. The actual products of an aggregated sector may vary from year to year, and along with them the inputs used to produce them. The shift within the food sector from canned foods to frozen foods has

changed the inputs of packaging materials, just as diet soft drinks have reduced the demand for sugar and syrups.

6. Increased imports to a region or nation mean that the inputs of the goods purchased (and their multiplier effects) are shifted to the place of production.

Incorporating technological change into input–output models is difficult and rarely attempted, as mentioned earlier (Rose 1984). To incorporate changing production technologies, rather than constant technical coefficients, to represent allotment of inputs, is typically done by extrapolating based on past changes in coefficients or by setting the coefficients proportional to some other variable, such as investment, to represent embodied technological change (Carter 1970). Technological change can also be estimated by sampling the most efficient producers in an area, providing an approximation of best-practice technology (Miernyk *et al.* 1970). However, technological change may affect the pattern of imports and interregional trade generally more than it does the input coefficients. Branch plants which produce standardized products tend to have rather stable inputs and sources in comparison with small firms (Emerson 1976).

The literature on input–output models, estimation, and updating is large and quite technical (Hewings and Jensen 1986). It is sufficient for our purposes to point out that all such models are an aggregation; they cannot fully measure the activities of individual firms (unless a firm dominates a sector in a region). The size of firms also matters; large firms typically generate larger multipliers via their linkages, even if small firms have larger overall employment (Meller and Marfán 1981).

The input–output model became, early on, a standard tool of regional analysis and innovative ways were proposed for its use (Isard 1960; Keeble 1967). In the context of regional policy, it allows for the identification of clusters of linked industries and of 'key sectors' which are often important for regional policy, including growth centres, a topic discussed at greater length in Chapter 3 (Beyers 1976). The creation of an industrial complex of closely linked industries remains a common policy objective, and its operationalization relies on input–output data (Czamanski 1974; Czamanski and Czamanski 1976, 1977; O'hUallachain 1984a).

More recent research has focused on the 'fundamental economic structure' of economies of all sizes (Jensen, West, and Hewings 1988). In studying the input–output structure of 10 regions in Queensland, Australia, they found that the common thread in economies of all sizes is a set of tertiary, urban service sectors which increase in importance with region size, much as the minimum requirements approach suggests.

The principal effect of technological change is to increase the complexity of fundamental economic structure (Hewings, Sonis, and Jensen 1988). The importance of urban services lends weight to the continued validity of central place theory as an organizing framework for regional economic structure (Berry *et al.* 1988).

Econometric models

The interrelationships among sectors and variables in a regional economy can be quite complex, and this complexity can be incorporated in econometric, or multiple-equation, models of an economy. Particularly for forecasting purposes, these models allow a researcher to estimate the impact of a change in any of the variables that affect the economy, such as employment, output, wages, prices, income, population, retail sales, construction activity, and so on. Figure 2.3 illustrates some of the typical interactions incorporated in these models, such as the effects of migration on labour supply and the effects of income, prices, and investment on industries in the region.

Regional econometric models are constructed specific to each region, and employ time-series data to estimate the hypothesized relationships by means of regression analysis (Pleeter 1980). These models can be very large; the Philadelphia Region Econometric Model contained 228 equations (Glickman 1971). Taylor (1982) has shown that such models need not be overly large or complex to be reasonably accurate, but they should contain an appropriate degree of 'simultaneity' or interaction among the components of the economy, especially local labour-market conditions that affect population, employment, and income.

A major question that nags econometric modellers is whether to employ a 'top-down' or 'bottom-up' approach (Klein and Glickman 1977; Bolton 1982a). In the 'top-down' approach, the regional model is designed as a satellite system, attached in some consistent way to a system for the national economy as a whole. Consequently, some regional variables are related to national macrovariables in a one-way specification that to a large extent reduces regional analysis to determining regional income elasticity and industry mix (Pleeter 1980: 21).

At the opposite extreme, the ultimate 'bottom-up' approach would be to establish models for each region in a nation and combine the various relationships into a national aggregate. This allows regional changes to have an explicit effect on national performance, which is desirable given

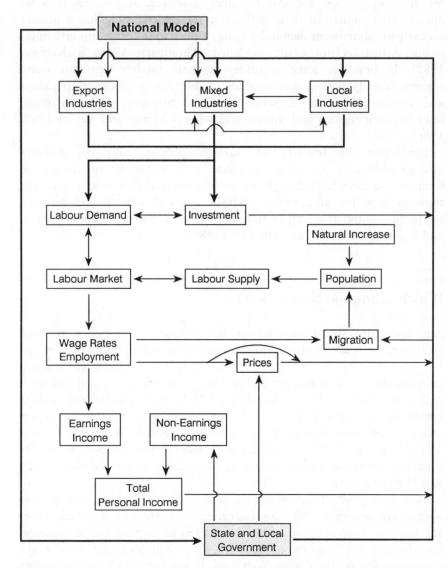

Figure 2.3 A typical regional econometric model. *Source:* Bolton (1985: 501 (fig. 3))

the actual interdependence. However, severe data problems at the regional level preclude the pure bottom-up approach from being feasible. Statistical coverage is typically relatively poor at the regional level, and there is unlikely to be adequate coverage of profits, trade, and consumption at a regional level (Klein and Glickman 1977). The resulting reliance on national variables, however, also makes regional models very similar to their national counterparts, including a greater concern for short-term demand impacts rather than long-term structural change which is not easily modelled (Thompson 1968; Richardson 1988). In practice, some relatively accurate models exist for many regions, usually with a combination of 'bottom-up' local relationships and 'top-down' equations. Similar concerns surround efforts to model links between regions and subregional areas (Charney and Taylor 1983, 1986).

Simulation of impacts of various policies and of different demographic and economic scenarios is a common application of econometric models. Although it has been covered only briefly here, the topic is a major area of research in regional science and regional economics. Summaries can be found in Bolton (1985), Glickman (1977), and Nijkamp, Rietveld, and Snickars (1986).

Which economic base is best?

Any industrial sector can act as the economic base of a regional economy. Even in the absence of an employing sector, a community can be supported by income from retired people's pensions, government unemployment payments, rents, royalties, or dividends earned outside a region (McNulty 1977; Tiebout 1962: 40). Some of this income is spent where the recipients reside, even if employment in a conventional export sector is non-existent since the pensioners are retired (Beyers 1979; Gillis 1987; Mulligan 1987). Government transfer programmes may be the dominant source of income in remote economies, such as Alaska (Knapp and Huskey 1988).

Although primary activities, such as mining, forestry, fishing, and agriculture, together with manufacturing, are customary economic bases for analysis, several other sectors have risen to prominence as potential foundations for regional economies. We will examine briefly the tourism, military, high technology and, in greater detail, service sectors as economic bases of regions.

Tourism as an economic base

Tourism has moved to 'centre stage' in economic development policy, especially in regions or nations where employment creation is a dominant objective. Tourism tends to be labour-intensive and shows little potential for substitution by capital in the production of tourist services (Gershuny and Miles 1983; Williams and Shaw 1988). Locations with special natural, cultural, and historical attractions have been able to attract revenue from tourists in the same way as production that is exported. Tourism is increasingly 'created' by construction of convention facilities and shopping outlets in urban locales which, along with hotels and restaurants, comprise the destination of most tourist expenditure (Patton 1985). It also is currently fashionable to promote 'hallmark events' to secure the attention of the tourism market for a short, defined period of time, in contrast to regular or seasonal attractions. Olympic Games and other infrequent sporting events, international expositions or world fairs, and Grand Prix motor races are international examples. Much the same is attempted at the local level, where annual festivals attract media attention not possible the remainder of the year (Hall 1989; Syme *et al.* 1989).

Tourism has some characteristics which make it unique. The touristic experience is a combination of everyday elements and extraordinary pleasurable experiences (Urry 1987: 19–20). Gray (1970) and Pearce (1987) categorize tourist preferences as either 'wanderlust', which includes urban pursuits such as shopping and seeing historic sights, and 'sunlust', which captures the popularity of island and other coastal resorts. Recreation participation and coastal landscapes are the principal images portrayed in advertising brochures of island nations, whereas history and art are more common among Old World countries in Europe as well as India and Japan (Dilley 1986).

Tourism is a problematic sector to identify, since there is rarely a 'tourism' industry defined in official statistics, and simple combinations of tourist attractions, hotels, and restaurants do not do a very good job of representing a 'tourism sector' (Stevens and Lahr 1988: 94). In spite of this, numerous analyses have found large impacts from individual tourist attractions, such as the Louisiana Superdome in New Orleans (Ragas *et al.* 1987). This facility, subsidized by the state of Louisiana, accounted for enough revenue to result in a 13.0 benefit/cost ratio over the 1975–84 period.

Tourism is an industry with particularly low entry requirements, as seen in the various 'Mom and Pop' hotels, shops, and restaurants found both in developed tourist centres and in smaller locales where national and international chain establishments see insufficient market to warrant their investment. In either setting, employment in the tourist industry is known to be female dominated, and to be largely part-time and seasonal (Cater 1988; Shaw and Williams 1988; UN Centre on Transnational

Corporations 1982). However, tourism provides entrepreneurial opportunities that can supplement income from the other jobs held by local residents. Seasonality varies from region to region, in large part along with the nature of the built environment. Large 'theme parks' such as the Disneylands are able to attract tourists year-round, while many mountain and beach resorts have only seasonal attraction. The tourist market is also differentiated according to social class, generation, gender, and household type (Richter 1989; Urry 1987: 21).

Whether the benefits of tourism as an economic base are equivalent to those of other sectors depends on the degree of linkage within or leakage from the regional economy (Sinclair and Sutcliffe 1988). Multipliers from tourism typically are greatly reduced by leakages, and multiplier values hover in the range of 0.25–0.5 (Archer 1976). These are relatively low multipliers, and they reflect perhaps more the anaemic nature of the economies which depend on tourism than they do the nature of tourism itself. In major urban areas, tourism multipliers may be much larger, especially during hallmark events.

Despite uncertainties over benefits, tourism is an alluring source of income to struggling countries such as Cuba, which is only in the late 1980s attempting to revive its tourism industry by permitting several international hotel chains to establish resorts along its beaches (McGuire 1989). Infrastructure shortages of all kinds, including transportation, sanitation, and energy, are needed to lure tourists (Cater 1988; Stabler 1988). 'The luxury hotel in the developing nation requires excessive monetary support, disproportionate water, energy, food, land, and construction materials – all items in scarce supply. In some island destinations, reefs are dynamited and fishing destroyed so that beaches can be widened and swimming improved' (Richter 1989: 185).

The dominance of the large hotel chains and international tour companies reduces the income which any tourist region actually receives (McKee 1988; UN Centre on Transnational Corporations 1982). Holiday Inn's parent firm, Holiday Corp., owned 1907 hotels in 1986 (Williams and Shaw 1988: 6). The concentrated 'enclave tourism' found in many island nations is also subject to competition from best-practice technology, in the sense that the mix and quality of tourist services (e.g. golf courses and 'nightlife' activities) expected by tourists tend to increase over time. Finally, the seasonality of tourism makes it an unstable economic base, even where it is the only base (McKee 1988). None the less, several countries rely on foreign tourism for over 20 per cent of their total export earnings, including Austria, the Bahamas, Fiji, Greece, Morocco, and Spain (UN Centre on Transnational Corporations 1982: 96).

Military spending as an economic base

Military spending is another common economic foundation on which many regions depend (Bolton 1966). Lovering and Boddy (1988) estimate that defence spending in the UK generates an indirect effect on employment and exports nearly equal to direct military effects. Military spending has two kinds of regional effects: those that ensue from the purchase of hardware for the military services, and those that follow from the presence of armed forces stationed in a locality (Todd 1980a: 116). Perhaps most attention has been paid to the effects of production of military goods, both because the industries in which production is concentrated are 'high tech', and because marked spatial concentrations are exhibited in this production (Breheny 1988). Since the 1960s, increases in defence spending have been slowest in personnel (the direct effect) and greatest in purchases of air and naval equipment and thus in those industries (Law 1983; Malecki 1984b; Wells 1987). Space-related industries overlap considerably with military activities, and generate an employment multiplier of 1.89–2.24, high by most standards (Greenwood, Hunt, and Pfalzgraff 1987).

Military expenditure varies markedly among a number of categories, only some of which comprise the 'regional technology base' (Malecki and Stark 1988; Wells 1987). The demand for engineers and scientists in these production sectors has a distinct spatial bias, for example in the UK towards the South-east, and in the USA towards the East and West Coasts (the 'defence perimeter'), which cumulatively reinforces previous investments and location decisions of military–industrial enterprises (Kunzmann 1988; Lovering 1985, 1988; Markusen 1986a, 1988; Markusen and Bloch 1985; Todd 1988a). Even subcontracting of components for complex systems tends to favour the same regions and reinforce the concentration of multiplier effects (Malecki 1984b).

Much of the defence industry is considered high technology, employing large numbers of engineers and scientists and involving large efforts in research and development (R&D). Indeed, military production is considered the core of high-technology industry in the UK (Hall *et al.* 1987). The relationship of military production to other high-technology sectors is a tenuous one, since military products have idiosyncratic characteristics that may have little 'spin-off' potential in consumer markets (Kaldor 1981).

The presence of a local military installation is a more local, and usually a smaller, kind of economic base (Tiebout 1962: 40–2). Local firms will depend indirectly on this 'export' income, and local civilians will be employed to provide support functions. However, leakages of multiplier effects are often very large, because of centralized purchasing of supplies and services from non-local sources (Erickson 1977). More commonly, leakages are ignored in estimates of economic effects, even in small regions where they are likely to be large (Rodriguez and Krienke 1982). Instead, a symbiotic relationship develops between the military and the community in such places (Lotchin 1984; Grime 1987). Military bases are unstable sources of income, despite large multiplier effects during boom times, and base cut-backs and closings are strongly resisted by government officials. Military priorities in high-technology policies are discussed in Chapter 5.

High-technology industry

High-technology sectors are one of the most frequently cited objectives of recent regional development policy. The allure of high tech is twofold. Because of their reliance on scientists and engineers, high-technology industries are inherently more innovative than other sectors and this, in turn, should be related to above-average rates of growth. In addition, the growth of high tech during times of decline in most other manufacturing sectors, lends at least partial empirical support to the theoretical argument (Thompson 1988a). Obviously, part of high tech's boom is a result of military production, a connection not always noted when high technology is sought.

High-technology industry is both misunderstood and probably overrated. Even over the long term, its probable direct employment generation is relatively low, with employment in high-tech firms unlikely to exceed 10 per cent of the US workforce, and maybe 15 per cent in a few states, such as California and Massachusetts. Service industries – of which only a few, such as computer software and information processing, are high tech – will account for the lion's share of future employment (Browne 1983; Riche, Hecker, and Burgan 1983).

At the same time, high-tech industries are the most probable source of innovations, of successful entrepreneurs, of new firms, and of new industries. In this indirect and more long-term route, high tech is an important employment generator. Industries newly created since the Second World War, such as electronics, computers, and biotechnology, now employ thousands of people. It is the job-creation potential of high technology which attracts the interest of policy-makers. Based on the experience of such places as 'Silicon Valley' in northern California and 'Route 128' surrounding Boston, many regions have attempted to re-create the dynamism of technological and entrepreneurial 'fever' (Miller and Coté 1987; Rogers and Larsen 1984). It is not at all clear that high tech is a dependable economic base, or one which can be 'created' in regions which lack agglomeration economies and other dynamics common to high tech centres (Malecki 1987a; Scott and Storper 1987). High tech is returned to in Chapter 5.

Diversification of the economic base

The question of the optimal basic sector for a regional economy ignores the fact that any specialization is vulnerable to unpredictable and perhaps abrupt changes in demand for a region's exports. This problem is especially acute for Third World nations which rely heavily if not exclusively on a primary commodity for which demand and market price might shift suddenly and dramatically. As a poignant example, a single corporation, the Coca Cola Company, reformulated its principal soft drink in 1985, bringing about a large drop in demand for Madagascar's vanilla bean crop. Madagascar produced at the time about 80 per cent of the world's vanilla beans, and 30 per cent of the world's annual crop evidently was purchased by Coca Cola for its soft drink. A subsequent reformulation of 'Classic Coke' brought about a partial revival of demand for vanilla, an ingredient that apparently is used in that product but not in 'New Coke' (Mufson 1985).

The solution to such vulnerability is to diversify the regional economy with additional industries (Conroy 1975; Lande 1982). Thompson (1965: 133–72) argues strongly that city (or region) size, on average, decreases cyclical instability, in large part because of greater industrial diversity, and Marchand (1986) provides some evidence for this argument. Conroy (1975) dismisses this as overly simplistic; cities and regions of the same size vary. Other simple relationships, such as the age of the region's capital stock, also do not uniformly account for regional cycles (Howland 1984b). More recently, it has become clear that industrial diversity alone is not the answer, but the actual mix of industries: some industries are better (more stable) than others (Browne 1978; Grossberg 1982; Howland 1984a).

A hypothetical region with three industries is shown in Fig. 2.4. Sector A is in a long-term decline; sector B is growing slowly, with large fluctuations; sector C is growing steadily. Together A and B comprise a steadily weakening economic base. When industry C is added, the region's economic prospects are much brighter. In short, an economy which contains a set of growing industries and firms (such as sector C in Fig. 2.4) is preferable to a group of stagnant or declining ones (e.g. sector B). The occupational structure of a region, based on its place within corporate hierarchies and spatial divisions of labour – whether it specializes in research, decision-making, or production – is also significant from a long-term perspective (Pedersen 1978; W Thompson 1987).

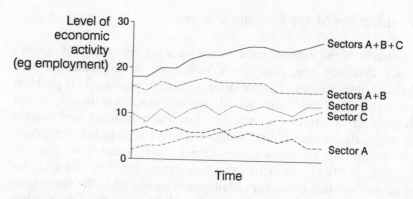

Figure 2.4 Schematic representation of industrial diversification

Industrial diversification of a regional economy has a long history as a policy recommendation, largely because it is a simple solution to an easily identified problem. The result of diversification, an increase in the number of sectors present in the region, is that any one sector has a smaller effect on overall change (growth or decline) in the region. However, it is also wishful thinking for many areas, most particularly poor nations which depend on a single primary sector for their export income. To diversify away from a mineral or agricultural commodity is difficult, if not impossible, for nations which do not have capital to invest or the ability to create a competitive industry in another sector.

Sectoral change and economic development

Part of the process of economic development entails the transition from one economic activity to another, as new industries arise and old ones fail to provide an adequate economic base for a region's population. The conventional way of presenting this change at an aggregate level is to consolidate economic activities into a small number of more general sectors.

The primary sector, consisting of extractive activities such as agriculture, mining, forestry, and fishing, comprises the traditional activities on which pre-industrial people relied, but which are now performed by a wide range of techniques, from highly mechanized capital-intensive methods to traditional labour-intensive systems. Secondary activities are manufacturing or processing sectors which, again, include both traditional and modern practices. Tertiary, or service, sectors have commonly been the residual category after primary and secondary activities are delineated.

More recently, services have been separated into those which involve a substantial information content (quaternary services) and those which do not (tertiary services). It is this portion of economic activity which is projected to grow most rapidly in the future, as primary and secondary sectors continue to decline in importance in relative employment (Fig. 2.5). For example, the knowledge-intensive or information-based sector

Labour force (%)

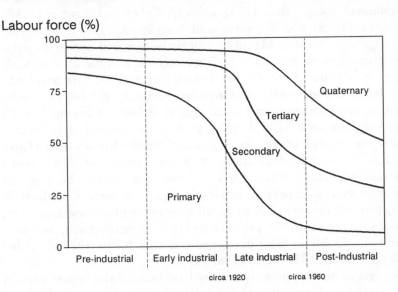

Figure 2.5 Transition of economic activity among broad sectors. *Source*: Abler (1975: 49 (Fig. 3.11))

was defined by Machlup (1962) as consisting of education, research, publishing, and broadcasting, who estimated it at 31 per cent of the US workforce in 1958. The 'information society', developed in the mid-1970s, is slightly different (Porat 1977), where information workers are defined as those engaged in creating or processing information, in three categories: (1) workers whose final product is information; (2) workers whose main activity is informational in nature, including information creators ('knowledge workers'), information transmitters, and information processors; (3) workers who operate information technologies (Katz 1988: 5). In addition, the diverse nature of the service sector has led some to propose a fifth 'quinary' sector which represents control functions and not routine information-processing activities (Harper 1982).

The shift among advanced economies from primary economic activities to manufacturing and, more recently, to services became part of the received and largely unquestioned wisdom since the 1930s, when Fisher (1933) noted the shift of employment towards secondary and tertiary production. Clark (1957) established the 'Clark–Fisher hypothesis' as a model of economic development that became wedded to policy prescriptions such as Rostow's (Ch. 1).

The sectoral shift remains embedded in many analyses that herald a 'post-industrial society' (Bell 1973), although Cohen and Zysman (1987), Singelmann (1978) and Browning and Singelmann (1978) are among those who urge caution. Manufacturing industries still appear to drive the economic growth process at both the regional and the national scale (Cohen and Zysman 1987; Gertler 1986a; Harris 1987; Park and Chan 1989). In the regional context, in addition, sectoral shifts fail to identify differences in occupations and in the spatial division of labour. The US South, long the most agricultural region in the country, has become virtually the average US region as primary activity has declined (Table 2.3). But it remains dominated by branch plants of firms based elsewhere (Malecki 1985a). Services have undoubtedly become the dominant sector of economic activity in many economies, at least in employment. If broadly defined as all non-goods-producing industries, they account for nearly 75 per cent of US total employment, and over 50 per cent of GNP in most developed countries (Illeris 1989: 1; Ochel and Wegner 1987: 105; Waite 1988: 1).

Throughout history, primary activity has absorbed the largest fraction of the labour force in primitive economies. Exports of staple commodities and, later, of primary manufactured products, are typical of frontier regions, whose markets are in other, often distant, regions (Hayter 1986). Simple processing, such as sawing logs into milled lumber, can evolve into further production, of plywood and veneers, and finally wood and paper products such as furniture and newsprint. Some extractive exports are more likely than others to promote desirable linkages and processing; coal mining rarely fosters local linkages. 'The export sector must grow and maintain flexibility, shifting commodities, markets and production techniques. . . . Lacking flexibility, regions may be forced into the staple trap, eking out a marginal existence by means of uncertain returns from the same old staples' (Gilmour 1975: 62). Regional circumstances are also related to the degree of internationalization and export performance of the region's firms (Hayter 1986).

Table 2.3 Percentage employment by industrial sector and region, USA, 1920–70

Industry sector and region	1920	1930	1940	1950	1960	1970
Primary						
USA	28.9	23.9	21.3	14.4	8.1	4.5
Northeast	10.1	8.0	6.9	5.0	2.7	1.7
North Central	27.6	22.3	21.0	15.0	9.1	5.3
South	49.4	42.1	35.4	23.2	11.8	5.9
West	28.7	23.2	19.4	12.6	7.8	5.1
Secondary						
USA	30.8	28.9	29.8	33.9	35.9	33.1
Northeast	43.6	37.9	39.7	41.8	42.4	35.9
North Central	31.5	32.6	31.0	36.2	38.8	36.1
South	18.7	20.0	21.6	26.9	31.0	32.0
West	25.2	23.9	23.3	27.4	31.2	26.4
Tertiary						
USA	40.3	47.1	49.0	51.6	56.1	62.4
Northeast	51.0	54.2	53.4	53.2	54.8	62.3
North Central	41.0	47.3	48.0	48.8	52.2	58.6
South	31.8	37.9	43.5	49.9	55.1	62.2
West	46.0	53.0	57.3	60.0	61.0	68.5
Producer services						
USA			4.6	4.8	6.6	8.2
Northeast			6.5	6.6	8.2	11.0
North Central			4.3	4.5	5.8	8.0
South			2.7	3.6	5.5	8.1
West			5.5	5.9	7.8	10.5

Source: Adapted from Singelmann (1981: 191 (table 7.8)).

The diversification embodied in the sectoral stages model suggests that some industries (the more advanced or newer ones) are 'better' for a region than are others, because demand for them is more likely to grow or to be stable. Certainly when examined historically, the shift of employment from agriculture to industry and, more recently, from manufacturing to services, appears to be part of the structural change of economic development. In advanced economies quaternary activities, related to information and knowledge-intensive activities, characterize the 'advanced industrial metropolis' (Gottmann 1961; Knight 1986).

If the Clark–Fisher model of stages is correct, then the historical progression seen in Europe and North America is the pattern to be expected elsewhere. However, in developing countries, this has not been occurring in the expected manner (Casetti and Pandit 1987; Pandit and Casetti 1989). Instead, the shift has been directly from agriculture to services, and manufacturing has remained at a relatively low level.

The service sector: engine of growth?

Criticisms of the basic/service division on which economic base analysis relies go back many years (Hoyt 1954; Massey 1973; Tiebout 1962; Watkins 1980). Most notable among the critics was Blumenfeld (1955), who pointed out that the economic base of metropolitan areas is very different from that of small towns, a distinction not captured by the model. In this view, the economic base of a region is maintained by three sets of factors:

1. A labour force of various skills, which in turn is attracted by local consumer services such as housing, schools, stores, and local transportation;
2. Business services, including transportation; and
3. Markets, local as well as regional.

These constitute the way in which an area is able to replace basic industries if they decline or fade away. In Blumenfeld's (1955: 131) words: 'It is thus the "secondary", "nonbasic" industries, both business and personal services, as well as "ancillary" manufacturing, which constitute the real and lasting strength of the metropolitan economy.' These enable the region to sustain, expand, and replace its 'primary' industries. This line of argument has been carried forward by Tiebout (1956), Thompson (1965), Hirschhorn (1979), Power (1988), and

Stanback *et al.* (1981), among others. Lewis (1972: 24) makes the point strongly: 'the increase in exports, and its effect on employment in the basic sectors, is a result rather than a cause of economic growth'.

Unlike manufactured goods, however, services are more difficult to define. Examples of services are easy to come by: a shoeshine, a heart transplant, a wedding (or divorce), a computer program, a tax consultation or audit, a speeding ticket, a memo to the files, a tennis lesson, a burial, a car wash, a crop dusting (Cohen and Zysman 1987: 51). But what do these have in common? Gershuny and Miles (1983: 3) identify the service industry as 'all those firms and employers whose major final output is some intangible or ephemeral commodity'. Cohen and Zysman (1987: 51–2) point out that not all intangibles are really services, because many have their end result in tangible goods, such as architectural services in buildings, product design in products, and shipping in stocks of goods. Non-storageability is another trait frequently attributed to services, but on-line data banks and computer tapes and discs belie this distinction. In many cases, products and services are interchangeable. A home washing machine (a product) can substitute for the use of a launderette (a service), and a frozen dinner can replace a restaurant meal (Quinn, Baruch, and Paquette 1987).

Such definitions, however, do not take us very far, and most accounts of the service sector lament the problems of delineation (Gershuny and Miles 1983; Marshall *et al.* 1988; Urry 1987), and each provides a different classification scheme. Services play a central role by facilitating and stimulating economic growth in all sectors. This *infrastructural role* of services, particularly for communication and information, is perhaps their most important function (Cohen and Zysman 1987; Riddle 1986). In retailing, for example, infrastructural services (transport and communications) link the producer with the market (Nusbaumer 1987a). However, such services are the 'borderline cases' between secondary and tertiary activity, along with public utilities, construction, and information transmission services (Illeris 1989: 16–18).

In order to simplify the variety of services, they are frequently divided into three general types (Marshall *et al.* 1988: 11–29):

1. Consumer services, including retail and household services;
2. Producer services, which also fall into three categories: (a) information processing, including R&D, consulting, engineering, legal services, and purchasing; (b) goods-related services, such as transport and waste disposal; and (c) personnel support, including welfare, cleaning, security, travel and accommodation;
3. Government services, including education, welfare, and defence.

It is perhaps more common to exclude government services or the public sector from the service sector, in order to separate public and private services (Waite 1988: 2).

Bailly *et al.* (1987) have proposed a new typology of sectors that incorporates the growing interdependence between services and production (Table 2.4). Services comprise the circulation, distribution, and regulation activities in this typology. Reiner and Wolpert (1981) and Wolch and Geiger (1986) note that a 'not-for-profit' sector (4.2 in Table 2.4) complements the services provided by government in the USA and accounts for substantial employment and multiplier effects. 'Mixed' activities, such as transport, telecommunications, delivery and postal services, banking, and insurance obviously serve both consumer and producer markets, thus raising definitional problems. Banking and insurance, for example, are typical 'producer' services, but their services (loans and coverages) and employees are typically more oriented towards final consumers (Allen 1988). Some services are easily classified, such as advertising, marketing, and R&D as producer services and recreation, education, health, welfare, and personal services as consumer services. It is financial and commercial activities which are the hardest to categorize, for they are *circulation* services, not intermediate or final demand services, and they depend on velocity of turnover, not an identifiable demand.

The level of service activity is a response to a set of demand and supply factors (Daniels 1982). The level and distribution of income, along with the relative propensity to consume services, affect the consumption of consumer or distribution services (Kellerman 1985; Stanback *et al.* 1981). Retailing establishments and jobs increased to help provide the desired goods and services, such as automobiles, refrigerators, washing machines, and meals away from home. The size of the government sector affects the degree to which employment is found in public administration and related regulation activities. Population growth, rising incomes, and increased participation of women in the labour force have increased demands for certain services, such as medical care, personal banking and financial services, and day care. Changes in the composition of the population also account for substantial portions of increased service sector growth, such as demand for nursing home and related health industry outputs which has grown with the ageing of the population.

Supply factors, such as mass production and new technology, have greatly increased the types and variety of services possible (Kellerman 1985; Waite 1988). This is especially evident in producer (circulation) services, where new financial services and global transactions have proliferated in recent years, but it is also a major factor in consumer services (Miles 1988). Airport growth has been another technologically induced circulation service on which employment growth has been based (Parsons 1984).

Table 2.4 Classification of economic activities

---

1. Manufacturing
   1.1   Use of natural resources
             agriculture; horticulture; silviculture; fishing;
             electricity, water, gas; mining
   1.2/1.3   Processing of natural resources and manufactured goods
   1.4   Construction and civil engineering

2. Circulation
   2.1   Physical flows and flows of persons
             wholesale trade; brokerage; transportation
   2.2   Information and communication flows
             information transmission (including post office);
             information processing
   2.3   Financial flows
             banking; insurance; financial companies

3. Distribution
   3.1   Health
   3.2   Education
   3.3   Retail trade
   3.4   Hotel and restaurant trade
   3.5   Repair of vehicles and consumer items
   3.6   Personal services
   3.7   Culture, sports, leisure
   3.8   Domestic services, other

4. Regulation
   4.1   Public: administration
   4.2   Private: organizations
             social welfare organizations; religious, social, cultural
             organizations; community services, common interest groups;
             private road works and sanitation
   4.3   International: diplomatic organizations

Sources: Bailly *et al.* (1987); Bailly, Maillat, and Coffey (1987).

At the same time, occupational structures have become more constrained partly as a result of information technology (Rajan 1987). Entry-level skills demands are higher, and this reduces the employment possibilities of disadvantaged groups (Cyert and Mowery 1987; Stanback 1987). In addition, job mobility to higher levels is less available than previously, or than was prevalent in manufacturing sectors. Management positions are more likely to be filled from the outside, by those with university credentials, than by those who have started within the organization at lower ranks (Rajan 1987: 213–28). Overall, job growth in services, while relatively high, may not be sufficient to replace jobs lost elsewhere (Gershuny and Miles 1983).

Job growth in services has been concentrated in relatively few sectors of the economy. In the USA, the greatest growth in employment from 1959 to 1985 was in wholesale and retail trade (12 million new jobs) and services (15 million new jobs), a category that includes repair, business services, hotels, and private medical care. These two major sectors grew from 50 per cent of total job growth in 1959–69 to 60 per cent of 1969–79 job growth to almost 99 per cent of 1979–85 job growth (Kutscher 1988: 26). When disaggregated further, it is seen that just four industries – eating and drinking places, other retail trade, business services, and medical services (encompassing doctor and dental services, hospitals, and other medical services) – have accounted for 43 per cent of the 1959–85 growth and 65 per cent of the more recent (1979–85) job growth (Kutscher 1988: 28).

In occupations, the largest increase has been among professional, technical, and kindred workers, and managers and administrators, and clerical workers. Rajan (1987: 6–8) notes that the main beneficiaries of new service jobs are likely to be female, part-time, and young employees, especially in clerical and sales occupations, and Townsend (1986) has shown this to be the case in the UK. Even in clerical jobs, however, the skill requirements imposed on new labour force entrants are much higher than in the past, frequently requiring some computer and other technological expertise (Ochel and Wegner 1987; Stanback 1987).

A final consideration on the growth of the service sector concerns the trend for government statisticians to include services in the value of the goods if they are performed internally, but to report them separately as services if they are purchased from outside the firm. Thus the nature of the contractual form used in procurement of intermediate inputs is crucial to whether the value of these inputs is included in services or goods-producing industries (Waite 1988: 2). This 'make or buy' decision has increasingly moved towards subcontracting or 'out-sourcing' as firms obtain data processing, management advice, transport, and even janitorial services from outside providers.

Producer services

Producer services, or business services, have been especially prominent in the recent rise of the service sector (Daniels 1987; Marshall *et al.* 1988). These range from routine services, such as window-cleaning and janitorial services to management consulting and R&D (sectors 2.2 and 2.3 in Table 2.4). In a 'make or buy' decision, the firm must decide whether the service can be better or more economically provided by the firm internally with its own employees, or procured from outside firms which may specialize in a particular type of service. The decision to have such services integrated within the firm or disintegrated is usually determined largely by the magnitude of the transaction cost involved in the arrangements with the providing firms (Scott 1988c). In effect, high transaction costs provide an incentive to internalize a service.

Internal or external 'sourcing' of services is but one of several *linkages* with other firms that influence the size of the regional multiplier. For example, Coffey and Polèse (1987a) surveyed a large number of Montreal firms to see which services are most often provided within the firm, purchased from local firms, or provided by distant firms or establishments. Several services are typically purchased rather than provided internally: consulting, personnel services, computer services, advertising, architecture, engineering and scientific services, and royalties/patents. If there is no local service infrastructure providing these services, then they are obtained from the firm's headquarters, even if located far away.

Business services and other producer services respond to different locational pressures and these reflect the variety found within the service sector generally. Coffey and Polèse (1987a) suggest that there are three types of orientation among business services: towards markets, towards human resources, and towards advanced corporate services. Firms attempt to minimize their overall costs of communication or transportation to these resources. Producer services required by firms in a region also reflect the level within the corporate hierarchy. Establishments performing corporate-level functions or activities exist in few locations – one in the case of the firm's headquarters – but a wide array of services are used. At the plant level, there are far fewer services demanded or utilized, but there are many more plants where these services are incorporated into the firm's activities.

In certain respects, producer services are the most important of services. They are frequently exported outside the region, thus constituting an economic base similar to the traditional role of manufacturing (Beyers and Alvine 1985). They are prominent in the dominant regional economies, and are concentrated in the seats of economic power (Borchert 1978). Together with corporate headquarters, non-profit institutions, and the public sector, non-consumer services account for roughly two-thirds of the labour force in cities such as New York (Noyelle and Stanback 1983). In fact, it is this counterbalancing of employment gain in producer services during a period of deindustrialization that has led to conclusions that they are the principal economic base of contemporary economies (Daniels 1983; Hirschhorn 1979; Illeris 1989; Stanback *et al.* 1981).

Producer services and management functions provided by firms internally account for the bulk of the *non-production* workers in manufacturing firms (Goddard 1978; Gudgin, Crum, and Bailey 1979; Marshall *et al.* 1988). Non-production activities include office occupations, such as clerical, information-processing, and other support activities, as well as administration and management functions, such as finance and purchasing. Classification problems in government reporting schemes render it difficult in all cases to assign these activities to the service sector and they are often classified into the manufacturing sector of the firm's primary product(s) (Marshall *et al.* 1988).

Producer services, although they are growing extensively in advanced economies, are not an equalizing element in the space-economy. In both Europe and in North America, for example, producer services display a distinct bias towards concentration in a very small number of major cities (Daniels 1982; Hepworth, Green, and Gillespie 1987; Howells 1988; Illeris 1989; Kirn 1987; Marshall *et al.* 1988; Noyelle and Stanback 1983). Hepworth (1986), in the Canadian context, shows that this outcome is a product of the centralization of control possible with modern computer networks. In particular, it is evident that high-level control and decision-making activities are centralizing even while more routine data-processing activities are decentralizing (Howells 1988; Marshall *et al.* 1988). The same tendency is visible in the white-collar occupations of manufacturing firms (Goddard 1978). Martin (1988) attributes much of the explanation of the 'North–South divide' in the UK to regional inequality in producer services, since only the South-east has above-average proportions of employment in banking, finance, insurance, and business services.

This should not be surprising, however, since the economic advantages of location in large cities affect the behaviour of both firms and individuals. From the firm's perspective, agglomeration economies revolve around the array of services and facilities available in large centres that permit them to purchase services and material inputs from specialized suppliers (O'hUallachain 1989; Scott 1988c). For individuals, the large labour markets found in major agglomerations increase their probability of suitable employment (Pascal and McCall 1980). That large urban economies continue to attract decision-making functions and multiplier effects is less surprising when viewed in the long term (Noyelle and Stanback 1983; Pred 1977; Thompson 1965). Noyelle and Stanback (1983) see a hierarchy of cities that is less an effect of city size and more a function of the array of office activities and producer services found in various places. Indeed, Kirk (1987) and Hoppes (1982) have found in the USA that stable and prosperous regional economies are more likely to be based on producer services. At the international scale, the disparity in service activities is even more pronounced, and many countries are simply left out of the picture in organizational location decisions as 'world cities' (Ch. 1) attract the bulk of corporate investment (Friedmann 1986a).

Can services be an economic base?

Proponents of services as a viable, if not a superior, economic base, typically cite the decline of manufacturing employment during the 1970s and 1980s and the rapid rise of service employment, particularly in producer services (Gillis 1987; Hall 1987; Marshall *et al.* 1988). This argument incorporates both the notion of immutable stages of economic development discussed above, and the deindustrialization argument evident in the recent data. Service jobs also come and go, perhaps at different rates than manufacturing jobs, and for many of the same reasons that affect manufacturing (Buck 1988). These include technical change (investment in new processes which reduce labour input), intensification (reorganization of work processes to speed up work), and rationalization (reduction of overall capacity by closing the least profitable locations in an organization) (Massey and Meegan 1982).

Kirk (1987) has asked whether business services are immune to the business cycle. If they are, then the cyclical nature of economic activity in manufacturing would be avoided at the local level. However, Kirk found that only a few services were not affected by recessions in the USA from 1973 to 1982. Computer services, including data processing and computer programming, showed no relationship to three national downturns. Advertising, mailing services, and building services were immune from two of three national cycles during the period. On the other hand, personnel supply services and credit reporting went through up-and-down cycles similar to the national economy. Relatively few business services are highly export-oriented (Groshen 1987). These include R&D laboratories, schools and educational services, hotels and lodging places, and direct mail advertising services. In general, however, business services in Cincinnati, Cleveland, Columbus, and Pittsburgh were not very export-oriented and imports of legal, computing, and engineering services were common (Groshen 1987). The evolving hierarchy of corporate activity embodies a high degree of specialization, with only a small number of urban regions containing control functions and high-level services (Hepworth 1986; Howells 1988; Noyelle and Stanback 1983). At the same time, services are closely correlated with growth of consumer and business markets (Kirn 1987).

Stabler and Howe (1988) suggest that services are commonly exported from even very small urban and non-metropolitan areas of the Canadian prairies. They conclude that service exports appear to be sufficiently related to regional growth to cast doubt on the dependent nature of service activity ascribed to it in export base formulations. This is a common thread among analysts in developed economies. For developing countries, however, it is less likely that they will be able to retain the multiplier effects locally – a point less often considered in developed contexts. Ownership by multinational firms and operations staffed by foreigners reduce the multiplier effect that remains within the host country (McKee 1988). In Beyers and Alvine's (1985) work, nearly all exporting service firms were located within the metropolitan agglomeration of Seattle. In general, the concentration, even if decentralizing somewhat from urban core to suburban periphery, of producer services is continuing. This trend 'reinforces spatial imbalance rather than ending it' (Morris 1988a: 757), suggesting that services as an economic base are unlikely to be a widespread basis for regional growth.

Service industries still depend on manufacturing in some way, whether or not they are as subordinate as the economic base model implies. The asymmetrical dependence of services on manufacturing was clearly found in a study of 26 countries (Park and Chan 1989). There are linkages between manufacturing and services which include the internal co-ordination of production and non-production activities within firms. Inter-firm linkages involve innovation as well as routine activities such as transportation and distribution. For the most part, innovation originates in manufacturing (Robson, Townsend, and Pavitt 1988; Scherer 1982). Innovations resulting from R&D flow to firms in a number of service sectors. The most prominent example in recent years is the effect of the computer sector on virtually every service sector, from architecture to medicine, typing, and retailing.

However, the relationship between manufacturing and services is truly symbiotic in the case of producer services (Park and Chan 1989). Producer service firms also play an important role in innovation by providing information and expertise that may not be available, especially in small manufacturing firms. MacPherson's (1988a) research in Toronto found significantly higher rates of innovation among manufacturing firms which utilized the services of local producer service firms. However, services may be possible as an economic base only in large urban regions such as Toronto where head offices and control are concentrated. The prominence of services as an employer has perhaps caused an ill-advised de-emphasis on manufacturing as the core of a prosperous economy. Spatial hierarchies and division of labour in services suggest that the 'better' services, especially producer services, are very different from routine, more footloose services (Coffey and Polèse 1989). Further, innovation, whether in the form of new products and services or new production processes, is a producer service that takes advantage, and may stimulate the development, of new products and infrastructures developed in other sectors.

Conclusion

The economic base of a region and flows of capital are fundamental concepts in regional economic development. Since no two regions are identical, it is not easy to define an optimal economic structure or to transfer successful aspects from one region to another. Multiplier effects shift capital, in the form of purchases and profits, from one firm to another and from one region to another. These shifts favour regions where economic power is centred and where innovations originate. Certain economic sectors will be dominant at any given time, however, thus lending credence to a general rendering of, if not a precise adherence to, the economic base concept.

The intricate linkages within and between firms, sectors, regions, and nations make attempts either to capture them with theories or to manipulate them by means of policies difficult. The following chapter discusses various attempts in both theoretical and policy contexts to come to terms with regional growth dynamics. It becomes clear that efforts to understand the dynamics of regional development must rely on the ability of both corporate hierarchies and innovation to alter abruptly the smooth flow and equilibrium of traditional approaches.

Chapter 3

Regional growth theory and policy: the burden of conventional wisdom

The concept of a multiplier and of causal connections in economic development, discussed in Chapters 1 and 2, have sparked a great deal of theoretical and empirical research. In addition, a wide array of policies based on the expectation of multiplier effects has been put into place in local, regional, and national contexts. These range from simple subsidies for private investment to massive public infrastructure projects in particular places; far less effort has gone towards attempting to restructure regional economies. In virtually all instances, the gains in backward areas have also been small quantitatively; they have also been distinct structurally and qualitatively. The problems of backward areas have been around for centuries and have been observed statistically for at least four decades. They are typically attributed to industrial structure: too heavy a reliance on agriculture or mining, too narrow a base of manufacturing, or a concentration in declining, rather than growing, sectors (Brown and Burrows 1977). The inclusion since the 1960s of decaying urban regions in the list of regions needing assistance posed new problems, since many explanations of regional growth rely on urban size and agglomeration as positive elements.

The explanation of regional economic growth, while often tied to measurement issues and discussion of the economic base, discussed in Chapter 2, is more frequently bound up in the standard models of economics (Hewings 1977; Richardson 1973). Indeed, the analysis of regional growth did not become widespread until Borts and Stein (1964) placed regional issues in the accepted and well-trod neoclassical economic framework of competitive equilibrium. These models, especially the neoclassical model of production, utilize the simplification of a regional production function with two inputs, capital and labour. (For excellent discussions, see Brookfield 1975: 28–32; Gertler 1984.) Although this simplifies economic activity to an analytically manageable form, it also neglects dynamic elements which do not easily collapse into capital and labour. This criticism is heard most loudly from the adherents of Marxism, who themselves focus on the capital–labour dichotomy.

The dynamics of regional economic change involve social, institutional, and political elements – as well as economic factors – that come together distinctly in each place. Technology is one of the principal forces behind regional dynamics, but is dealt with least satisfactorily by either conventional or Marxist approaches. Technology is behind the capability of multinational or transnational corporations to co-ordinate global operations; it is central to the competitive environment which firms and, increasingly, regions and nations must confront. Technology, along with capital and labour, is utilized differently by each firm or enterprise, as Chapter 4 will show. At the national and regional levels, where the aggregate impacts of firms' actions are felt, technology is more difficult for economic theory to capture and account for.

Policies to alter the disorderly realities of regional development have been advanced primarily since the 1950s, when regional growth became an issue in most advanced countries. The persistent backwardness of some regions led to proposals to establish 'growth poles' which would utilize the growth dynamics of certain carefully selected sectors in regions where a catalyst was needed. Growth centre policy, one of the most elaborately developed and widely used concepts in regional policy, however, failed to counteract market forces which continued to favour the large, attractive regions. This was especially true in Third World countries, where growth centres were complementary to modernization efforts in newly independent nations. Notably, the networks of large corporations channelled multiplier effects along routes not anticipated by prevailing theory.

This chapter reviews the theories and policies which have addressed regional development. The striking commonality in principles and strategies among advanced and developing countries, and the failures of such policies to deter market forces, is a central message. In many respects, economic activity refuses to become anything but 'uneven'. This unevenness has been attributed to cumulative causation and agglomeration of capital and technical progress, entwined in a complex process of regional growth and development.

Production functions and equilibrium: the neoclassical model

The production function approach to the study and measurement of economic growth is a direct outgrowth of its origins in neoclassical equilibrium economics (Brookfield 1975: 28–32). In this framework, all production (Q) is produced by two inputs, or *factors of production*, capital (K) and labour (L):

$$Q = f(K, L) \qquad [3.1]$$

Typically, simple mathematical functions, particularly the Cobb-Douglas function, are utilized because of their ease of manipulation (Borts and Stein 1964). More complex formulations are reviewed by Andersson and Kuenne (1986).

In the Cobb–Douglas form,

$$Q = K^a L^b \qquad [3.2]$$

Returns to scale are often assumed to be constant, resulting in the condition that $a + b = 1$, meaning that an equal amount of production results from increasing amounts of either input at all levels. Returns to scale can also be assumed to be increasing (or decreasing), such that larger (or smaller) amounts of production result from an increase in an input. Other conditions enter the world of the aggregate production function: identical products and production functions in all firms and locations, and a homogeneous capital stock, with the same technology and productivity assumed in all places.

The two-factor model can be used to measure productivity of capital (output per unit of capital) and of labour (output per unit of labour) in an economy. By adding technology to the model, the rate of growth of output (Q) over time can be attributed to the productivity of capital (K) and of labour (L):

$$q = ak + bl + ct \qquad [3.3]$$

where q, k, l, and t are growth rates of output, capital, labour, and technology, c is the level of technical progress, and a and b are elasticities of substitution with respect to capital and labour, respectively. The sum of $a + b$ (<0, =0, or >0) indicates whether there are decreasing, constant, or increasing returns to scale. Hahn and Matthews (1964: 825–53) and Kennedy and Thirlwall (1972) review comprehensively neoclassical views of technical progress.

Neoclassical theory treats capital as a flexible and malleable means of production. In particular, capital is able to assume new forms and incorporate new technologies as well as to change location without restriction (Gertler 1984). Since capital and labour are substitutable, and productivity is measured as output per unit of labour, labour-saving investment is a favoured means of increasing output. The focus on capital also results in a concentration on capital investment as the best means of increasing output, a conclusion that has profoundly affected development policy. Productivity growth resulting from capital investment was commonly used to justify large-scale production projects in developing nations, and appeared to affirm the inexorable nature of labour-saving machinery as the road to economic growth.

In the neoclassical model, labour is treated as a homogeneous input as well, with no recognition of varying skills or capabilities, and differences in the growth of output 'are due ultimately to variations between regions in the growth of their labour force' (McCombie 1988a: 270). A result of labour-saving investment is a 'de-skilling' of work as tasks are simplified or relegated to machines, often then requiring labour with no particular skill (Braverman 1974; Hirschhorn 1984; Leontief 1983; Massey 1984, Shaiken 1984). The issue of labour skills will be returned to in Chapter 4.

The regional production function was adopted from the national scale to the regional scale with few if any modifications (Richardson 1973: 22–9). The model in its pure aggregative formulation yields neat, precise predictions at the regional as well as at the national level. For example, in its most restrictive form – assuming full employment, perfect competition, one homogeneous commodity produced in all regions, zero transport costs, regionally identical production functions exhibiting constant returns to scale, a fixed supply of labour, and no technical progress – it can be shown that the wage (marginal product of labour) is a direct function, and the return to capital (marginal product of capital) is an inverse function, of the capital/labour ratio (K/L).

The mobility of labour and capital

The neoclassical model incorporates a theory of factor mobility as well as a theory of growth. Given identical production functions in all regions, labour will flow from low-wage to high-wage regions, and capital will flow in the opposite direction (since low returns to capital imply high wages, and high returns are obtained in low-wage regions).

These flows continue until returns (wages) are equal in each region. Simultaneously, capital (investments) will go to where the return (profit) is highest, which will be where labour costs (wages) are lowest (since the price of all output is assumed constant wherever it is produced). The equilibrium is the same as that brought about by migration (Greenwood 1985; Richardson 1973: 89–103).

In the standard and simplest neoclassical model, equilibrium is possible between two (or more) regions through the mobility of capital and labour, the 'factors of production'. Capital, in the form of new investment, moves in the direction of the location where it can receive the highest return; likewise, labour moves from low-wage regions and towards higher-wage regions. This process ends in an equilibrium where all regions provide equal returns to both capital and labour. Factor mobility remains a key element in nearly all enhancements of the neoclassical model (Carlberg 1981; Clark, Gertler, and Whiteman 1986; Ghali, Akiyama, and Fujiwara 1981).

The neoclassical paradigm has strong assumptions, but also has clear and simple consequences: as capital is accumulated and the labour force grows in a region, output increases. In this setting, a region can grow only if labour or capital flow there from other regions (Anderson 1976). Richardson (1971: 95) summarizes the implications:

> The growth potential of a city depends on its ability to create and attract from outside the productive resources needed for growth as well as on its ability to produce the goods and services in demand in regional, national and international markets. The growth of the city, as of the region or nation, is determined by its rate of population growth, its rate of capital investment and its rate of technological progress, widely interpreted. . . . To grow fast, a city must obtain productive factors from outside, and act as a magnet for migrants, outside capital, non-local managerial and skilled technical personnel, and innovation.

According to the neoclassical conventional wisdom, the process of regional growth will be associated ultimately with a *convergence* in regional per capita incomes. In the neoclassical model, in effect, 'growth is essentially a reallocative process' (Borts and Stein 1964: 106; McCombie 1988a). The evidence through the 1970s seemed to confirm such ideas in North America (Hewings 1977: 148–51; Newman 1984), the UK (Diamond and Spence 1983), and Spain (Cuadrado Roura 1982). The 1980s brought a return of widening differences in regional incomes in Japan (Abe and Alden 1988), the UK (Martin 1988), and the USA (Hansen 1988) (Table 3.1). Thus, the convergence theory has seemed weaker than it did during the 1970s. In addition, Tam and Persky (1982) have found increasing inequality *within* regions of the USA at the same time that interregional convergence has taken place. In a comprehensive examination, Williamson (1980) shows that long-term convergence in the USA is in doubt, despite apparent convergence between 1929 and 1950, in large part because of substantial regional differences in technological progress, a view which may well represent the global state of affairs as well (Krugman 1979).

Table 3.1 Per capita incomes in regions of the USA as a percentage of the US average, 1929–86

Region	1929	1940	1950	1959	1969	1973	1979	1986
Mideast	138	133	116	114	113	110	106	113
Far West	129	132	121	119	115	110	114	112
New England	125	128	108	109	110	106	104	117
Great Lakes	114	112	111	107	105	104	104	99
Rocky Mountain	85	89	99	95	89	95	96	90
Plains	82	81	97	92	93	101	99	96
Southwest	68	71	87	92	93	101	99	96
Southeast	53	58	69	74	80	84	85	87

Source: Hansen (1988: 110 (table 1)).

The conflict between interregional equity and national efficiency has been a thorny issue in regional theory and policy for some time. In general, the most rapid rates of growth occur when economic activity has concentrated in large urban regions, maximizing national efficiency, a process which leads to inequality among regions. An inverted U-shaped curve, which describes this increasing divergence followed by convergence, is an expected pattern during the course of economic development (Williamson 1965). During periods of rapid national growth, the gap between rich and poor areas can become quite large, as Lakshmanan and Hua (1987) show occurred in China from 1979 to 1984. Mao's China had done little to narrow the urban–rural gap and inequalities. A 'new Great Wall' between town and country – the urban and rural sectors – has kept rural and urban sectors very separate in the push for growth (Aguinier 1988; Zweig 1987).

Although there is serious reason to question the empirical inverted-U pattern as a general model or as a necessary evil (Gilbert and Goodman 1976; Hollier 1988; Therkildsen 1981), it continues to be part of the conventional wisdom in development studies (Alonso 1980b; Mera 1974; Renaud 1973). The efficiency–equity trade-off may not be a policy conflict over the long term (Alonso 1968; Richardson 1979). Concentration, especially in urban areas in lagging regions, may provide a suitable balance (Todd 1980b).

Borts and Stein (1964) sparked a flurry of equilibrium-oriented regional economic research, especially in the USA, which attempted to account for migration (interregional labour mobility) in terms of response to differentials in interregional rates of return to labour (wage levels) (Gober-Myers 1978; Greenwood 1975). Recent research has shifted from this approach primarily by adding amenities as a further attraction (in addition to higher wages). Migration to nice (high amenity) areas, however, can increase labour supply and decrease wages, leading to employment increases (Knapp and Graves 1989).

The interplay between migration and regional growth is complex and includes a variety of issues, such as selectivity of migrants, whereby the most informed and most productive have the highest propensity to move. As discussed above, migration has traditionally been considered an equilibrating process, but substantial evidence to the contrary has weakened appreciably its ties to the neoclassical paradigm (Clark and Ballard 1981; Knapp and Graves 1989). Migration is a response to much more than simply income levels; the possibility of an income gain is a necessary but by no means sufficient condition for migration. Mobility costs, spatial frictions, non-economic resistances to migration, occupation, job opportunities, and family ties all enter into a migration decision (Gleave and Palmer 1979; Greenwood 1973, 1985; Isserman *et al.* 1986). In addition to migration, other factors, such as fertility, are part of the picture of population and labour force change in a region (Alonso 1980a; Ettlinger 1988).

Persistent interregional wage differentials, which ought to have been eliminated by migration, have also nagged at such equilibrium views, and considerable effort has gone into attempting to account for them (Isserman *et al.* 1986: 569–74). Historically, differences in technology and its correlates (industry mix, human capital, and occupational differences) were thought to account for wage differences (Batra and Scully 1972). The allure of equilibrium in the neoclassical model is so strong that valiant efforts have gone into determinations of 'real' wages, adjusted for cost of living and characteristics of the labour force in each region (Stamas 1981). When such variables are taken into account, a negligible differential is found (Gerking and Weirick 1983).

Some facts remain poorly accounted for by conventional methods. Persky (1978) found that capital–labour ratios were quite high in some sectors, not in the high-wage US North where capital is expected to be substituted for relatively costly labour, but in the US South, where wages are persistently low. Hansen (1980) suggests that capital-intensive branch plants may be able to employ low-wage workers because of their standardized production processes. When different occupational groups are examined, not all groups are experiencing convergence of wages across regions. Although the pay of professional workers tends to be nearly equal regardless of location, wages of blue-collar workers are least similar across regions (Goldfarb and Yezer 1976, 1987).

The study of capital movement, in contrast to the study of labour mobility, is more perplexing. Since capital flows are much more difficult to measure (Persky and Klein 1975; Richardson 1973: 103–13), capital investment flows are commonly measured by manufacturing location decisions which represent only a subset of capital investment decisions. By virtually any measure used, however, capital is not available equally across space (Gertler 1984). The locational decisions of both people and of firms rely on information which is often far from perfect (thus violating one of the key assumptions of the neoclassical theory). Workers are not equally aware of wages in all places, and may migrate for non-economic reasons. Regional differentials in the return to capital is a poor explanation of the interregional mobility of capital. Investors have incomplete information about such differentials, and consumer tastes and preferences may also vary from region to region, affecting profitability (Courchene and Melvin 1988). In addition, capital is not perfectly mobile, for two other reasons. First, investments are often 'lumpy', whereby they must be made in given units that may be relatively large (e.g. a paper mill or a car assembly plant of minimum efficient size). Second, 'industrial inertia' suggests that once firms have made an investment at a location, the advantages of remaining there are greater than from moving elsewhere.

The theoretical objections to the neoclassical model are quite serious. It neglects some essential characteristics of regional economies. 'From other branches of economic analysis, we know that certain phenomena are of some importance: agglomeration economies in location and urbanization, transport costs, interdependence of location decisions, metropolitan–regional relationships. But these have no role in the neoclassical system' (Richardson 1973: 27). In addition, regional production structures are assumed to be identical, despite substantial differences across regions in products, capital and labour inputs, and efficiency (Beeson and Hustad 1989; Clark, Gertler, and Whiteman 1986; Gertler 1984; Lande 1978; Luger and Evans 1988). And, significantly, technical progress is accorded surprisingly little attention given recognition of its importance.

In sum, the sameness across regions assumed in the neoclassical model presumes a very simplistic economy where multilocational firms with an internal division of functions and labour – where a head office and a branch plant, with their different functions, are in different locations – do not exist. Spatial segmentation of product and of labour and capital markets is omitted from the neoclassical world (Clark, Gertler, and Whiteman 1986; Richardson 1973). Moreover, most research on migration ignores the role of large organizations – firms and governments – in generating flows of migrants (McKay and Whitelaw 1977).

Technology, productivity, and economic growth

Technical progress or technological change initially posed little trouble for economists who admired the explanations derived from the production function model. Within any given region, capital can be induced to stay by increasing its return or profitability at the expense of labour. That is, capital investment which replaces labour allows profits to rise by reducing labour costs. This new capital–labour combination provides an incentive to invest in new technology; technology other than the labour-saving kind is not a part of the standard model.

A great deal of research at the national level has attempted to understand the contributions of factors to variations in the rate of economic growth (Denison 1967, 1985; Jorgenson 1988; OECD 1980b; Odagiri 1985; Sato 1978). Productivity growth is considered to be the principal means of economic growth, by which a region or a nation can increase the level of income and well-being of its population (Baumol 1989; Kendrick 1977). Overall, studies have found that capital investment growth accounts for less than half of the measured increase in output. The balance is attributable mainly to improvements in the quality of labour, from education, health, training, and experience. Indeed, the related concept of 'human capital' to describe investments in education and other improvements in labour skills persists (Becker 1964; McCrackin 1984; Nussbaum *et al.* 1988). In part because of differences in education and skills, national economies vary greatly in productivity growth (Table 3.2). Among the industrial nations, productivity growth has been highest since 1960 in Japan, lowest in the USA. Within Europe, France and Italy have had higher growth rates than West Germany or the UK. These differences also result from industry mix, since national productivity growth is a composite of the changes taking place across many sectors (McUsic 1987). But the principal cause of slow productivity growth is the failure of firms in some countries – notably the USA since 1960 – to take advantage of new technological ideas and to incorporate them into production (Baily and Chakrabarti 1988).

Table 3.2 Growth in manufacturing productivity in selected countries, 1960–84 (percentage change in output per hour, annual rate)

	USA	Canada	Japan	France	W.Germany	Italy	UK
1960–84	2.7	3.4	8.8	5.6	4.8	5.6	3.5
1960–73	3.2	4.7	10.5	6.5	5.9	7.3	4.3
1973–80	1.2	1.6	7.0	4.6	3.8	3.7	1.0
1980–82	4.0	2.4	6.8	4.7	3.1	3.5	5.3
1982–84	5.8	5.2	7.3	4.6	4.7	4.4	5.3

Source: McUsic (1987: 10 (Table 9)).

From these aggregate studies, however, a large residual continues to emerge beyond that which can be explained by increases in the quantity and quality of capital and labour. This residual, labelled 'technical progress' by Solow (1957), was disaggregated by Denison (1967) into three components:

1. Resource shifts (from agriculture to industry) and economies of scale;
2. Advances in technical, managerial, and organizational knowledge, and
3. A further 'residual productivity'.

Scott (1981) considers investment – both capital investment and investment in human capital – to account for nearly all of the 'technical progress' found in earlier studies. This may help in measurement terms, but it fails to deal with the many ways in which productivity can improve without measurable investment, such as learning by doing and learning by using. Nelson (1981) sees technological advance as quite separate from capital growth and rising educational attainments, and all three sources as complementary 'ingredients' of productivity growth. Soviet research reaches a similar conclusion to that in the West: scientific and technical progress is the chief factor underlying growth in labour productivity in socialist economies (Kapustin 1980).

Elaboration of the neoclassical model permits sources of productivity growth to be distinguished. McCombie (1988a: 270), for example, includes three main types of productivity growth in addition to exogenous technical progress and capital deepening:

1. Growth of productivity due to the improved interregional allocation of resources;
2. That due to the improved allocation of resources between industries within a given region (i.e. the improved intraregional allocation of resources); and
3. The rate of technical progress in addition to ct (in Eq. 3.3) due to the diffusion of innovations caused by differences in the level of technology to which the regions have access.

It is far more common for neoclassical analyses simply to assume that new techniques are freely available to all and largely to disregard the actual mechanisms of technical change.

One realistic enhancement of the neoclassical model is the 'vintage' approach, which acknowledges that not all machinery is alike. Technology is 'embodied' in capital equipment, and as progress is made, newer machines (newer *vintages*) make use of the most current knowledge and techniques of production, but these remain fixed at the level of productivity and labour intensity of their vintage (Salter 1966). The vintage concept appeared to allow for a portion of productivity growth that could not be accounted for by capital and labour inputs alone (Thirlwall 1972).

Vintage models measure cumulative capital investment, but capital which is fixed or invested in a particular year remains distinct from capital fixed in earlier and later years, for the duration of its productive life. The stock of productive capital, or the means of production, at any point in time, is the sum of invested capital of various different vintages. Associated with each vintage are separate productivities and labour intensities and, thus, the aggregate productivity of total invested capital and the implied demand for labour are determined (Gertler 1988a). The principal simplification imposed by vintage models – despite their apparent complexity – is that the diversity across firms, establishments, and regions, of production technologies is simply reduced to spatial variations in the mix of different vintages of capital. For example, Varaiya and Wiseman (1981) show that the regions of the USA vary significantly in the age of their manufacturing capacity (Table 3.3). Further, the technology of older machinery is much more labour-intensive, suggesting that plants with older capital stock will experience large employment losses as the least competitive facilities are closed (Varaiya and Wiseman 1981: 441). This indeed has been the finding of research on plant closings (Bluestone and Harrison 1982; Howland 1988a; Massey and Meegan 1982).

Table 3.3 Median age in years of manufacturing machinery, by region of the USA, 1966 and 1976

Region	1966	1976
Northeast	6.23	7.24
North Central	5.95	7.18
South	5.67	6.41
West	5.77	6.65
All regions	5.87	6.82

Source: Varaiya and Wiseman (1981: 440 (Table 1)).

Aggregate data can be deceptive, as Howland's (1984b) study illustrates. She found that states in the USA with newer capital experienced more severe recessions, in contrast to the logic of the vintage approach. Howland accounted for this finding by citing the greater proportion in such states of new, small firms, which are more likely to close during an economic downturn. If only large firms are considered, states with older capital stock were found to have had harsher recessions. Similarly, Carlsson (1981) has shown that the aggregate productivity data conceal a great deal of variation among individual plants.

In studies where the production function allows for technological change to be included, in the form of a residual as in Eq. 3.3, this form of technical progress changes only over time, not across regions, so it simply accounts for higher productivity as time goes on, and does not account at all for productivity differences among regions. A great deal of evidence, in addition to casual observation, verifies that regional technologies differ in ways beyond varying regional rates of adoption of new vintages of technology (Blackley 1986; McCombie 1988a). In particular, the vintage approach permits new technology to affect only the newest machinery in which it is embodied, and to have no effect on the efficiency of machines previously installed (McCombie 1988a: 279).

Gertler (1988a: 155) believes that the vintage approach, while an improvement over the neoclassical model, is far too simplifying: 'Clearly, this top-down reductionist view of technological change in space does not do justice to the actual diversity of local production and technology histories, which are themselves shaped by local firms and their interactions with labour in employment relations that are locally defined.' Gertler also points out that technologies are introduced and developed in many places, and even as they diffuse – unevenly – they are modified by each firm as they are adopted. The interplay between workers and firms in any location determines the shape, quantity, and the overall locational pattern of new investment over time (Gertler 1988a). The ultimate mix of technologies and the speed of the adoption of successful new technologies, then, are indeterminate (Storper 1988). There also are rigidities which prevent the full impact of new vintages of machinery from taking effect for some time (Johansson and Strömquist 1981).

More recent views of productivity have focused on the dynamics of technical change at the level of the firm (Carlsson 1981; Dosi 1988; Gold 1979; Griliches 1984; Nelson 1981). Thomas (1986) has discussed these ideas in a regional context. Rather than to begin with the premises of an equilibrium of identical firms, companies are considered to vary – as they do in reality – in organization, and in their knowledge about and access to technology (Nelson 1981). Research and development (R&D) efforts persistently appear as important indicators of general progressiveness at the firm level in both neoclassical formulations and in 'evolutionary' theories of economic change (Dosi 1988; Nelson and Winter 1982). Chapter 5 considers R&D in greater detail.

Productivity in services
The growth of services, discussed in Chapter 2, poses a particular problem in the context of productivity growth. Services are typically thought to be less productive and less amenable to technological change than manufacturing industries (Baumol 1967). This is a result of the fact that productivity is typically measured as output per unit of labour, a criterion that works against many services, such as education, medical care, and cultural activities, which are characteristically very labour-intensive. Recent research has stressed the diversity of services, from very capital-intensive ones, such as radio and television broadcasting, electric utilities, and pipeline transportation, to the most labour-intensive, retail and wholesale trade (Kutscher and Mark 1983).

The service sector has been greatly affected by the rapid diffusion of information-based technology and computers (Ch. 5). Computer information services are particularly important to the service sector because many service industries (e.g. banking, finance, health, transportation, utilities, communications, and business services) require substantial information handling (Quinn, Baruch, and Paquette 1987; Waite 1988). This trend is significant because it radically changes the skills necessary for such jobs (Cyert and Mowery 1987; Guile and Quinn 1988; Stanback 1987). The expected measurable rise in productivity from this capital investment, however, has not accompanied this increase in capital intensity. However, when services are divided into progressive and stagnant subsectors, progressive sectors – particularly knowledge production – 'contributes strongly to productivity growth' (Baumol, Blackman, and Wolff 1989: 159). Stagnant sectors, all conventional sectors of distribution of goods and services, seem to follow the usual problem of labour-intensive activity.

In addition, the intangible nature of services makes the measurement of productivity (output/worker) more difficult than in manufacturing where products are standardized. Services involve 'person-processing' in addition to information-processing, changing the state of people through educational, medical, or social services (Gershuny and Miles 1983: 137–8). The output of services is hard to measure, as in, for example, an education or a surgical operation, because quality is an important dimension (Waite 1988: 13). Baumol, Blackman, and Wolff (1989) suggest a new approach to measuring productivity in services, based on increased productive capacity. It remains true, however, that productivity measurement is problematic (Baumol, Blackman, and Wolff 1989: 225–50).

Agglomeration and regional economic growth

Part of the efficiency of large cities and their regions is a result of *agglomeration economies*. Agglomeration economies – a term used to refer to advantages such as higher productivity and profits accruing to firms located in large urban areas (except perhaps the very largest) or in concentrations of other firms – are also used to account for wage differentials and other technological variations (Carlino 1982; Kawashima 1975; Louri 1988; Meyer 1977; Moomaw 1981, 1983). For the most part, this research has confirmed the presence of agglomeration advantages, although these may vary by industry (Moomaw 1988; Sveikauskas, Gowdy, and Funk 1988). Productivity advantages are greater in non-manufacturing than in manufacturing sectors (Moomaw 1983). Such advantages, in turn, attract new investment and new technology to large cities, and wages are expected to be higher in urban areas or densely populated regions because productivity is higher (Beeson 1987). This is a variant of the process of circular and cumulative causation reviewed in Chapter 2.

One reason for higher productivity in large cities and developed regions is that different occupational groups are found there. Large cities attract educated and skilled workers to a disproportionate degree (Burns and Healy 1978; Klaassen 1987; Sveikauskas 1979). To a great extent, this is due to the spatial division of labour in large firms, who concentrate managerial and administrative staff in particular cities and regions, especially large urban regions. Chapter 6 will return to the allocation of various types of company facilities and its implications.

When the two-factor neoclassical model is extended to include technology or technical knowledge as a factor of production, the result may be disequilibrium rather than equilibrium. Siebert (1969) argues that higher output in a region leads to more investment in R&D, a hypothesis borne out by analysis of recent American data (Parisi 1989). The resulting innovation and knowledge offer initial advantages to that region. This leads, in turn, to higher output and more R&D. Even though technical knowledge is mobile, it tends to be polarized at the point of research or a small number of other points in space. The 'polarizing incidence of technical progress' also leads to polarization of capital and labour because of the initial advantage – profits and higher wages – obtained from innovation. Siebert (1969: 40) continues: 'This process of polarization is increased if we take into account . . . economies of scale. The existence of these internal and agglomeration economies will intensify the polarizing effects. . . . The more mobile capital and labour, the stronger are the polarizing effects which technical knowledge induces with respect to these determinants.'

Such a hypothesis is difficult to test empirically, and Siebert (1969), Richardson (1973), and others were forced by data constraints to cut down their more general models of regional growth to much simpler 'reduced form' models that are scarcely different from neoclassical formulations (McCombie 1988a). In essence, however, such cumulative causation processes are reinforced by diffusion of new technologies, which are adopted first in larger, richer cities and regions. Agglomeration is a powerful concept, which captures the idea that spatial change is conditioned on the past and accumulated spatial patterns. Knowledge, skills and capital, once acquired, do not vanish, but become an endogenous source of future endowments.

Disequilibrium models of regional growth and development

Alternatives to the equilibrium processes inherent in the neoclassical model are based explicitly or implicitly on the unbalanced growth in the circular–cumulative causation model. The process of circular and cumulative causation attempts to include a fuller set of the changes that take place within a region as growth takes place and reinforces prior growth, among them polarization and agglomeration economies. Attributable initially to Myrdal (1957), more complex conceptualizations have been developed by Thompson (1965), Kaldor (1970), and Pred (1977) (Ch. 2).

Cumulative causation suggests that if a region gains some initial advantage, new growth and multiplier effects will tend to concentrate in this already expanding region, rather than in other regions. The effect goes beyond the multiplier effect. Differences in regional productivity growth are predominantly due to differences in output growth, which allow some regions to benefit more than others from economies of scale and agglomeration. Growth becomes self-reinforcing with strong endogenous forces tending to increase regional differences in productivity growth, which may persist for a long time (McCombie 1988b). In more formal presentations, this Verdoorn–Kaldor 'law' leads to cumulative causation: increased investment occurs in faster growing regions, thereby reinforcing their higher growth (Casetti 1984; Swales 1983; Thirlwall 1983). Dosi (1984: 171) suggests that the Verdoorn–Kaldor relation captures the effect of technical change that is not embodied in capital equipment, as will be the case in more innovative sectors.

To a large extent, cumulative causation is a natural corollary of agglomeration economies (Gore 1984: 49). 'Large urban regions are expected, ceteris paribus, to have higher rates of innovation, more rapid adoption of innovations, and higher proportions of skilled workers than smaller places' even if technological change is not endogenous to the process (Malecki and Varaiya 1986: 633; Meyer 1977).

Formal models of disequilibrium regional growth largely keep to the rudimentary capital–labour constituents of economic activity, but structure them in a way that leads to outcomes other than regional equilibrium. Notable efforts at developing such a regional growth theory include Richardson (1973) and von Böventer (1975). Virtually all such efforts strive to incorporate agglomeration economies and other trappings of cumulative causation.

Clark, Gertler, and Whiteman (1986) are sceptical about cumulative causation, largely because they take it literally, citing investment flows out of large cities and into peripheral areas as evidence counter to agglomeration. As with neoclassical flows of capital and labour, there is more than a grain of truth in models which account for cumulative processes favouring large regions. Indeed, Scott's (1988b, c) analyses of agglomeration economies suggest little decline in their influence. The advantages to both workers and firms from agglomeration are substantial. Proximity between employers and workers facilitates specialization and lowers training costs for firms. Employers have access to pools of labour in times of growth, and the costs and time of search for both parties are lower (Kim 1987; Pascal and McCall 1980).

In the sense that they are demand-oriented, the non-equilibrium approaches are 'post-Keynesian', with growth responding to the rate of expansion of demand (McCombie 1988b). The Verdoorn–Kaldor 'law', for example, concerns the relationship between the growth rate of labour productivity and the rate of growth of output. A region's growth rate depends on four main factors: the income elasticity and price elasticities of demand for its exports, (as the economic base model posits), the rate of inflation, and the Verdoorn effect, which assumes rates of investment and productivity growth occur where output growth is greatest – a version of cumulative causation (Thirlwall 1974: 7). This particular disequilibrium model also concludes that manufacturing is the engine of growth, in concurrence with the 'manufacturing matters' school in the controversy over the role of the service sector discussed in Chapter 2 (Thirlwall 1983).

Among the strong arguments in favour of cumulative causation models was a need to allow for increasing returns due to scale and agglomeration economies in rich regions. Many researchers during the 1970s attempted to account for the observed ability of established regions to remain centres of production and economic capacity (von Böventer 1975; Dixon and Thirlwall 1975a, b; Pred 1977; Richardson 1973; Thirlwall 1974). Most accounts rely on agglomeration economies, which are present to a greater degree in large cities and which have several functions in regional growth: to boost the rate of technical progress and productivity, to attract industry and capital into a region, to influence the migration decisions of households, and to improve the intraregional spatial structure (Richardson 1973: 175–96). However, in operational terms, the effect of urban size in these models became primarily one of innovation diffusion down the urban hierarchy (Berry 1972; Richardson 1973). Agglomeration economies, rather than technology, become the driving force behind regional growth in these models.

Marx, dependency, and technology

Although technology is included in the main bodies of explanation of economic change – including Marxian thought, dependency theory, and the neoclassical model – it is played down relative to capital and labour. In each of the non-neoclassical views on economic activity, capital accumulation is both motive and process, and it may in analysis be easy to forget the adaptations to changes in labour, technology, and markets, especially those represented by product innovation (Sayer 1985). In Marxist formulations, for example, profit rates are regarded as possessing an inevitable tendency to decrease over time, because of the endemic contradictions of capitalism as identified by Marx (Gertler 1988a: 154). These contradictions include the capital deepening of new labour-saving technology, which permits higher profits in the short run, and industrialization of rural areas and Third World countries where labour power is lower (Carney 1980). However, in detailed empirical work on Canada, Webber and Tonkin (1987, 1988a, b) have found that this 'tendency' for labour replacement is by no means certain.

Largely through the efforts of Michael Webber (1987a, b; Webber and Foot 1988), a Marxist model of production can be analysed in a manner similar to that of neoclassical formulations. In an excellent series of empirical studies, Webber and Tonkin (1987, 1988a, b) show that individual sectors in Canada differ markedly in the types of technical changes they underwent in response to declining rates of profit. The food industry was able to stave off profit decline largely by decreasing turnover times in production and by exploiting the technical advances of supplier industries. The clothing industry experienced little technical change, but was characterized by labour exploitation. Textile firms, by contrast, generally incorporated increases in the technical composition of capital through capital investment. Knitting firms were somewhere in between these two extremes. Finally, in the wood, furniture, and paper industries, the interactions between technical changes and market relations were more complex, and involved a combination of capital and labour adjustments. As these studies illustrate and as Chapter 5 will show, profit and growth are not constant or steady, but follow a pattern of 'long waves' of cyclical growth and decline. Technology is a key element of the resilience of capitalism that is not captured by a focus solely on labour.

Dependency theory, unequal exchange, and underdevelopment are among the alternative explanations to modernization and development. A central process is the geographical transfer of value from peripheral areas to the core as the basis of uneven development (Chilcote and Johnson 1983; Hadjimichalis 1987; Soja 1984). More importantly, this transfer occurs because an economy's ties to the international economic system are stronger than within the domestic economy (Caporaso and

Zare 1981). Wilber (1984) provides a useful compendium of the primary sources on dependency and underdevelopment. In a distinct terminology, geographical transfer of value is the same process as leakage of multiplier effects, channelled along paths governed by the organizational structure and spatial division of labour within and among firms. The largest surplus values (multiplier effects) end up and are reinvested in control or core regions rather than in peripheral regions. Unimpassioned descriptions of branch plant economies have found virtually the same 'transfer of value' within advanced economies (Watts 1981). Braudel (1982) and Landes (1980) suggest that such transfers were commonplace by the sixteenth century, and placed Europe at the hub of a far-reaching network of trade and of an international division of labour.

Dependency and underdevelopment, concepts developed on a global scale to describe relations among nations, are thus applicable in a regional setting as well (Moulaert and Salinas 1982). Marelli (1983: 69) concludes that such transfers in Italy 'are just a necessary consequence of this (probably unavoidable) lack of homogeneity' among regions. The distinct vocabularies of Marxist and neoclassical analyses obscure the fact that the processes portrayed are identical. Parallel research on the same topics may fail to cite studies written in the other perspective and jargon (Bourne 1988). It is perhaps preferable not to assert that a particular paradigm is most appropriate, but to employ an eclectic approach and local knowledge (Brown 1988).

The division of labour is seen by Marxists as a means of effecting the geographical transfer of profits, and results from the need of capitalist firms (often termed 'capitals') to create and augment surplus value. Tasks are routinized to extricate additional surplus value out of production labour; new technologies are created by specialists for the purpose of reducing the labour content of production (Hadjimichalis 1987: 139–68; Mandel 1975; Massey 1984; Morgan and Sayer 1988). The literature on this topic is large and complex; good summaries of this literature and its arguments at the international scale are found in Weaver and Jameson (1981) and the various editions of Wilber (1984); Moulaert and Salinas (1982) provide regional examples.

For many underdeveloped countries, unequal exchange began with colonialism, although trade, multipliers, and accumulation unfolded in Asia apart from European capitalism (Braudel 1982: 581–94). The plantation system of agriculture, perfected under colonialism, concentrated effort and infrastructure on a few export crops, such as rubber, palm-oil, sugar, coffee, or tea. Traditional subsistence farming was neglected, and productivity efforts were channelled towards the export commodities (Braudel 1982: 272–80). The extreme specialization had severe impacts. Some commodities, such as rubber, have been substantially replaced by cheaper and higher-quality synthetic substances. Others are subject to the vagaries of perhaps a single large corporation (Simpson 1987: 84–94). Recall the example of the Coca Cola Company's reformulation of its principal soft drink, immediately bringing about a drop in demand for Madagascar's vanilla bean crop.

Development as a temporal process

Urban and regional processes of change differ both in the speed with which they take place and the dimensions and actors which influence change. Regarding speed, Wegener (1986) classifies processes as fast, medium-speed, or slow. *Fast processes* include the mobility of people and goods within existing structures and communication channels, such as relocations of firms and changes of workplace by workers, both of which in turn affect housing and infrastructure location.

Socio-economic or technological change are *medium-speed processes*, which occur outside or 'above' the local urban area or region. Relocations and daily movements take place within the longer life cycles of people, households, products, firms, and technologies. Their speed and direction are determined either by biological processes such as ageing or by economic cycles outside a given region; they may have impacts on, but are largely unaffected by what is going on in, a region. Thus, it is nearly impossible to reverse their direction. The impacts of technological and socio-economic change themselves do not change the physical structure of a city or region, but they change the activities performed within it.

Finally, *slow processes* are exemplified by construction. The physical structure of a city displays a remarkable stability over time because of the long lifetime of buildings, the large investment involved, and the long delays between the first decision to invest and completion. The most rigid elements of the regional fabric are transportation lines, such as major highways and railroads, which usually remain in place over centuries, even if the transport equipment using them is replaced several times. Therefore decisions on the location of transport infrastructure may be considered as virtually irreversible (Wegener 1986: 438–9; Baumol 1981).

In addition to the different rates of change, development also includes structural dimensions and embedded and inertial structures. The geological analogy of layers of sediment piled on top of one another illustrates just one way in which change over time may be envisioned (Massey 1984: 117–18). As Dosi (1984: 298) says, 'Through time, history becomes structure, while at each point in time structure shapes history.' The study of time within economic development has been largely simplistic thus far, appealing to concepts such as layers or general processes that often do not involve real people, firms, or institutions.

In an important contribution, Storper (1988) has distinguished between events, structures, and processes. Events, which are typically 'small', are the outcomes of structured, but not fully determined, situations. They are the results of choices and strategies by individuals and actors (human agents) undertaken within structured circumstances. Sequences and collections of these small events produce large processes, and large ('global') outcomes of small events may not be fully predictable from the specific events themselves. The large processes thus generated may prompt reproduction or change in big structures (Storper 1988). An example may be found in long-term structural change which, based on Kaldor's (1970) views on disequilibrium, results from two types of smaller changes which take place within firms: productivity increases from new processes and the growth of new industries through the creation of new products (Furtado 1980: 204; Thomas 1975: 13).

Another view of the process of regional economic change assumes that changes need not be smooth, but may be quite abrupt. Major 'logistical revolutions', such as the Industrial Revolution, caused large and rather sudden changes in the means by which work is done and movement is carried out (Andersson 1986; Leontief 1983). These abrupt changes or 'catastrophes' alter the relationships among variables. Casetti (1981), for example, interprets recent, rather abrupt, shifts from north to south in the USA as evidence of catastrophic change. Gertler (1986b), on the other hand, reviews the empirical evidence and concludes that this shift was not so abrupt, but had been taking place more or less steadily for some time before it attracted the notice of observers. Similarly, Nijkamp (1982) and Nijkamp and Schubert (1985) suggest that catastrophe-type models can best incorporate the exogenous and endogenous factors, such as infrastructure and innovation, that precipitate urban and regional change.

The view of catastrophes in economic change has not been accepted by all observers. Short-term, evolutionary adjustments may be closer to reality, and are the mechanism invoked by contemporary 'evolutionary' theories of economic change outside the regional context, as Chapter 4 will show. In this context, Clark, Gertler, and Whiteman (1986) have proposed an 'adjustment model' of regional production which is cast in the same capital–labour mould as the neoclassical model. 'Allocations' of resources via the invisible hand mechanism in the neoclassical model are replaced by short-run adjustments in which firms are constrained in the extent to which they are able to alter prices and to substitute capital for labour. In this situation, fixed costs can continue to rise and profits diminish.

The prime variable over which firms have control is labour costs, if only through changes in location, since locations vary not only in wage levels but also in the degree to which the quantity of labour can be adjusted through lay-offs or changes in the length of the working week (Ch. 6). Local economies are 'differentiated along the basic dimension of economic uncertainty' concerning the degree to which workers are willing to absorb economic fluctuations (Clark, Gertler, and Whiteman 1986: 35–8). To a considerable extent, these differences among locales are primarily related to differences in the portion of the spatial division of labour found in an area.

Regional policy

Policies intended to alter geographically uneven development were begun only since the Great Depression of the 1930s, in both advanced countries and in Third World countries. These were grounded primarily on demand-based Keynesian and economic base theory as the appropriate rationale for government intervention in development (Higgins 1981; Leven 1985). The experience of underdeveloped and advanced countries, while certainly different in several respects, proceeded quite similarly as the same theories and policies were applied and confronted the same forces of the world capitalist system in all places.

The widespread unemployment of the 1930s prompted experimentation with subsidies for manufacturing to create jobs. This has been a fundamental policy mechanism since the 1936 Balancing Agriculture with Industry (BAWI) Act in Mississippi in the USA (Cobb 1982). In addition, construction projects for infrastructure in lagging areas became a common mechanism to provide jobs, if only temporarily (Leven 1985; Savoie 1986). In the decade immediately after the Second World War, the preferred policies relied on large-scale 'big push' investments, capital accumulation, and overall national growth (Brookfield 1975; Streeten 1979b). This emphasis on export industries had the effect of benefiting owners of capital more than workers (Leven 1985).

Explicit policies to alter the regional distribution of population and economic activity have only a brief history, largely dating from the 1960s (Sundquist 1975). Concentration is a much greater problem in Europe, where capital cities dominate several national economies, and in Third World countries, where urban primacy is an acute dilemma. The net result, however, is that few countries have successfully dispersed population as a result of policies to restrict economic growth in some places (OECD 1977). Most interest in population dispersal among advanced countries diminished during the early 1980s coinciding with the identification of 'counterurbanization', net migration away from established urban areas (Vining and Kontuly 1978). In truth, Richardson's (1973) earlier observation of the spatial pattern of recent economic change as 'decentralized concentrated dispersion' was quite accurate. To a large degree, population and economic activity dispersed only to a fairly small number of concentrations, large urban areas, where growth has occurred primarily in a decentralized fashion in the suburban and exurban periphery.

A prominent dilemma in regional policy has been the question of 'people prosperity or place prosperity' (Agnew 1984). It has arisen largely from the conceptualization of people and places as factors of production: labour and land. If the goal is to help people in a place, then policies aimed at regions are only indirect, and they may indeed benefit the rich rather than the poor in such regions (Richardson 1979: 168). Others have proposed that people in backward regions could best be helped by subsidizing their outmigration to other regions (Hansen 1973). The 'people versus place' choice arises repeatedly as policies attempt at considerable cost to 'create' jobs (Chisholm 1987). Regional 'triage', favouring investment in some places at the expense of others, is fairly pervasive in regional policies, including growth pole and growth centre policies (Agnew 1984).

Growth poles and growth centres

Regional policy became a widespread phenomenon during the 1960s, applied to solve persistent regional inequalities found in the advanced economies during a period of growth and planning newly independent states in the Third World. Among the most influential policies was that based on Perroux's (1955) 'growth pole' concept. Based on aspatial economic reasoning, Perroux merely suggested that variations among industries in their degree of motive power and differences in the strength of their linkages to other sectors made it important to select some industries as 'poles of growth'. The selection of sectors was based on input–output data that identified not only the sectors on which other sectors relied for inputs, but also those which sold a great deal of their product to other sectors. The steel industry was the most frequently chosen as a growth pole, especially in developing countries, where this sector also served as an import substitution mechanism (Hansen 1967).

Hirschman (1958) and Boudeville (1966, 1968) extended Perroux's concept and suggested that a propulsive industry, identified by its input–output multiplier, can have a significant effect on its region through backward (upstream) and forward (downstream) linkages. In Boudeville's (1966: 11) words, 'A regional growth-pole is a set of expanding industries located in an urban area and inducing further development of economic activity in its sphere of influence.' Following Darwent (1969), most authors began to follow the distinction between *growth poles*, set in the economic space of input–output relations after Perroux (1955), and *growth centres*, propulsive industries set in a regional context; Todd (1974) and others use *development poles* and *development centres*, respectively, for the same concepts; Moseley (1974) and Polenske (1988) provide a summary of the categorizations; Friedmann and Weaver (1979) and Gore (1984) place growth centre policy into its historical context.

The emphasis of these spatial concepts was on the location of growth, and frequently their application bears little resemblance to Perroux's initial ideas (Brookfield 1975: 105–8). Agglomeration economies were a key element of growth centre policies, in both an industrial and a spatial sense (Moseley 1974). For example, industrial complexes attempt to incorporate highly linked industries within a relatively small geographical area. The creation of industrial complexes as a planning tool, thus mustering together productive forces in a particular location, has been especially prominent in Canada and in the Soviet Union (Bandman 1980; Czamanski and Czamanski 1977; Karaska and Linge 1978; Luttrell 1972; Norcliffe and Kotseff 1978).

However, the appeal of growth poles was largely in their ability to justify unbalanced growth policies which, if pursued, make it possible to turn around persistently lagging regions, reinforced by the ideas of Myrdal (1957) and Hirschman (1958). Closely related were expectations that input–output linkages should form in the vicinity of growth centres, as part of a general optimism concerning the unfolding of agglomeration economies (Moseley 1974). Much of the expected benefit of growth centres was also tied to optimism concerning the process of 'trickle-down' or 'spread' effects, based on Hirschman (1958) and Myrdal (1957). Higher incomes and economic growth were envisioned to emanate outwards from a growth centre over time, gradually affecting the entire hinterland of the centre. Similarity between this process and spatial diffusion is evident, and the theories of the latter seemed to lend further support to growth centre policies (Berry 1972; Moseley 1974; Richardson 1973). Indeed, growth centres paralleled a related body of research from the 1960s on *modernization*, which was once influential but is now in disrepute even among its former advocates, who applied them to the context of newly independent African nations (Riddell 1970, 1981a, 1987; Soja 1968, 1979).

So enthusiastic was the social science community about the concept that, even where no growth centre policy was in place, 'spontaneous' or 'natural' growth centres were identified (Alonso and Medrich 1972; Harner and Haynes 1975; Ironside and Williams 1980). As Higgins (1983: 8) stated it:

> The greatest attraction of all was that application of the growth pole doctrine seemed to require so little in the way of intervention. All we needed to do was lure some propulsive industries (and in the euphoria of the time, all lurable industries seemed more propulsive than repulsive) into urban centres of retarded regions. Then, according to the Boudeville revisionist doctrine which all of us siezed, we would sit back and let the market generate spread effects to the peripheral region.

Gore (1984: 262–3) makes the same point about the political attraction of growth pole policies: '. . . from a political point of view, the idea that if resources are spatially concentrated in particular localities and regions, the "benefits of development" will eventually trickle down into other areas is an essential way of maintaining support in the neglected areas'.

However, Myrdal and Hirschman also identified the process of 'backwash' or 'polarization' which refers to the tendency for factors of production to be drawn from the periphery to the centre. Taken together, 'spread' and 'backwash' represent simultaneous processes and, depending on which is dominant and after a necessary time-lag, growth will occur in the hinterland (Gaile 1980; Richardson 1976). This expectation of a temporal pattern could be added to Gertler's (1988a) list of 'inevitable' processes in Chapter 1. By contrast, empirical evidence suggests that backwash processes dominate in practice, and that the regional effect of growth centres is limited to a rather small area and, at best, only after a long time (Gilbert 1975; Richardson 1976).

However, even in developed countries, growth centres and other regional policies fought an uphill battle against the slippery slope of capitalist tendencies. The evaluation of regional policies and assessment of their success are difficult tasks at best, because there are many different – and often conflicting – criteria on which they can be evaluated. Diamond and Spence (1983), for example, suggest several criteria:

1. *Labour* criteria, including a decrease in unemployment or an increase in employment;
2. *Capital* criteria, the value of investment or the number of firms locating in a region; and
3. Environmental or *social welfare* criteria, primarily the value of infrastructure (transportation, communications, electricity, water, and social services) provided.

In addition, other indicators of economic performance may be more important, such as regional levels of output, productivity, profits, and wages. There may be indirect effects and impacts as well (Folmer and Nijkamp 1985; Higgins 1988). Total employment, for example, masks the variations that may be present between male and female employment, full-time and part-time jobs, and wage levels – the quality and status of employment, rather than the number of jobs (Martin and Hodge 1983; Townsend 1986). Similarly, investment measures, such as the number of firms moving to or locating in a region, often do not take into account subsequent closures. Infrastructure can have uneven impacts as well, benefiting some industries or individuals more than others.

Some examples of regional policy in developed economies

Growth pole policies have been applied in dozens of countries, documented through the reports of the UN Research Institute for Social Development (Kuklinski 1972, 1975b, 1978, 1981; Kuklinski and Petrella 1972), the UN Centre for Regional Development (Gana 1981), and elsewhere (Hansen 1972; Lo and Salih 1978; Sundquist 1975). A great deal of effort was devoted in all countries to technical details of such a policy, such as how much concentration is needed; how many centres; how are they to be selected and how promoted? (Morrill 1973; Moseley 1974: 50–1).

The national experiences of a number of countries provides a backdrop for evaluating the effectiveness and evolution of regional policies.

The French experience

France presented an early experiment of Perroux's growth pole concept. A set of *métropoles d'équilibre* were identified in an elaborate attempt to decentralize growth away from the Paris region (Dunford 1988: 231–95; Hansen 1967; Moseley 1974: 41–50). Marseille, the second largest French city, would seem to have been a likely candidate for a successful growth centre, one of several internationally competitive industrial complexes envisoned (Dunford 1988: 249). However, the plan to create in nearby Fos-sur-Mer an integrated port and industrial zone to rival Rotterdam did not quite succeed. Instead, what emerged was something quite similar to the Third World model of a dualistic economy, with multi-plant organizations importing their requirements and exporting their output from the region (Kinsey 1978).

However, things were learned from the French experience with regional policy which presaged issues for later policies. Regional development aid varied depending on location within France, generally being greater in the South and West (Dunford 1988). Efforts were focused on large firms, rather than on small and medium-size enterprises, and were centred on the number, rather than the type, of jobs created, which were frequently unskilled and poorly paid. Finally, policies were directed at manufacturing at a time when services were becoming the lead sector in the economy (Prud'homme 1974).

Firms began in the 1970s to move towards areas where there were reservoirs of unskilled workers, ignoring regional policies. As Hansen (1987: 11–12) describes French regional development,

> the model that prevailed during the 1960s was essentially top-down in nature. It was based on large-scale, spatially-concentrated industrial and infrastructure investments, with decision-making largely in the hands of large industrial oligopolies and financial institutions. The prevailing conceptual base for regional planning was growth pole theory, whose influence eventually extended throughout the world. Although a considerable amount of industrial decentralization took place in this context, the quality of the decentralized jobs left much to be desired. In retrospect, it is apparent that externally-induced growth typically did not provide a solid basis for sustained regional or local development.

Consequently, French regional theory and policy have shifted 'away from top-down models in favour of endogenous development at the regional and local levels' (Hansen 1987: 5). Although it took many years for this conclusion to be reached, the French experience was paralleled elsewhere.

The Italian experience

The Italian case provides a clear look at a situation where regional policy failed to have any discernible beneficial effect at all. The southern portion of Italy, the Mezzogiorno, has long been less industrial, and has much lower incomes and higher rates of unemployment than northern Italy, where industrial agglomerations such as Milan and Turin are found. A 'big push' level of 40 per cent of total national industrial investment was determined to be necessary to develop the South (Cao-Pinna 1974).

Both through subsidies to private firms and through investments made by the Italian government through its large state holding companies and semi-public firms in oil, petrochemicals, and steel, 80 per cent of all new investment was required to be located in the South (Sundquist 1975: 179–87). These were large, capital-intensive plants, which had relatively little subsequent effect on employment or income in the region (Amin 1985; Dunford 1988: 169).

However, local linkages were minimal and multiplier effects in the Mezzogiorno were small (Dunford 1986; Martinelli 1985; Rodgers 1979). 'The entrepreneurs who had been relied on to establish satellite plants, it turned out, did not exist' (Sundquist 1975: 181). What employment they provided tended to be mainly unskilled labour specializing in a limited array of production activities (Amin 1985). Italian industrial growth, and the development of producer services, remains primarily in the North.

Regional policy in the USA

American regional policy was at its height during President Lyndon Johnson's 'new society' years of the late 1960s. During that period, emphasis was on the Appalachian Regional Commission and its valiant efforts to spread development to that backward region (Rothblatt 1971). Infrastructure investment was undertaken, especially highway construction in the mountainous region, but other policies, such as subsidies for industrial development, had little effect (Martin 1979; Miernyk 1980).

Although some infrastructure and land use policies predated formal regional policy, regional policy in the USA has always been secondary both in priority and in its effects relative to broader sectoral policies (Cumberland 1973). Consequently, the direct effect of federal policies on industrial location have been slight (Rees 1980). The *implicit* or *indirect* regional effects of these other policies, on the other hand, have been enormous. The policies include tax structure and macroeconomic policy, which favours some sectors and regions, social security, which stimulated migration to the sunbelt, and defence spending, which is highly biased geographically (Bolton 1982b; King and Clark 1978; Malecki 1982; OECD 1980a; Rees 1987; Vaughan 1977). Because of the absence of direct benefits from regional policy, many conclude (in neoclassical fashion) that market and institutional forces, especially migration and investment – factor mobility – are the greatest causes of patterns of regional economic change in the USA (*Business Week* 1976; Newman 1984; Weinstein, Gross, and Rees 1985).

In the absence of regional policy at the national (or federal) level in the USA, the individual states and localities have taken many uncoordinated initiatives to promote economic development (Haider 1986; Osborne 1988). The decentralization of policy, perhaps a spontaneous move towards bottom-up policy, is rather widespread among the developed countries, as Fox Przeworski's (1986) survey of practices in the OECD countries in the mid-1980s illustrates. A consistent emphasis in American research is the prominence of economic and demographic variables, and the absence of influence of policy variables, in accounting for spatial patterns of economic growth (e.g. Carlino and Mills 1987).

Regional policy in the United Kingdom

British regional policy has centred around the subsidization of industrial location in designated 'assisted areas' located primarily in the north of England, Scotland, and Wales. Many assessments of British regional policy have been positive. Ashcroft (1982), in reviewing the large literature on earlier attempts to assess the effects of regional policies in West Germany, Ireland, the Netherlands, as well as the UK, concluded that regional policy indeed appeared to have had positive effects on investment and employment. However, these benefits appeared to have occurred primarily by the movement of mobile, multi-plant firms rather than through any local or indigenous economic activity (Ashcroft 1982; Moore and Rhodes 1976). Measurement problems, however, plague such evaluations, and Buck and Atkins (1983), for example, conclude that the benefits of such policies are usually overstated.

In addition to the possible overstatement of direct benefits of regional policy, the indirect effects – linkages and multiplier effects – have probably been more limited, as in the case of Scotland, often cited as a region where 'regional policy has less obviously failed in this region than in others' (Damesick and Wood 1987: 261; Randall 1987). Although in Scotland 'regional policy contributed significantly to the movement of industry', the closure rates of firms were also much higher than elsewhere in the UK (Diamond and Spence 1983: 118–23). However, in part because of policy bias towards investment (ie subsidies for capital investment), the policy cost per job generated was quite high (Diamond and Spence 1983).

The British example illustrates that in many respects it is complex – if not impossible – to sort out the cause and effect of regional policies. In part this is attributable to the variety of methods and data used in the evaluation of regional policies (Bartels, Nicol, and van Duijn 1982). Whether it is regional policy, for example, rather than simply the availability of low-wage labour, that has attracted firms and investment to assisted areas is not easily sorted out. There is now ample evidence that firms locate in peripheral areas for advantages related to unskilled labour, including cost, control, and reproduction, rather than because of policy benefits or other attractions (Cobb 1982; Erickson 1976; Massey 1979b; Morgan and Sayer 1985). Table 3.4 illustrates the overwhelming importance of labour to the location decisions of mobile plants in the UK during the height of regional policy.

The Regional Studies Association (1983) concluded its report on regional problems in the UK with the recommendation that some form of regional preferential assistance is needed in order to reduce the disparities between the Greater South-east and the North and West. Martin (1988: 408) is less sure of this, concluding that 'regional policy did little to reduce or eliminate the unemployment differential between the "industrial periphery" and the prosperous "south and east"'. For a variety of political and economic reasons, there was a gradual but inexorable evolution during the 1980s toward policies which relied on market forces rather than intervention (Chisholm 1987; Damesick and Wood 1987; Martin 1988; Martin and Hodge 1983).

Canadian experience
Canadian regional policy was initiated in the late 1950s primarily to reduce regional economic disparities, and specifically to address the problems of the Atlantic region, a traditionally backward area reliant on primary activities (agriculture, fishing, forestry, and mining). Disparities

in incomes and employment opportunities across Canada's 10 provinces are of long standing (Ray and Brewis 1976; Raynauld 1988). A bewildering array of policies, ranging from subsidies and industrial incentives to growth centres, has been applied in Canada, with minimal effect (Britton 1988a).

Table 3.4 Factors influencing the location decision of interregional moves in the UK, 1964–67

Reason for location	% of all respondents		Outstanding single factor
	Major reason	Minor reason	
Availability of labour at new location	72	20	20
Expectation that industrial development certificate available	48	18	2
Accessibility to markets or supplies	39	21	9
Availability of government inducements	39	7	7
Assistance from local authorities	36	30	3
Access to firm's plants or old location	32	18	7
Access to specified transport facilities	31	20	2
Good amenities and environment	29	41	1
Availability of non-government factory	28	5	6
Special characteristics of site	20	17	3
Other factors	12	2	3
No single outstanding factor			38

Source: Ashcroft (1978), cited in Diamond and Spence (1983: 57).

A strong federal role, implemented through the Department of Regional Economic Expansion (DREE), has been abandoned in favour of co-operation among federal, provincial, and local authorities. Discussions of Canadian regional policy have always been long on procedures, such as who should do what, and short on results (Brewis 1969; Firestone 1974; Savoie 1986). The current tenor of Canadians is to decentralize economic policy and planning to provincial and local governments, rather than to depend on federal policy (Coffey and Polèse 1987c). As a very 'open' economy because of its proximity to the USA,

Canada has circumstances which make it particularly vulnerable to international influences, such as technological change and global corporations, and Canadian policies have yet to come to grips with these facts (Britton 1988a, b).

Regional development in Germany and central Europe

Regional development is decentralized in West Germany much as it is in the USA. However, as in the USA, the impact of regional development programmes 'is overshadowed by sectoral development programs, many of which have regional economic implications' (Romsa, Blenman, and Nipper 1989: 51). Too many growth centres were identified, and assisted areas covered 45 per cent of the country, so that any regional policy was overextended.

Likewise, in Austria, Berentsen (1978) detected little impact of regional development policy relative to the effect of free market forces. In a larger study of Austria, the Federal Republic of Germany, the German Democratic Republic, and Switzerland, Berentsen (1987) found small decreases in regional inequalities which policies in all countries had attempted to reduce. However, regional inequalities persisted even as industry dispersed, suggesting that industrial employment was not the appropriate policy goal for the late 1980s and 1990s.

Under central planning, the countries of Eastern Europe experienced polarization as pronounced, if not more so, as in capitalist economies (Berentsen 1981; Fuchs and Demko 1979; Kowalski 1986). Berentsen (1981) and Fuchs and Demko (1979) attribute this result to a political concern for efficiency over regional equity in investment location, embodied in large industrial complexes in a small number of locations and industrial investment in established centres. A degree of regional convergence which took place in Hungary during the 1960s and 1970s was brought to an end by new policies in the mid-1980s which are based on a greater spatial division of labour and regional specialization (Bartke and Lackó 1986). It remains to be seen if regional disparity will widen as capitalism takes hold in Eastern Europe during the 1990s.

Regional development in the Third World

Regional policies in underdeveloped countries were frequently only parts of larger national policies for modernization. Growth pole policy provided an efficiency-oriented spatial policy, with popular elements such as industrialization, disguised as an equity strategy through processes such as diffusion (Gore 1984: 113–17; Richardson 1979). The success of such policies relies on the presence of an urban system or hierarchy to channel growth to secondary cities, typically in lagging regions, where incomes and development are low (Hansen 1971; Rondinelli 1983). The focus on the urban hierarchy as a channel of economic activity and a constraint on policies is a recurring observation (Friedmann 1966; Pred 1976; Richardson and Townroe 1986).

However, it is difficult – and it may be impossible – to alter the powerful forces which attract capital and labour to primate cities in developing countries (Clapp and Richardson 1984; El-Shakhs 1982a), or to large urban areas in any country (Ettlinger 1981; Sheppard 1982). In primate cities and other cities of the Third World, both colonialism and the penetration of capitalism have stimulated a flow of migration from rural areas. In the colonial era, such cities had opportunities to earn money for European goods and other 'bright lights' attractions, and individuals had a far greater chance of employment and education than in rural areas (Riddell 1981b; Tolosa 1981; Yap 1977). Primate cities remain the first recipients of technological progress, reinforcing advantages previously accumulated there. Multinational corporations need communications and agglomeration advantages (including access to government offices and agencies) (Richardson and Townroe 1986: 668).

In part because of urban primacy, McGee (1971: 31) suggests, 'the form of the urbanization process in the Third World may appear to be the same as that which characterized the West, . . . however, the city as an inducer of change appears to be absent from the urbanization process in the Third World'. This results both from dualism within primate cities (Weitz 1986: 37–9) and from the tendency of multinational corporations to focus their activities in such cities, at times in conjunction with subcontracting of manufacturing with cheap labour in nearby peripheral areas (Coraggio 1975). The urban–rural contrast has persisted with little evidence of integrated regional development. In some newly industrializing countries (NICs), such as Korea and Taiwan, industry has decentralized from large cities in search of lower labour costs (Meyer 1986). However, peripheral industrial growth has not challenged the hegemony of the primate city in these countries in control and financial functions, and the periphery remains dependent on the urban centres economically as well as in other dimensions. It may also be the case that peripheral nations are characterized by primacy to a greater extent, but the evidence on this is mixed (Frey, Dietz, and Marte 1986).

Primacy and a backward agricultural sector have been among the difficulties facing regional policy in Africa. As we shall see in later chapters, the orientation of conventional regional policies largely ignored the indigenous knowledge and skills of the local population in favour of large-scale projects and export industry. Indeed, to a large degree, policies to reduce regional disparities worked in opposition to the concentrated urban structures and economic flows established by the relatively brief colonial period, as Moudoud (1989) vividly demonstrates in the case of Tunisia.

The Kenyan experience

Kenya was an early 'laboratory' for spatial policies of modernization based on assumptions of diffusion and spread effects (Soja 1968). Theories of modernization were rather quickly replaced in Kenya by frameworks of dependency and underdevelopment (Soja 1979; Szentes 1971). This occurred elsewhere in the world as empirical knowledge showed that spread effects did not prevail (Friedmann and Weaver 1979). For example, the spread effects of agricultural innovations were minimal in most instances, largely because they focused on progressive farmers, who could afford them, and these initial advantages were strengthened by successive innovations over time (Rogers 1976; Röling, Ascroft, and Chege 1976). Brown (1981) shows, via examples in both Kenya and Mexico, that initial decisions regarding organization and infrastructure (for processing, storage, and transportation) have a major influence on the spatial spread of cash crops and other innovations. In this as in most technological settings, the early adopters experience a windfall profit or advantage that is not available to later adopters, reinforcing cumulative causation (Brown 1981: 270–7).

Kenya represents a Third World country where policy changes have taken place in recent years as well. 'Bottom-up' strategies gradually evolved, as the national urban development strategy shifted from nine growth centres in 1969 to a four-tier hierarchy of 33 cities and towns in 1984 (Evans 1989). The two chief instruments of Kenyan policy, as elsewhere, were investments in physical infrastructure and industrial location incentives to induce industries to locate outside of Nairobi and Mombasa. Instead of continuing in this approach, however, Kenya has moved away from concerns about the urban settlement pattern and the form of the urban system as a major element in the diffusion of economic growth and development impulses, and toward rural areas. According to Evans (1989: 262),

The new paradigm is based on the proposition that rising rural incomes increase demand not only for inputs for agricultural production but also for consumer goods and services, much of which can be supplied by small-scale enterprises in intermediate-size towns and smaller villages. This increase in demand in turn spurs the growth of non-farm activities, and hence the creation of new employment opportunities for the growing rural labour force. The diffusion of non-farm manufacturing, retailing networks and marketing channels in smaller urban centres closer to farmers, in turn helps to raise agricultural productivity, hence boosting rural incomes. . . . The main objective of the new urban policy is to support the larger goal of accelerated economic development, in part by promoting greater productivity in agriculture, and in part by spurring the growth of non-farm activities and employment opportunities in smaller towns and rural areas. The form or pattern of the urban settlement system is no longer the focus of urban policy, but largely the residual of macroeconomic policies.

Richardson and Townroe (1986: 671) believe that such a system is particularly appropriate for agricultural regions.

Spatial planning in Nigeria

Nigeria is another African country where modernization through spatial planning and urban and regional development were attempted on a large scale. The disparity between the country's growing cities and populated rural areas posed particular problems for planning (Ajaegbu 1976). Gana (1981: 203) sums up the result in the Nigerian setting, where 'growth axes' were attempted: 'Rather than transform the rural economies, the industrial growth centres became "enclave" economies specializing in import substitution industries. . . . Worse still is the fact that these so-called growth centres have been strongly linked with external economic entities whose "centres of gravity" lie in the nerve centres of the capitalist world.'

Mabogunje (1978) contends that the failure of growth centre policies in Nigeria was primarily attributable to an emphasis on large-scale industry and a failure to incorporate small and medium-size enterprises which would have greater linkage effects in towns and villages. Throughout Africa, industrialization has typically meant large operations by foreign firms, often in joint ventures with government. The dualistic contrast between traditional technology and modern, Western technology, which is typically large and inflexible, created an increasing dependence on Western inputs, managers, and technologies, with a consequent leakage of multiplier effects out of the country (Barker *et al.* 1986). Under these conditions, the development of regions and rural areas does not take place (Moudoud 1989).

Toward indigenous potential

Other examples would reinforce the conclusion of those presented so far: that the results of most regional planning efforts have been rather dismal. This has been a consequence of the structure of large organizations and the failure of 'spread' and 'trickle-down' actually to work in practice (Kowalski 1979; Pred 1976; Stöhr 1981). In developing countries, in particular, the expectations of spread effects were far too optimistic. Conventional 'top-down' or 'from above' regional policies, such as growth centres, rooted in neoclassical economic theory, entailed an optimism concerning regional convergence and a blindness to the leakages of economic effects that flow up the urban hierarchy (Stöhr and Taylor 1981). An even greater failure in practice was that national plans were often simply disaggregated by planners and established as regional policy (Kuklinski 1975a). Indeed, the large-project orientation of growth centre policies and the influence of external aid agencies reinforced concentrated investments and industrial projects over dispersed rural development schemes based on agriculture. Top-down strategies, like growth centres, emphasized an urban and industrial orientation.

By contrast, 'bottom-up' or 'from below' policies attempted to integrate the rural economy of backward regions better, often by focusing on processing of locally produced agricultural goods. In the Third World setting, 'bottom-up' strategies are tied to a 'basic needs' approach to development, wherein food, shelter, education, health, and other basic human needs are set as a priority in development planning (Dube 1988; Emmerij 1981; Lee 1981; Nagamine 1981; Streeten 1979a). The discussion of bottom-up strategies shades into arguments about appropriate national government structures (Richardson and Townroe 1986: 671). The fundamental changes in social and political structures bound with a bottom-up strategy make its adoption rather remote in actuality (Lee 1981). In another variant on the bottom-up strategy, Friedmann has proposed an 'agropolitan' development strategy, which links agricultural and resource-based industries in rural regions (Friedmann and Douglass 1978; Friedmann and Weaver 1979: 193–207;

Richardson 1978). It avoids the urban and industrial bias of growth centres and most other regional policies.

A minority view of growth poles, but one stressed by Perroux himself, sees technological innovation as the propulsive force behind the disequilibrium of growth poles (Erickson 1972; Hermansen 1972; Lasuen 1969; Thomas 1972a, b). 'Lead firms' or 'growth firms' are those whose rate of product growth or productivity growth is above average, or whose market share is increasing (Erickson 1972: 428). The effects of the lead firm's innovativeness and growth are then spread to other firms through its linkages with them. Indeed, there is no reason to expect that these linkages will be confined to the region of the lead firm (Erickson 1972: 434–7), as ample empirical evidence on linkages has made clear.

Thomas (1972a) suggests that there is a great similarity between Perroux's (1955) economic 'field of forces' and the linkages described by Hirschman (1958), and that innovation is a key element in the structural change which takes place in firms, industries, and regions (Thomas 1972b, 1986). The essence of economic growth at the national, regional, and local levels is *structural change* (Ch. 1). Unfortunately, the technology dimension has not been a major element of most attempts to develop growth centres in practice (Savoie 1986). Technology arose only later as a policy focus quite distinct from regional policy.

The empirical experience of growth centre policies has shown that researchers and policy-makers did not foresee the proliferation of interregional linkages, especially in the context of multilocational firms. Growth centre policies often resulted in '"cathedrals in the desert" – industrial complexes whose developmental impulses are real enough, but which are channelled to linked firms and industries hundreds of miles away' (Moseley 1974: 5).

A few urban centres – 'world cities' and those linked to them (Ch. 1) – have benefited most despite regional policies. As Stöhr and Tödtling (1977: 35) put it, 'Backwash effects – social, cultural, and political as well as economic – have prevailed, and the determinants of change have become vested in a few functional and geographical centres, on the impulses of which the rest of the system has increasingly become dependent.'

As this chapter has shown, development policies were at first largely oriented towards neoclassical economics, export-base theory, large-scale organizations, and formal institutions. During the 1980s development policies emerged to embrace a much greater orientation on local and informal mechanisms and on stimulating private investment and entrepreneurship (Marelli 1985). As Chapter 8 will elaborate, not all areas are equally prone to success in new firm formation. It remains a process heavily biased towards areas where entrepreneurship has occurred previously, and tends to reinforce the growth of places and regions where education and opportunity are greatest. It is not difficult to conclude that regional growth based on entrepreneurship could have the same concentrated pattern as that based on large-scale enterprises.

These sorts of characteristics are more appropriately put into the context of the power of large corporations and their channelling of growth and multiplier effects (Aydalot 1981; Pred 1977; Stöhr 1982). There are also a number of side-effects of traditional economic development mechanisms outlined by Stöhr (1982) in the particular context of peripheral areas, although they are likely to ensue in other regions as well. These include: accelerated introduction of external value systems, a trend towards sectoral monostructure, a reduction of the region's share of key functions, such as R&D and decision-making, increased import propensity, dependence on external economic decision-making, and reduction of intraregional functional relations. Although some of these appear to be related to the 'branch plant economy' problem (Watts 1981), they are complex processes related to the peripheralization of places (Ch. 1).

Linkages, multilocational firms, and technology

The current view of growth poles is that they are not 'dead' but that too much was expected of them, especially when the complexity of linkages is taken into account (Polenske 1988). As Boisier (1981: 79) puts it: 'In many cases or regions, the problem will not only be one of introducing destabilizers or poles within the regional system, but a problem of simultaneously creating a system of economic linkages and one of spatial linkages.'

The issue of industrial linkage was only beginning to be heeded when Moseley (1974) compiled his survey of growth centres. At that time, it was clear that few, if any, local linkages occurred among firms in the regions studied (Moseley 1974: 132–6). A few years later, Thwaites (1978) and Watts (1981) were able to document rather persuasively that linkages are few from the branch establishments of multi-plant firms located in peripheral regions. Pred (1976), on the other hand, was even more convinced that the spatial structure of large, multilocational organizations serves to channel the multiplier effects of corporate activity out of peripheral regions and towards the main urban centres. Growth transmission, in the presence of non-local linkages – especially of what we now call producer services – takes place upwards, rather than downwards, through the urban hierarchy. The structure of inter-firm linkages is quite complex, and comprises formal subcontracting arrangements as well as other inter-firm transactions (Fredriksson and Lindmark 1979; Hoare 1985; Scott 1983a, 1984). In Third World settings, the lack of efficient transport systems, which in many countries has never quite approached the final stage of the model in Fig. 1.3, reduces linkages with other places even further. Virtually everywhere, the structure of inter-firm linkages favours large cities and developed 'information-intensive' regions. The survey research of Montreal firms by Coffey and Polèse (1987a, b) found that the producer services on which firms depend are highly concentrated in large cities (over 300 000 population). Consequently, several services are typically purchased rather than provided by firms internally: consulting, personnel services, computer services, advertising, and engineering.

The similarity between the development experiences of advanced and underdeveloped countries is striking, despite obvious differences in their standards of living. Positive development is a product of the formation and maintenance of inter-firm linkages and continued entrepreneurship. Where these structures are lacking, development is more an aspiration than a reality. The critical position of technology in development is not a recent discovery (Hagen 1963). Technology has increased over the past century, and where pre-industrial craftsmen could formerly understand, copy, and improve technologies, this now requires not only literacy, but also knowledge of science and engineering. As a consequence, underdeveloped countries now must turn to the advanced countries for their capital equipment, thereby losing the benefits – multiplier effects, backward linkages, externalities – of the capital goods industries (Headrick 1988: 11).

The mechanistic view of technological change – that new technologies begin and are adopted first in large cities, which then benefit from economic growth – ignores the more complex reality found empirically. Pred (1966, 1977), Siebert (1969), and Thompson (1965, 1968) focused several years ago on the internal generation of innovation found in large urban regions through R&D efforts. This generation of new innovations may be 'growth-inducing' (Pred 1977), or not, depending on the structure of corporations and their allocation of activities.

Technological change does not benefit all places equally. Process innovations, those embodied in machinery, have become those least likely to advance the places where they are 'adopted'. Branch manufacturing plants are common locations of such new technology, and the benefits to their rural regions may be slight or even negative. Thwaites (1983) found employment losses were common in the locations of such early process innovations. By contrast, product innovation, the endeavours found in corporate R&D facilities, is commonly associated with employment gains. The separation of process changes from product innovations within large firms, in particular, casts spatial and industrial complexity into the 'steady' process found in models of economic growth (Sayer 1985).

Coffey and Polèse (1984, 1985) have suggested that the process of local development involves several steps: (1) the emergence of local entrepreneurship; (2) the growth and expansion of local enterprises; (3) the maintenance of local enterprises under local control; and (4) the attainment of an autonomous local control structure and of a local business sector. The generation of regional creative capacity (Perrin 1974) is one of the most vibrant concepts of contemporary development thinking, for it relies on the everyday activities and contacts of ordinary people. This framework is the core of economic development within a region, large or small. Linkages and entrepreneurship are intertwined, each depending on the other.

Conclusion

Macroeconomic forces, such as the oil crises of the 1970s, in addition to political changes related to supply-side economics, forced revision of regional development policies away from earlier redistributive policies. The current focus in the advanced economies is on eliminating bottlenecks in the supply of capital, skills, land, and other factors (Chisholm 1987; Marelli 1985). This efficiency orientation which grew during the 1980s has the consequence that 'the traditional policy tradeoff between equity and efficiency is more relevant now than before' (Bartels and van Duijn 1982: 105).

In response to the failure of conventional theory and policies to generate regional growth in peripheral regions, local or 'bottom-up' development strategies have been proposed to attempt to make local development independent of global processes (Riddell 1985; Stöhr 1981; Stöhr and Taylor 1981). The *indigenous potential* of a region and its technological capability have emerged as major elements in regional economic growth. This indigenous potential is largely embodied in innovative potential, local skills, and entrepreneurship (CURDS 1979; Ewers and Wettmann 1980; Pajestka 1979; Sweeney 1987). A great deal of the indigenous potential is reflected in the technological capability of a region, a multifaceted concept to which Chapter 4 is devoted.

From the point of view of regional policy, few new initiatives have been submitted. At one level, the political sentiment has swung towards greater reliance on 'the market' rather than on government or state action, a shift which has become the hallmark of the 1980s and 1990s. In the Third World, the World Bank recently has shifted its emphasis away from large-scale projects to an increased focus on private industry, if not quite small-scale entrepreneurship (Cook and Kirkpatrick 1988; Hemming and Mansour 1988; Pfeffermann and Weigel 1988). The rather abrupt shift from a public sector to private sector emphasis is a key change of the 1980s, but it is too soon to tell if benefits from such a policy will spread more widely than did those from the emphasis on large-scale public projects in the past. The policy choices made in the USSR and Eastern Europe during the 1990s will be important ones in this regard.

A second major shift is a prominent decentralization of policy initiatives to the local level (Albrechts *et al.* 1989; El-Shakhs 1982b; Hanson 1983; OECD 1987). There is a similar 'balancing' issue here, to produce 'hybrid' development strategies, not 'top-down' or 'bottom-up' policies alone (Weitz 1986: 81). Albrechts *et al.* (1989: 4) suggest that this is a necessary – if long unrecognized – aspect of the regional development process. A top-down approach on its own '. . . runs the risk of failing to fully utilize local (historically evolved and accumulated) potential, whilst an overemphasis on a bottom-up approach can deny – or at least underestimate – the importance of linking local conditions with broader structural tendencies'.

This chapter has shown that regional policies, based initially on the orthodoxy of neoclassical economics, were not up to the task assigned to them. Convergence of regional incomes, spread of economic growth to rural areas and backward regions, and formation of linkages within these regions were among the Utopian expectations. Neoclassical economics, even less applicable to regional situations than to national ones, remains a powerful force because of the dominance of neoclassicists in academic economics. Cumulative causation, while not universally subscribed to by theorists, has advantages over neoclassical processes. It may be expressed more as social than as spatial inequality, however, as the spatial division of labour has made production more discontinuous over space (Gertler 1986b: 81).

The evolution of regional planning and policy involved little more at times than observations of reality – and often not enough of those. These observations have led to a cognizance that economic growth* is possible, but mainly in a small number of dominant locations. These locations contain an array of decision-making functions of powerful corporations as well as a collection of upper-level service functions that serve these corporations. At middle levels of the urban hierarchy, even relatively large urban regions, which were significant during the industrial era, saw that business services and decision-making were not part of their capability (Noyelle and Stanback 1983; Stanback and Noyelle 1982). Having been relegated to production activities, but excluded from decisions, control, and research, many of these regions began a slide downwards in economic sovereignty (Beauregard 1989).

The worst conditions of all are found in peripheral areas in the Third World. 'Third World countries cannot break the structures of underdevelopment by relying on the impulses generated in the heartlands of world capitalism' (Mittelman 1988: 177). Some NICs appear to be emerging successfully from the predicament of the rest of the underdeveloped world through exporting. As Chapter 4 will suggest, the reasons for this have to do with technological capability. This capability is able to stretch beyond a dependence on giant firms and their dominance to generate indigenous industry and economic activity, frequently based on local and traditional techniques.

Chapter 4

Technological capability: the core of economic development

This chapter continues to examine regional economic change from the perspective of places, regions, and nations. Now we turn to the capability and potential of places to control as well as to respond to economic change. By directing attention to local or indigenous capability, however, we see that microeconomic behaviour – the decisions and actions of individual people and firms – is more critical to the process of change than are larger, 'top-down' processes. As Dosi (1988: 1131) puts it, 'Innovative activities present – to different degrees – firm-specific, local, and cumulative features.' In particular, the ability of people and of organizations to experiment with,' to learn, and to become proficient in new technologies has a significance beyond that captured in past, or current, theoretical or policy milieux.

Rosenberg (1982: 8) describes the local variation in technological capabilities:

> One of the most compelling facts of history is that there have been enormous differences in the capacity of different societies to generate technical innovations that are suitable to their economic needs. Moreover, there has also been extreme variability in the willingness and ease with which societies have adopted and utilized technological innovations developed elsewhere. And, in addition, individual societies have themselves changed markedly over the course of their own separate histories in the extent and intensity of their technological dynamism. Clearly, the reasons for these differences, which are not well understood, are tied in numerous and complex ways to the functioning of the larger social systems, their institutions, values, and incentive structures.

This chapter examines the process of economic development as an outcome of the manner in which technology is generated, acquired, and utilized. The source of technology may be endogenous (or internal), generated primarily through research and development (R&D), a topic which Chapter 5 examines in greater detail. Alternatively, it can be obtained from exogenous (or external) sources – other people, other firms, other regions. As we shall see, if external sources are relied upon, the successful integration of new ideas and techniques demands an internal capability to assimilate external knowledge. This competence to discover, select, adopt, utilize, learn, and improve new technology is a key determinant of economic success of firms, of their employees, and consequently of the regions in which they are located.

It is clear that there is a great deal to be learned from the experience of Third World nations, as some succeed in developing a technological capability while others do not. Consequently, there is an emphasis in this chapter on the Third World, both rapidly developing and underdeveloped countries, relative to previous chapters, because of the special significance of the topic to such nations (Fransman 1986a, b; Fransman and King 1984; McIntyre and Papp 1986). The notion of technological capability provides an operational, if complex, concept for assessing the development situation of regions and nations.

The linear model of technological change

Technological change is a complex process whose workings are only partially understood (Gilpin 1975; Rosenberg 1982). The complexity lies partially in the diverse collection of phenomena which can be called 'innovation'. Bienaymé (1986: 139) distinguishes between:

1. Product innovations;
2. Innovations destined to resolve, circumvent, or eliminate a technical difficulty in manufacture or to improve services;
3. Innovation for the purpose of saving inputs (e.g. energy conservation, automation, robotization); and
4. Innovation to improve the conditions of work (security or safety measures).

These very different entities have made generalization difficult.

The 'linear model' of innovation (Fig. 4.1) provides the conventional wisdom which underlies most policy thinking about technology and economic development (Malecki 1990). This model suggests that the sequence from research through development to production and marketing is the standard or predominant path of innovation in both firms and national economies (Buswell 1983; Kline and Rosenberg 1986; Marquis 1988; Ronayne 1984). Mitchell (1989: 107) terms the linear process the 'bucket brigade': 'Someone in the research lab comes up with an idea. Then it is passed on to the engineering department, which converts it into a design. Next, manufacturing gets specifications from engineering and figures out how to make the thing. At last, responsibility for the finished product is dumped on marketing.' Companies have abandoned the linear approach as slow and inefficient. Kline and Rosenberg (1986) maintain that it was never such a simple process in any firm.

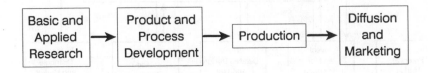

Figure 4.1 The linear model of technological change

If the linear model is assumed to prevail, the policy implications are also straightforward. If the level of R&D is increased, for example, a corresponding increase in technological innovation should follow. Therefore, to continue the argument one step further, government science and industrial policy must include measures aimed at achieving an appropriate balance between basic and applied research, since it is basic research from which innovation ultimately flows (Ronayne 1984: 64–5).

Contemporary economists stress the linear model's shortcomings, especially in light of the diversity of activities that make up the innovation process, the variation across industry lines, and the apparent disorderliness of the innovation process in reality (Dosi 1988; Kline and Rosenberg 1986). Kline and Rosenberg (1986) depict the process of innovation as shown in Fig. 4.2. Numerous feedbacks take place as solutions are sought to production problems, as new products are developed to meet a customer's needs, and as learning and process innovation take place during production (Casson 1987: 21–2; Teece 1988). Market-driven innovation and rapidly changing technology suggest to corporate managers that the linear model is too slow and fraught with organizational barriers which contribute to failure (Frey 1989: 7). R&D by firms can be thought to serve two functions: learning as well as the pursuit of product and process innovation (Cohen and Levinthal 1989).

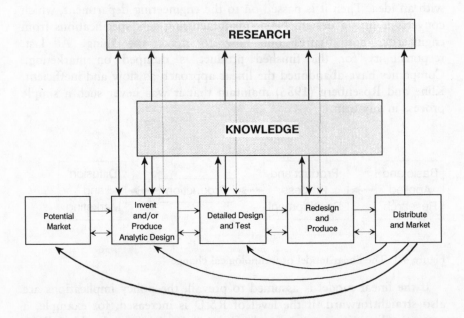

Figure 4.2 Feedbacks and interactions in the innovation process. *Source*: Kline and Rosenberg (1986: 290 (Fig. 3))

Both product and process innovation necessitate adjustments in an economy and among economies (Kuznets 1972). To a great extent, new product innovations are the greatest source of new jobs, since this type of innovation allows the creation of new firms and new industries to provide the new products. Computers, electronics, and biotechnology are prominent current examples of the job-creating effects of product innovation. Successive generations of innovations have permitted new firms to enter these industries and have created jobs in old firms. This process of 'creative destruction' occurs more in some settings than in others. Scherer (1982) found that in the USA, 74 per cent of firms' R&D efforts are aimed at new products, and only 26 per cent at internal processes. In Poland, by contrast, 48 of 86 innovations (56%) concerned process technology (Poznanski 1980).

Despite its simplicity, elements of the linear model of technological change remain appropriate for understanding the dynamic nature of technology within economic activity. Particularly when viewed within the framework of market competition, product and process innovation and competition over time highlight flows of knowledge that originate in scientific research and manifest themselves in products, processes, and new economic activity. This section discusses in turn each of the components of the linear model, with the aim being to show the possible direct paths to regional growth and development. Clearly, the route will differ among firms and is often more circuitous and indirect, because of the feedback loops which are present (Fig. 4.2). However, it is valuable none the less to assess the potential impact of each component.

Basic and applied research

The starting-point of technological innovation in the linear model of technology is basic research. Basic, or scientific, research is primarily the purview of universities and government research institutes, although some is conducted at the central laboratories of large firms. The feedbacks within the innovation process and the frequent and intermittent need for new scientific knowledge have led to a focus on the links between industry and academic institutions (Mogee 1980b; OECD 1981b). These links are also used as a justification for increasing local or regional R&D funding: the expectation is that firms will form these links in the area, and innovations and production will take place locally as well. Howells (1986), reviewing the evolution and variation of university–industry links, shows that there is little real connection on which regional economic development could be based. The most important ties are often made across long distances, and with a fairly small number of prestigious institutions.

Applied research is the typical entry point of industrial firms into the innovation process, since they commonly undertake relatively little fundamental or basic research. At the stage of applied research, however, the capabilities of many firms match or exceed those of many universities and research institutes. In the applied research stage, potential products are already envisioned, although considerable effort – more focused in nature than in basic research – might still be necessary. Drawing the line between basic and applied research in practice may be quite difficult; Rosenberg (1982: 149) considers it 'a hopeless quest'.

The biotechnology industry illustrates well the place of applied research in the technological innovation process (Kenney 1986). Biotechnology has followed rather closely the linear sequence from university-based scientific research through to the marketing of innovative new products. The example of biotechnology also exemplifies the entry of large firms into the innovation process. Firms primarily, but not only, from the pharmaceutical industry, have attempted both to influence the direction of research and to draw upon the entrepreneurial energies of small firms and university scientists through equity investments, joint ventures, and contracts with university researchers. As commercial products become available, both small and large firms are likely to have a part in the expanding market (Kenney 1986). The location of applied research is largely a puzzle, governed by the location choices of scientists and engineers and of large firms and their R&D labs. These choices tend to favour large urban areas, a point to which Chapter 6 will return.

Product and process development

Product development is the last major stage in the linear model of innovation, and is that towards which the bulk of industrial R&D is oriented. Product development and refinement customarily occur within product-line labs of large firms, rather than in central research facilities (Steele 1975). Interaction and 'coupling' with marketing and with manufacturing become critical as product design, production engineering, and market acceptance all must be assessed and incorporated into decision-making. These tasks recur intermittently throughout a product's 'life', and perhaps more within firms which specialize and innovate as part of their routine activity (Sabel *et al.* 1987).

Process innovation is typically distinguished from product innovation, as if they are truly separate. In fact, 'the same technical advance may be viewed as a product innovation by its producers and as a process innovation by those using it' (Baldwin and Scott 1987: 144). The firm originating the innovation may be at either the producing or using level, and diffusion may occur upstream to producers, downstream to users, or horizontally to competitors of the innovator. The vast majority of process innovations in the service sector, in particular, are product innovations from the manufacturing sectors (Robson, Townsend, and Pavitt 1988; Scherer 1982).

Economic competitiveness is a product of the ability of firms to operate at the *best-practice* level in their industry (Le Heron 1973). Production technology (machinery, equipment, management practices) ranges from the best currently known and available to the worst. The spectrum from best-practice to worst-practice technology is largely a function of the age or vintage of capital equipment employed. Newer vintages will incorporate or embody newer concepts, techniques, and knowledge which tend to give an advantage to firms – and regions – where newer technology is employed (Salter 1966; Varaiya and Wiseman 1981).

In less industrialized countries, much if not most technological change consists of the *adaptation* of imported technology to the local environment and factor supplies. 'Upgrading' traditional technology retains more local knowledge than does 'descaling' or 'scaling down' modern technology, which takes modern technology as its starting-point (James 1989). For example, scaling down a plant to smaller output and incorporation of manual rather than automated processes are common adaptations by Indian firms (Desai 1984). Price distortions and supply constraints provoked by government policies may be one reason for input modification. R&D efforts may involve the need to use different raw materials, to scale down to a smaller plant size, to diversify the product mix, to use simpler, more universal, less automated, and lower capacity machinery, and to stretch out the capacity of existing equipment (Teitel 1981). These adaptations must still take as their starting-point the technologies originating outside the local setting and intended for other economic and political conditions. In centrally planned economies, such as Poland, where supply constraints are often severe, innovations to overcome supply shortages or a lack of hard currency are especially common, and are applied to indigenous as well as foreign technologies (Poznanski 1980).

The purchase of machinery or capital goods from industrial countries by firms in less developed countries does not by itself contribute to economic development. There must also be considerable effort in learning the new technology in order to master it, to improve it, and to develop the capability to produce machinery. These activities, and especially the development of a design capability, require both investment in R&D and the presence of (initially the nurturing of) suppliers and subcontractors for production of machinery (Chudnovsky 1986; Enos and Park 1988; Fransman 1986c; Pack 1982).

Technically progressive firms obtain knowledge from customers and suppliers as well as generate it internally. A customer may request the imitation of an imported machine, but simplified and improved in various ways (Fransman 1984: 45). The development and nurturing of links to outside information networks go well beyond what is encompassed by R&D (Brugger and Stuckey 1987; Håkansson 1989; Sweeney 1987). In the American context, Miller and Coté (1987) suggest that subcontracting and purchasing from local firms is one of the best means for technology to diffuse widely in a region. Once again, Fong (1986: 192–95) recognizes that linkages, particularly from large establishments to small firms, ensure a rapid flow of technological know-how among various industries, reminiscent of industrial districts.

Production and the diffusion of process innovations

The diffusion of process innovation and the resulting distribution of best-practice (and older) technologies is a critical determinant of regional economic competitiveness. To a large extent, however, it remains somewhat a mystery as to why investment in new technologies varies from region to region. The simple and attractive finding, that the largest firms adopt earliest, has numerous proponents and supporters (Baldwin and Scott 1987: 129; Ettlie and Rubenstein 1987; Ewers 1986; Mansfield *et al.* 1977; Oakey, Thwaites, and Nash 1980; Rees, Briggs, and Oakey 1984; Rossini and Porter 1987). This finding is often used to justify the linear model of innovation, since large firms also do most of the R&D in any industry. However, Ray (1969) sees no such clear pattern. He notes that there is no definitive evidence that large companies have always been 'leaders in innovation and in the adoption of new techniques. The leading role which they often play in research and development, their generally more sophisticated management set-up, and their easier access to new capital are likely to give them a lead over smaller firms . . . in other cases it has been the opposite way round' (Ray 1969: 83). Rather than size of firm, Ray attributes differences in 'the attitude of management' is likely to have the greatest impact on the application of new techniques (Ray 1969: 83).

For regions, the conclusion often drawn from the association between firm size and diffusion is that R&D leads to general economic competitiveness (Meyer-Krahmer 1985). For developing countries, indigenous R&D is a major hurdle, since it demands managerial and technical skills not often found in local firms. The view that innovation is a continuous, non-linear process thrusts complications into this argument. Innovations – especially process innovations – do not remain constant during their diffusion (Gold 1979; Rosenberg 1976, 1982). Rather, in large part because most process innovations are the product innovations developed by supplier firms, improvements and modifications alter the characteristics of the innovation over time. Thus, adoption may lag because of the expectation, based on past experience, that newer and better innovations will come along in the near future (Rosenberg 1976).

A number of other factors contribute to delays in adoption, and again, regional variations are present. Adaptation and implementation within each adopting firm can greatly affect the rate of adoption of new technologies (Fawkes and Jacques 1987; Voss 1985). Perhaps even more important are access to information and channels of supply for the innovation (Ganz 1980) – factors which vary markedly from place to place (Brown 1981; Sweeney 1987). Information may be withheld or proprietary to an innovator, or it may be slow to accumulate as firms adopt and gain experience with new technology. The availability of 'expert information', often provided by third parties, such as consultants, government agents, and early users, may affect the speed of diffusion (Mantel and Rosegger 1987). The slowness of diffusion is often noted, and is part of the reason why equilibrium rarely characterizes economic change (Nelson 1981; Mowery 1988b; Ray 1989; Soete and Turner 1984). These institutional factors provide elements of a more realistic model of the diffusion of technological innovation, even if they are much more complex than what we have relied on in the past (Abraham and Hayward 1984).

One of the key pieces to the diffusion of innovations in industry is simply the organizational structure – the division of labour – of large firms. As Oakey, Thwaites, and Nash (1980, 1982) have shown in the UK, process innovations are implemented earliest in branch plants of large, multilocational firms. If a process is still under development, it is likely to remain in the South-east, where most R&D in the firms takes place; once it is readied for standardized, large-volume production, peripheral locations are more likely. A similar regional division of labour is evident in US high-technology industries (Glasmeier 1986; Malecki 1985a).

The diffusion of product innovations

The diffusion of product innovations is perhaps somewhat different. New products typically replace old ones but must be both manufactured and marketed. In British research, product innovations were 'adopted' earliest (i.e. produced) mainly in plants where R&D was done. In general, this can be thought of as typical of the pattern expected in the product/innovation cycle as the product moves from the innovation phase to the growth phase where volume production is the priority. The combination of product and process innovation at this point relies on interaction between R&D and production within the firm (Utterback

1979). The implications of this geographical pattern for regional development are clear (Malecki 1981b). Technological capability for new product R&D remains concentrated in some regions, where early production and associated new employment also take place (Schmenner 1982; Thwaites 1983). Process innovations, which usually result in employment declines as new capital equipment reduces the demand for labour, are nevertheless related to product innovations. Product innovations, especially radical ones, usually require substantial process changes as well (Ettlie and Rubenstein 1987). It is large firms, although not the very largest, which tend to dominate in the introduction of radical product innovation.

The technological capability or 'know-how' embodied in a technical labour force – whether employed in private or government-controlled enterprises – also provides a firm, region or nation with the skills necessary to evaluate new technologies as they emerge in other places. A key criterion in the evaluation process is the level of *best-practice technology*: the best products, the most efficient production methods, the most innovative markets. Technology does not remain static; new products substitute for old ones and new techniques allow improvements in production processes. The continual transformation of product and process makes it imperative that information acquisition concerning new techniques and actions of other producers be a routine part of business activity. As new technology and products are learned, acquired, evaluated, and improved upon, a firm or region comes to know about best-practice technology and to be prepared for new and substitute products.

The interaction between product and process innovation reinforces the organizing concept of best practice. As goods and services flow between producers and users a form of 'learning by interacting' occurs (Lundvall 1988). Firms which are active at product innovation tend to demand innovative components from their suppliers, thus reinforcing the concept of the 'user as innovator' (Foxall and Johnston 1987; Johne 1986; von Hippel 1979, 1988). Firms which are abreast of one dimension of technology (e.g. *machine technology*) tend to be competent in *procedural* or *organizational technology* as well. *Knowledge technology* integrates these dimensions within a firm and provides for the observed feedback which weakens the linear model of innovation (Shrivastana and Souder 1987). The amalgam of technology provides the impetus for coupling or integration of different corporate functions which is widely viewed as a critical element of the R&D strategy of firms. A variant on the three types of technology above are Casson's (1987) three forms of know-how: technical, marketing, and managerial.

Knowledge as conceived here also enhances the ability of firms to utilize profitably strategies for technology acquisition from outside the firm, through licensing, joint ventures, equity participation in other firms, and outright acquisition of innovative firms (Friar and Horwitch 1986). As Casson (1987: 12) describes:

> Know-how depends in turn on 'know-that' and 'know-who', and upon being 'known-of'. Know-that is the factual knowledge that underpins all successful problem-solving. Know-who is the knowledge of who is able to supply missing information and, more generally, of who is willing to buy or sell a resource that is not regularly traded on an organized market. Being known of means having a reputation that makes other people willing to offer information and, more generally, to become a trading partner.

For example, a reputation for product quality is invaluable in marketing a product, and a reputation for integrity and sound judgement is crucial in procuring finance for production (Casson 1987). The biotechnology example discussed earlier illustrates how large firms, which must tend a broad portfolio of products and technologies, seek out such technological opportunities. Graham (1986b) stresses that firms which acquire technology from outside must have substantial in-house R&D and it must be broader-based in order to integrate internal and external technology. In essence, firms – and regions – follow technological trajectories which build on existing strengths and which have difficulty dealing with new paradigms (Dosi 1982, 1984; Nelson and Winter 1982).

The integration between corporate R&D and other information and innovation-related functions typically has a geographical impact: R&D takes place at or near corporate headquarters, in order to communicate with corporate marketing, financial, and strategy functions. Within multinational enterprises, R&D and strategic activities may be retained largely in the 'home country' where it is headquartered. Subsidiaries elsewhere, then, possess only a 'truncated' structure oriented towards routine production and marketing (Hayter 1982; Etemad and Séguin Dulude 1986a). Consequently, the level of knowledge and information on the state of the art in products, materials, markets, and production processes is not available to these plants until and unless corporate headquarters decides to transfer them to other establishments.

The impact of the product cycle on regional development

Perhaps the dominant model in recent years for understanding the nature of technological change and regional economic development is the *product cycle* model and its corollaries, the profit cycle, the innovation cycle, and the manufacturing process cycle (Hayes and Wheelwright 1988; Krumme and Hayter 1975; Markusen 1985; Seninger 1985; Suarez-Villa 1984; Thomas 1975; Vernon 1966). These models describe the typical pattern of a product's development and production during its 'life', from R&D to market success to ultimate decline and replacement by new products (Fig. 4.3). Product development peaks early in the commercial life of a product, whereupon process innovation begins to take precedence as the firm prepares for

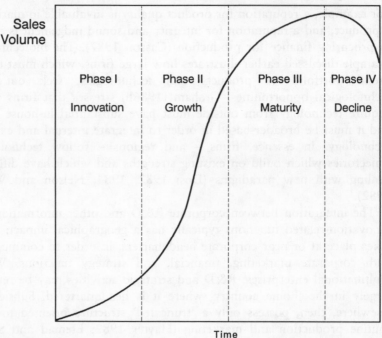

Figure 4.3 The product life cycle model

large-volume production (Utterback 1979). Other characteristics, especially those related to labour and management, also vary over time, in accordance with the innovativeness of the product and the technical expertise needed to produce it. In the innovation phase, as the product is begun to be marketed, firms require skilled labour, such as scientists and engineers, for refinements and improvements. The standardization or mature phase, by contrast, is characterized by shifts of production to low-cost, and especially low-wage, locations (Abernathy and Utterback 1978; Utterback and Abernathy 1975).

The product life cycle remains a fundamental concept in marketing, where it is used to generalize about circumstances across stages for a product class or product brand (Baker 1983; Onkvisit and Shaw 1986). The product cycle was also used by Vernon (1966) to describe the investment behaviour of multinational firms in the 1950s and 1960s, when shifts to foreign markets generally coincided with a decline in a product's innovativeness. Beginning in the mid-1960s, when infrastructure and management innovations began to permit low wages, rather than markets, to draw multinational investment, the product life cycle took on a more rigid interpretation, corresponding to the spatial division of labour. Some locations were preferred for R&D and new product manufacture; other sites – and entire countries – seemed to be chosen only for routine manufacture of mature products (Fröbel, Heinrichs, and Kreye 1980; Kolm 1988; Seninger 1985). Chapter 6 will return to the location decision. In terms of regional technological capability, however, routine production regions may operate far from global best practice. In other situations, state-of-the-art production processes and machinery are used, but these nearly always originate elsewhere and provide little opportunity for imitation or local control.

Markusen (1985) has suggested a *profit cycle* model which is analogous to the product life cycle and which clarifies several of its points (Fig. 4.4). Although Markusen presents her cycle model in the context of an industry (rather than a product) life cycle, the mechanisms are similar. As Fig. 4.4 shows, whereas profits are negative during the initial birth and design stage (I) of a product, the profit level which accrues to an innovation is greatest shortly after its introduction. This 'super profit' stage (II) derives from the monopoly position held by the innovator. Relatively soon afterwards (stage III), competitors enter the market, and the profit per unit drops sharply, as output approaches saturation. The fourth stage is uncertain. In some industries firms will be able to earn profits, whereas in others competition eliminates all but the lowest-cost producers. Indeed, negative profits prevail in the final stage (V) of an industry, when large corporations disinvest and the sector becomes the domain of small firms (Markusen 1985: 27–38).

Markusen believes that industry structure parallels these changes (Dosi 1984; Onkvisit and Shaw 1986). A large number of competitors gives way to an oligopolistic market structure, a small group of firms which dominate production in an industrial sector. Industrial sectors can

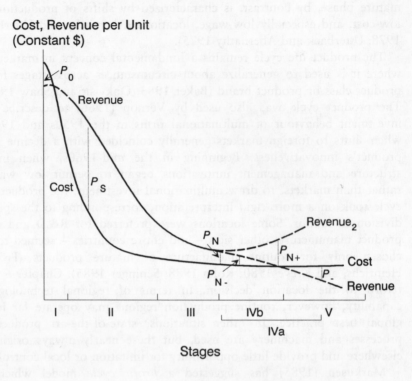

Figure 4.4 The profit life cycle model. *Source*: after Markusen (1985: 28 (Fig. 3.1))

then be classified according to their stage in the profit cycle (Table 4.1). Porter (1980), while acknowledging that the evolution of industries is more complex than this, also characterizes industries as 'emerging', 'mature', and 'declining', in addition to fragmented and global.

Generally, production-based industrial sectors follow an S-shaped pattern (Cooper and Schendel 1976; Utterback 1987). The analogy is a convenience to which few (e.g. Storper and Walker 1989) would expect strict adherence (Arnold 1985: 205–17; van Duijn 1983).

Table 4.1 Examples of industries in various stages of the profit cycle

Stage	Industrial sector
II	Frozen fish
	Wineries
	Aluminium
	Pharmaceuticals
	Semiconductors
	Computers
III	Soyabeans
	Cigarettes
	Canned fish
	Knit textiles
	Women's suits
	Motorcars
IV	Sawmills
	Steel
	Cotton textiles
	Women's dresses
	Brewing
	Shoes

Source: Markusen (1985).

The Abernathy–Utterback model of innovation (Fig. 4.5) in industry is complementary, but focuses on the types of innovation that take place at different points in the life of a product or firm (Utterback and Abernathy 1975; Utterback 1979). Product innovation is the core of the firm's activities in the early stages, because products are more easily patented and thus secure monopoly profits. The need for this technical activity wanes relatively quickly as the product is standardized for large-scale production, although it does not end, as Abernathy and Clark (1985) show in the case of the motor car industry, which remained innovative over many years. Process innovation, by contrast, is critical over a fairly long period: first to attain some degree of standardization in production which can be achieved with workers of lower skill than those in R&D who develop new products.

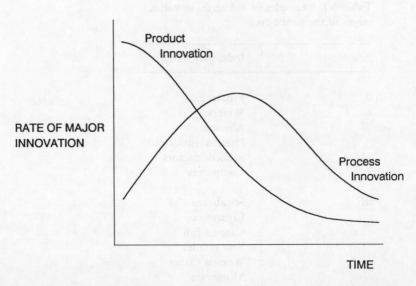

Figure 4.5 The Abernathy–Utterback model of product and process innovation.
Source: Abernathy and Utterback (1978: 40)

For products with short lives or with small production volumes, some modifications to the Abernathy–Utterback schema occur. On the one hand, the organization of batch production utilizing complex automation technologies has become commonplace. As Fig. 4.6 shows, these exhibit a high degree of knowledge complexity typically found only in continuous process production such as chemicals. On the other hand, even for products whose ultimate production volume may be large, process innovation must take place simultaneously with product innovation, because there is a need for continuous integrated innovation (Camagni 1988).

Dosi (1982, 1984, 1988) and Nelson and Winter (1977, 1982) continue this logic through the concept of 'technological paradigms', which define the ease of achievement of innovations and improvements. Technological opportunities vary from industry to industry. In 'oligopolistic maturity', technical changes are closely related to production, and technology is normally embodied into capital equipment, in contrast to the technology of new innovating firms which is 'people-embodied' in a community or informal network (Dosi 1984: 193–4; Shearman and Burrell 1987). The lack of structure and standardization in new industries shifts dramatically towards a more limited and limiting environment as a sector's life cycle progresses (Macdonald 1985).

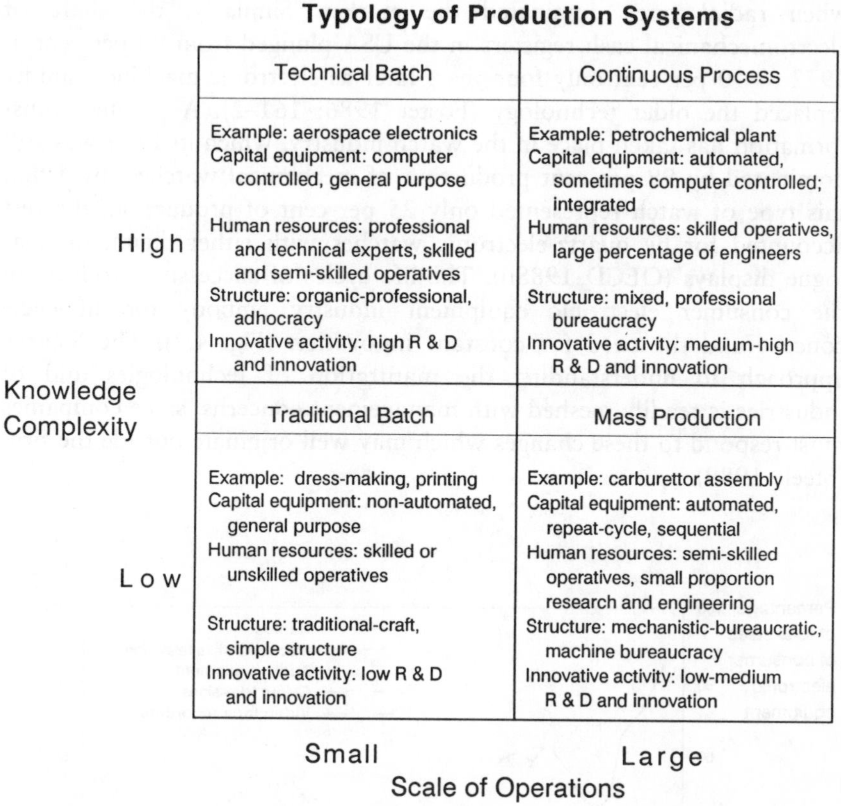

Typology of Production Systems

	Technical Batch	Continuous Process
High	Example: aerospace electronics Capital equipment: computer controlled, general purpose Human resources: professional and technical experts, skilled and semi-skilled operatives Structure: organic-professional, adhocracy Innovative activity: high R & D and innovation	Example: petrochemical plant Capital equipment: automated, sometimes computer controlled; integrated Human resources: skilled operatives, large percentage of engineers Structure: mixed, professional bureaucracy Innovative activity: medium-high R & D and innovation
	Traditional Batch	Mass Production
Low	Example: dress-making, printing Capital equipment: non-automated, general purpose Human resources: skilled or unskilled operatives Structure: traditional-craft, simple structure Innovative activity: low R & D and innovation	Example: carburettor assembly Capital equipment: automated, repeat-cycle, sequential Human resources: semi-skilled operatives, small proportion research and engineering Structure: mechanistic-bureaucratic, machine bureaucracy Innovative activity: low-medium R & D and innovation

Knowledge Complexity (vertical axis)

Small — Large
Scale of Operations

Figure 4.6 A typology of production systems. *Source*: Hull and Collins (1987: 788 (Fig. 1))

The technological opportunities present in an industry are often portrayed as an 'S-curve' that interprets technical performance, relative to the effort or expenditure on R&D, as increasing only slowly at first, exploding, followed by gradual maturation and ultimately decline (Ergas 1987; Foster 1986). In this view, the upper limit or technological ceiling of technical performance is the natural limit of a given technology (Ray

1984). Only if there is discontinuity, usually a shift to a very different technological base – a new or more efficient raw material or a more reliable manufacturing process – can a new product (and its producer) supplant an old one. Such a change can occur quite suddenly, as occurred when radial tyres replaced cross-ply (bias-ply) tyres in the 1970s. Cross-ply tyres held a 70 per cent share of the tyre market in 1977; this fell dramatically to less than 10 per cent only three years later when radial tyres dominated the market. Similarly, the share of electromechanical cash registers in the USA plunged from 90 per cent in 1972 to 10 per cent only four years later as electronic machines rapidly replaced the older technology (Foster 1986: 161–2). A parallel transformation has taken place in the watch industry, which in 1974 was still dominated by 99 per cent production of mechanical watches. By 1986, this type of watch represented only 25 per cent of production, the rest accounted for by quartz-electronic watches with either digital or analogue displays (OECD 1988b). The life cycles of successive products in the consumer electronic equipment industry embody the life-cycle concept and the need for constant innovation (Fig. 4.7). The S-curve approach to understanding the maturation of technologies and of industries is readily meshed with management concerns, since companies must respond to these changes which may well originate outside the firm (Steele 1989).

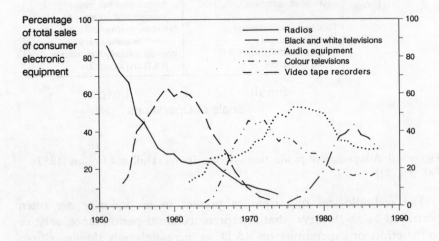

Figure 4.7 The life cycles of products in the consumer electronics industry: production composition of principal electronic equipment. *Source*: Vickery (1989: 116 (Fig. 1))

Generally, firms choose to compete either in new products, where innovativeness and newness are paramount for success, or in older, established products, for which low price and large volume predominate (Porter 1980). A *process life cycle* also highlights the difference between one-of-a-kind and low-volume products and those produced in large, standardized quantities (Hayes and Wheelwright 1988). A distinct contrast in the basis for competitiveness is present. Low-volume products, whether truly new and innovative or serving a small market niche, require attention to flexibility and quality. Volume production, on the other hand, stresses dependability, uniformity, and low cost. If a product or industry follows the life cycle from low to high volume, a 'shake-out' period ensues where competition shifts rather suddenly from product innovation to price, and from skilled labour-intensive to capital-intensive production, resulting in the demise of firms unable to make this transition (Olleros 1986).

A related *manufacturing process cycle* model has been proposed by Suarez-Villa (1984), who usefully contrasts developed and less developed countries in terms of organizational, labour, technology, and spatial outcomes over several stages. Managerial skills and non-routine activities are the norm in the early stages, and such activities concentrate in developed countries, as contemporary trade theories suggest (Krugman 1979; Nelson and Norman 1977). Regional specialization by stage of the product cycle is a common conclusion: innovative activities and new products made by skilled workers are found in some (usually large urban) regions. At the other end of the spectrum, mature products produced by unskilled labour gravitate towards regions with low labour costs (Norton and Rees 1979). The details of any firm or industrial sector are more complex than this dichotomy suggests (Dicken 1986; Morgan and Sayer 1988). However, it remains true that contrasts in development are largely a result of differences in labour skills utilized or employed in different regions.

Although firms and industries differ greatly in their specific responses to competition and technological change, perhaps because of 'corporate culture' as well as the firm's own 'technological trajectory', there are notable commonalities among competing firms (Dosi 1984, 1988; Malecki 1986c; Morgan and Sayer 1988). Despite its apparent linear nature, the advantages of the product cycle model are several: it emphasizes the labour as well as the capital needs of firms related to products at different phases; it emphasizes the ebb and flow of innovative activity observed in many industries; and it emphasizes that the location of economic activity varies with the type of activity undertaken (Utterback 1979). The volume of production largely defines the degree of innovativeness, non-routineness, and flexibility: small-volume products embody these qualities; high-volume products concentrate on cost and economies of scale (Hayes and Wheelwright 1988: 433). The product life cycle also represents the best-practice product technology in an industry. New products will be made only by those firms able to identify the combinations of inputs and markets that permit 'super profits'.

Flynn (1988) extends product, process, and technology life-cycle models to develop a *skill-training life cycle* in which 'products, production processes, and technologies are seen as dynamic phenomena whose skill and training requirements change as they evolve' (Flynn 1988: 7). This approach looks directly at the local and regional labour force impacts of technological change. One of the more innovative features of Flynn's framework is a concern with the provision of labour training. Although this typically is not a concern of regional analysts, how firms meet new, high-skill job requirements is a challenge equal to that of reallocating displaced workers (Ch. 9). The size of the labour pool needed for a particular skill varies across the life cycle, and thus the difficulty of finding workers to fill positions (which are at first ill-defined) is a predicament seldom explored in regional analysis. For example, if new skills are needed and are unavailable in the external labour market, firms commonly rely on training by equipment manufacturers. The labour market in the Lowell, Massachusetts area underwent a transformation from textile town to high technology over the past several decades, and worker training shifted to the local schools and technical institutes, which were somewhat constrained by a lack of suitable equipment for such training (Flynn 1988).

The product cycle model, especially its regional variant, has undergone a great deal of criticism in recent years (DeBresson and Lampel 1985; Steiner 1987; Taylor 1986). These criticisms basically address two elements. First, products rarely are constant and unchanging, even when in mass production and somewhat standardized. The Kline–Rosenberg portrayal of innovation embodies this dynamism. Improvements in products are made, often in back-and-forth competition between rival firms, and production techniques are enhanced.

Second, as has been stressed, not all products fit the pattern of achieving mass production where low cost and high volume are the norm. The standard typology of production systems stems from the work of Woodward (1965), who distinguished among batch, mass, and continuous-process production. Custom, or one-of-a-kind, production is often added to this list. Reprogrammable machines now permit changes in products to be made readily and routinely, so that batch production is highly automated, rather than reliant on traditional craft skills (Hull and Collins 1987). Figure 4.6 illustrates the types of production systems based on scale of operation and knowledge complexity. Many products, especially custom and small-batch products, never attain large volumes (DeBresson and Lampel 1985; Hull and Collins 1987; Kim and Lee 1987). Generally, product differentiation, a widespread corporate strategy, seems to contradict the homogenization of the mature stage of the product cycle (Porter 1980; Taylor 1987). The geographical dimensions of the possible shift to a new 'flexible' production are taken up in Chapter 6. The related extensions to industry life cycles and regional life cycles (Norton and Rees 1979), if not strongly related to the complexity of decision-making and resource allocation within firms and to contrasts in labour skill or knowledge, can vastly oversimplify descriptions of geographical outcomes (Taylor 1987).

It has been observed that the pace of technological change has become more rapid in recent years, especially when viewed in terms of the life of products (Goldman 1982). Short product life cycles are a direct consequence of product differentiation strategies, the purpose of which are to render older products obsolete (Baker 1983). As product life cycles 'shorten', illustrated in Fig. 4.8, only an innovative firm is able to reap large profits, since there is too little time for imitators to

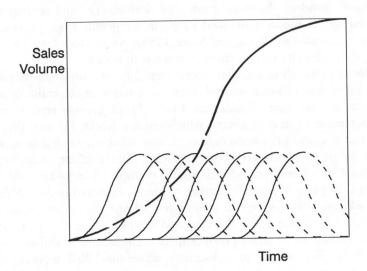

Figure 4.8 A succession of short life cycles

attain large-scale production before the product is replaced by the next 'generation' of technology. Since technology is cumulative within firms and regions, the innovative firm or country today is, other things being equal, among the most likely to be innovative tomorrow (Dosi 1984: 223). Fewer products fit the classic life-cycle pattern of long production runs and large volumes. In this setting, technology does not flow down or diffuse to less innovative countries; rather, relative advantages and disadvantages remain more or less constant. This cumulative 'technology gap' view of the international economy has many adherents (Dosi 1984: 219–25; Kaplinsky 1984b; Krugman 1979).

Third, product life cycles, when viewed in isolation, assume that products are independent of one another, although a product may be improved in order to prolong its 'life' (Fig. 4.9a). Other strategies by which declining industries can evade a decline in sales include extending the high-volume mature phase, usually done by line- or brand-extension, producing variations on successful products, and adopting new technology, which can generate a new, higher level of sales (Ballance and Sinclair 1983: 189–96; Hammermesh and Silk 1979; Porter 1980). The 'de-maturing' of products has been identified in the case of motor cars since the 1970s (Clark 1983), and televisions (Morgan and Sayer 1988; Rosenbloom and Abernathy 1982). Products also build upon one another and are interconnected in technological systems, families, or clusters of technologies. These might evolve as successive products following each other in rapid succession (Fig. 4.9b). This means that each 'new' product benefits from the knowledge and experience developed for its predecessors and its producer profits from previously generated externalities (Perez and Soete 1988). As product lives become shorter, their ultimate sales volume also tends to shrink.

Product cycles illustrate that some elements of corporate strategy make a great deal of sense. Indeed, they are a staple in the training of a generation of managers (Henderson 1984). Profit growth results from (1) a succession of new products, which require R&D, (2) attention to cost reduction, and (3) identification of new markets, especially niches ignored by other producers (Porter 1983a). The clothing industry, a low-technology sector, provides a useful example. The fashion design activities take place in a small number of places, such as London, Milan, New York, and Paris, where at most a few of any design are produced by skilled craftspeople under the direction of the fashion designers. If a design is chosen to be mass produced, its manufacture is shifted away from the fashion centres to a location where unskilled workers can perform the production tasks, such as cutting, sewing, and finishing. Routine activities for mass production of garments is thus commonly done by women in low-wage countries, especially in Asia. Indeed, some producers specialize on fashion and style, others on staple products for which style is less important than price (Steed 1978, 1981).

A   Extending a Product's Life Through Product Improvements

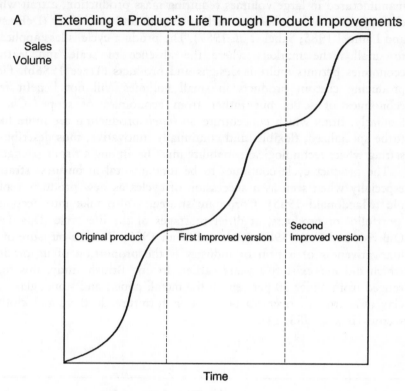

B   Successive Life Cycles of Low-Volume Products

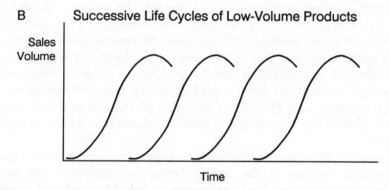

Figure 4.9 (A) Extension of a product's life; (B) successive life cycles of replacement products

The product cycle and its counterparts, while they do not provide universal applicability, capture the skill and knowledge differences among economic activities and types of products. Not all products are manufactured in large volumes requiring mass production, a trait which characterizes entire industries such as aircraft and machinery (DeBresson and Lampel 1985; Sabel *et al.* 1987). The product cycle is less applicable to small niche markets where the absence of scale and learning economies permits multiple designs and products (Teece 1986b). Firms producing custom products in small volumes will not benefit from economies of scale, but rather from economies of scope (Ch. 6). Similarly, firms which concentrate on *batch* production are more likely to be specialized, flexible, and continually innovative, thus describing a setting where technological capability must be among a firm's strengths.

The product cycle continues to be fundamental in business strategy, especially when seen as a succession of cycles as new products replace old (Macdonald 1985). Corporate strategies also take into account a 'portfolio' of products, at different stages of the life cycle (Day 1975; Onkvisit and Shaw 1986; Rink and Swan 1979). Thus, a measure of the innovativeness of a firm or industry is the proportion of its products which did not exist five years earlier. In one British study, this figure ranged from under 30 per cent in the metal, paper, and stone, glass, and clay industries to over 50 per cent in furniture, leather, and clothing sectors (Baker 1983: 14).

Technological change in the Third World

Research is a luxury affordable only by wealthy nations. The product life cycle conjecture that innovation begins in countries with markets for new products and services highlights both the contrast between rich and poor nations and the direction of technological dependence. R&D can be performed only by people who have attained a relatively high level of education, typically at least a post-secondary technical education. At higher levels of state-of-the-art research, a graduate degree (and usually a doctorate) is the entry ticket to research laboratories where such work is done.

Table 4.2 illustrates the disparity in educational opportunities found among the nations of the world. Tertiary education – that in all post-secondary schools and universities – includes vocational schools, adult education programmes, two-year community colleges, and distance

Table 4.2 Percentage of age group enrolled in primary, secondary, and tertiary education, 1985, selected countries

Country	Primary	Secondary	Tertiary
Ethiopia	36	12	1
Mozambique	84	7	0
Pakistan	47	17	5
Kenya	94	20	1
Haiti	78	18	1
Ghana	66	39	2
Indonesia	118	39	7
Malaysia	99	53	6
Thailand	97	30	20
Bolivia	91	37	20
Egypt	85	62	23
Brazil	104	35	11
Hong Kong	105	69	13
Portugal	112	47	13
Poland	101	78	17
Algeria	94	51	6
Saudi Arabia	69	42	11
Spain	104	91	27
Italy	98	75	26
United Kingdom	101	89	22
Netherlands	95	102	31
Australia	106	95	28
Federal Republic of Germany	96	74	30
Japan	102	96	30
Canada	105	103	55
Sweden	98	83	38
USA	101	99	57

Data are expressed as the percentage in each category of education relative to the population aged 6–11, 12–17, and 20–24, respectively. Thus figures higher than 100% result in some countries where school ages differ or where late entry is common.

Source: World Bank (1988: 280–1 (Table 30)).

education and correspondence courses. These data illustrate vividly the fact that primary education and, thus, literacy rates may be rather high in many countries, but the opportunity for education beyond a rudimentary level is still far from widespread. University education remains the province of élites in many, if not most, Third World countries, and the prevalence of study in the humanities, rather than in science and engineering, is also associated with these patterns. 'To create modern industries and businesses, entrepreneurs also needed information about foreign machines, technical processes, and business practices, information which was not forthcoming from the educational system' (Headrick 1988: 352). The gap between education and industry is yet another example of the dualistic nature of science and technology in less developed countries (Bhalla 1979; Stewart 1978).

The concentration of technical change in a few rich countries further reinforces disparities between rich and poor countries. By far the largest fraction of global R&D expenditure is located in developed countries, as is the greatest employment of scientists and engineers (Table 4.3). (Much of this is directed towards defence and space research which, it can be

Table 4.3 Scientists and engineers, 1985, selected countries

	Number of scientists and engineers (thousands)	Scientists and engineers	
		Per million population	Engaged in R&D (1980) (per million pop.)
Africa	1 623	3 451	91
Asia	32 670	11 686	272
Latin America and the Caribbean	4 746	11 759	252
Europe	37 369	48 600	1 732
North America	33 247	126 200	2 678
Oceania	1 105	48 213	1 483
Developed countries	81 247	70 452	2 984
Developing countries	29 513	8 263	127

Source: Unesco (1988, Tables 5.1, 5.2).

argued, has little benefit for either poor or rich nations.) Consequently, 'a fundamental distinction between rich and poor countries is that poor countries are for the most part *recipients* of technology developed in rich countries, while rich countries, as a block, generate their own technology' (Stewart 1978: 274). A further implication of this imbalance in research origin is found in modern trade theories within economics, which stress R&D and innovation as the sources of economic growth, resulting in a growing lag in incomes between innovators and followers (Krugman 1979; Nelson and Norman 1977). This gap in technological capacity, largely represented by scientists, engineers, and technical personnel, translates into technological dependence on the regions or nations from which technology and knowledge are generated (Stewart 1978: 116–40).

Perhaps even more importantly, the concentration of research in developed countries prompts other negative consequences. For example, agricultural research in wealthy, labour-short countries with temperate climates, such as the USA, will be concerned with economizing on labour and on developing mechanized equipment. Large, prosperous farmers may be able to adopt these innovations, but they will be well beyond the reach or situation of impoverished peasant cultivators (Griffin 1978). It also reinforces the dualism present between élites (wealthy farmers) and the bulk of the population who are poor and landless.

The inappropriate nature of technology in underdeveloped countries, then, goes beyond this simple example of a capital-intensity bias in agricultural research. The concern with temperate crops has also led to substitutes for tropical crops, as in the replacement of sugar cane by sugar beets after considerable research effort. Synthetic fibres and plastics have markedly reduced demand for tropical exports of rubber, cotton, and silk (Griffin 1978). The current furore over tropical oils, such as palm and coconut, contributing to heart disease will likely see their replacement by temperate substitutes such as sunflower, corn, and soyabean oils, again harming Third World economies.

When research is conducted in underdeveloped nations, it has often imitated scientific research in industrial nations, where most scientific research is conducted (Fig. 4.10). The consideration of which problems deserve attention is established in the West, and methods of investigation are learned in advanced countries' universities and laboratories (Juma and Ojwang 1989; Rahman 1979; Reddy 1979). Figure 4.11 illustrates the limited interaction of teachers, scientists and engineers with rural and urban poor in such dual societies (Reddy 1979). It is accepted that some connection with international scientific communities is important for underdeveloped countries, especially for monitoring research and keeping abreast of new developments (Shahidullah 1985; Wijesekera 1979). However, such a concern can neglect the needs of poor nations. Griffin (1978) cites as an example of such a scientific élite the fact that India conducts research on atomic energy while many households still use animal dung as their major source of fuel.

Traditional agricultural technologies do not respond quickly, if at all, to new techniques emanating from R&D. Most such agricultural

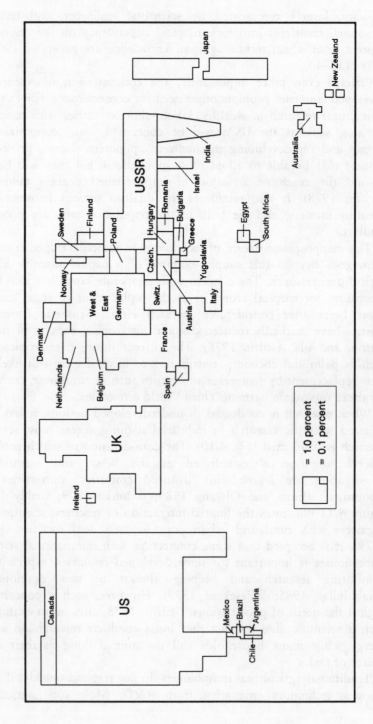

Figure 4.10 Scientific authorship by country. *Source:* de Souza (1985: 138) (Fig.1)

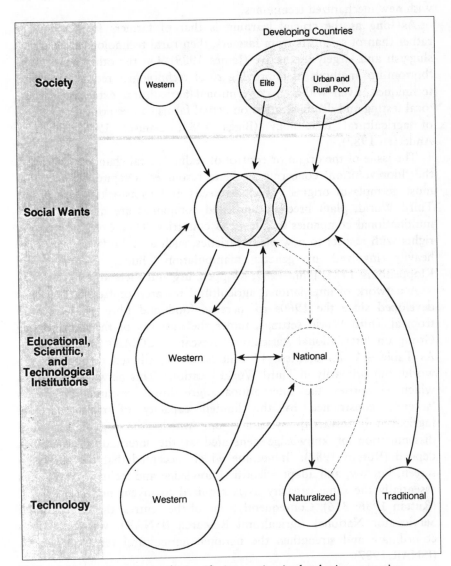

Figure 4.11 Dual structure of scientific interaction in developing countries.
Source: Reddy (1979: 97 (Fig. 3))

innovations take place via 'top-down' technology transfer, whereby
highly trained engineers, scientists, agronomists, and others develop
technologies in laboratories and experiment stations and then attempt to
transfer them to farmer 'clients' (James 1989). These innovations are
frequently costly, and therefore out of reach of the poor, and may even

harm the rural poor if they allow wealthy landowners to replace labour with new mechanized techniques.

As long as the crucial learning is that of farmers from scientists, rather than of scientists from farmers, then rural technological change is sluggish and largely ineffective (James 1989). On the other hand, most 'bottom-up' methods of improving rural technologies retain traditional techniques. Even within a conventional agricultural extension setting, local testing and demonstration on actual farms can be critical to success of agricultural technology (Biggs 1983; Gamser 1988; Pinstrup-Andersen 1982).

The issue of the origin or control of technological change is critical in the 'biorevolution' concerning genetic resources. Despite the fact that most germplasm origins – such as seeds and rootstocks – are in the Third World, plant breeders and seed companies are most frequently multinational companies in advanced countries. They exercise property rights with respect to new plant varieties, such as hybrids, and are now heavily involved in genetic manipulation (Buttel, Kenney, and Kloppenburg 1985; Juma 1989; Kloppenburg and Kleiman 1987).

A network of international agricultural research centres (IARCs) has developed since the 1960s to increase food and fibre production in tropical Third World settings, under the auspices of the Consultative Group on International Agricultural Research (CGIAR) (Ruttan 1986). As Table 4.4 shows, these 14 centres are distributed throughout the world, but primarily in Third World locations. This network of centres which comprises the international agricultural research system is 'severely constrained by the limited capacity of many national [agricultural research] systems', on which the adaptation and dissemination of knowledge generated at the international institutes depend (Ruttan 1982). 'It became widely accepted that the ability to screen, borrow, and adapt scientific knowledge and technology requires essentially the same capacity as is required to invent new technology' (Ruttan 1986: 316). Consequently, one of the centres, the International Service for National Agricultural Research (ISNAR), was created to co-ordinate and strengthen the national agricultural research systems (ISNAR 1987).

The success of the CGIAR centres and their interaction with national research systems has led to the development of a larger set of over 60 international agricultural research 'networks' outside the CGIAR. Three official IARCs have grown out of networks or were established specifically to promote networking among agricultural researchers (Plucknett, Smith, and Ozgediz 1990). The largest of the networks, the International Network on Genetic Enhancement of Rice (INGER) has led to the development of rice varieties in 43 different countries. These centres and networks are an attempt to break down the dualism that pervades research in and for the Third World. It is not clear that they have fully succeeded, and the dominance of the international research institutions and the involvement of large corporations have led to concern over the control of agriculture and of biotechnology in the Third World (Juma 1989).

Table 4.4 International agricultural research centres, 1985 (centres supported by the Consultative Group for International Agricultural Research)

Centre	Acronym	Year established	Location
International Rice Research Institute	IRRI	1960	Los Banos, Philippines
Centro Internacional de Mejoramiento de Maiz y Trigo	CIMMYT	1966	Mexico City, Mexico
International Institute of Tropical Agriculture	IITA	1967	Ibadan, Nigeria
Centro Internacional de Agricultura Tropical	CIAT	1968	Cali, Colombia
Centro Internacional de la Papa	CIP	1971	Lima, Peru
Asian Vegetable Research and Development Centre	AVRDC	1972	Shanhua, Taiwan
West African Rice Development Association	WARDA	1971	Monrovia, Liberia
International Crops Research Institute for the Semi-Arid Tropics	ICRISAT	1972	Hyderabad, India
International Laboratory for Research on Animal Diseases	ILRAD	1973	Nairobi, Kenya
International Board for Plant Genetic Resources	IBPGR	1974	Rome, Italy
International Livestock Centre for Africa	ILCA	1974	Addis Ababa, Ethiopia
International Fertilizer Development Center	IFDC	1974	Muscle Shoals, Ala., USA
International Centre for Agricultural Research in the Dry Areas	ICARDA	1976	Aleppo, Syria
International Centre for Research on Agro-Forestry	ICRAF	1978	Nairobi, Kenya
International Service for National Agricultural Research	ISNAR	1980	The Hague, Netherlands

Sources: Plucknett, Smith, and Ozgediz (1990); Ruttan (1986); CGIAR.

Production in the Third World

Production in the Third World often has little connection with formal R&D activities. Two cases may be examined as evidence of the constraints on technological capability. Malaysia has a great deal of investment by multinational corporations, but is struggling to attain technological proficiency. Tanzania, like many African countries, suffers from a lack of local skills.

The level of technology in a region or nation goes beyond R&D to encompass the stock of knowledge within firms concerning what is being made and how things are made, and to whom they can be sold. Fong's (1986) findings concerning the inability of Malaysian firms to seek out export opportunities are pertinent. Firms in the plastics, textile, and veneer and plywood industries, for example, are not major players in international markets. Fong suggests that 'besides the upgrading of technology, Malaysian planners ought to seriously consider means of "broadening the horizon" of Malaysian firms' (Fong 1986: 191).

Best practice or competitiveness can be assessed at the scale of national industries, as in the case of Malaysia, where manufacturing is considered 'the principal sector that could propel the country towards greater heights in economic organization and achievement, enabling the nation to attain the status of an advanced Newly Industrialized Country (NIC) by the end of this decade' (Fong 1986: vii). Fong (1986) examined five sectors: electronics, textiles, petrochemicals and plastics, iron and steel, and wood-based products, comparing 300 Malaysian firms with modern Japanese firms in those sectors. He found that a large technology gap exists between the industrial technology of the two countries (Fong 1986: 184–7). In a comparison of productivity, Malaysian sectors ranged from one-tenth the level of Japanese firms in iron and steel to one-half in electronics, Malaysia's leading export sector. In nearly all industries, this was attributable to the much higher capital intensity of technology employed by Japanese firms. There is considerable competition from neighbouring Singapore, where R&D in the electronics and computer sectors is increasingly concentrated, while labour-intensive operations have been shifted to Malaysia, where R&D is uncommon (Fong 1986: 74–6; Hakam and Chang 1988).

Fong (1986) emphasizes the causal linkages which lead from technology to productivity to competitiveness to performance and finally to profitability. This variant on the linear model of technological change is fairly accurate for process technology, the focus of Fong's analysis. 'By encouraging existing firms to modernize their equipment, and by promoting new ones which will utilize the latest technology, the productivity and competitiveness of Malaysian industry may be upgraded' (Fong 1986: 190). Singh (1988: 84) is less optimistic: 'The 'know-how' and 'know-why' of modern technology remain beyond the reach of Singapore and Malaysia; much of the more advanced activities are supervised by foreign technologists without greatly enhancing local capabilities.'

In Tanzania, industrialization has taken a slower route, and its shortcomings can be related to shortages of skills and technology. Imports of capital goods – mechanical equipment – in 1972 consisted of such machines as generators and boilers, tractors, metalworking machines, textile machinery, gas and liquid heaters and coolers, and lifting and loading equipment. Very little in the way of machine tools was imported, and only 10 per cent of the total capital goods imports were for industrial purposes; the balance was for agriculture and relatively simple machinery (Barker *et al*. 1986: 118–19).

The level of technological dependence in Tanzania is extreme. In importing technology, the country had too few nationals qualified in branches of engineering, economics, production management, laws of international trade and patent rights, and had very little access to global information on the sources and comparative costs of technology. The factories that are highly automated use little local skilled labour (Barker *et al*. 1986: 120–34). But the story is best told in the context of routine repair. Most maintenance engineers are expatriates, and there is a particular shortage of electricians.

> Therefore, companies have either service contracts with the equipment supplier or with an independent engineering service abroad, who in urgent cases send an expert from Nairobi or from Europe. We observed the case of a one week downtime of an entire factory because the fault (a broken electrical relay controlling the automatic boiler) could not be identified. An expert from another enterprise had to be called in to do the fault finding. The component could not be repaired, nor was it available in Tanzania or Kenya. Hence it had to be ordered from Europe. The whole plant and all the workers were idle for one week, because the boiler system is the key plant for the factory (Barker *et al*. 1986: 137–40).

Skills related to technological capability and R&D are rare; of 26 firms studied by Barker *et al*. indigenous design work was done only by 4 establishments. There is generally little concern with worker skills. Most companies seemed, to be more *concerned about the training costs and the labour costs* involved than about the skills required. A general fear by managers is related to the fact that formally trained managers are in great demand by other enterprises. Therefore any effort contributing towards more formal training would increase the labour turnover of formally trained staff, and be a loss for the sponsoring enterprise (Barker *et al.* 1986: 144) [emphasis in original].

Learning technology

Technological innovation is a learning process, or several processes, which overlap and feed back on one another (Rosenberg 1982). To appreciate technological capability,

> an important distinction must be made regarding the depth of understanding involved in the transformation of inputs into outputs. To take an example, while it might be possible for a firm to make an exact replica of another firm's product without an understanding of the fundamental principles underlying the product's working, to change the product will require a greater degree of understanding; the bigger the change, the greater the amount of knowledge that will be required (Fransman 1984: 33–4).

There are several forms of learning of new technology, and these form a hierarchy that usefully serves to indicate technological capacity itself as well as, more indirectly, the degree of dependence on external sources of technology (Bell 1984). Not only is R&D effort a source of technological improvement; many are by-products of production itself. Nobel laureate Kenneth Arrow's (1962) seminal work on *learning by doing* prompted recognition for relatively hidden forms of technological change. Arrow was concerned to explain how productivity increases as a result of production; 'learning by doing' represents that phenomenon. However, as Bell (1984: 189) points out, doing-based learning arises passively, is virtually automatic as 'doing' occurs, and it is costless, a free by-product from carrying out production. Rosenberg (1982: 120–40) suggests that experience-based learning ('learning by using') involves not just steady improvements in productivity but also incremental increases in understanding of design and performance of a product and the machinery with which it is produced. 'Learning by using', like 'learning by doing', receives no direct expenditure (Rosenberg 1982: 121) and is easily overlooked.

Particularly in the context of underdeveloped countries for which any expenditure on technical change is atypical, greater attention has been paid to learning. For many industries in such countries, technology – old and new – is seen only from the acquisition of machinery in a branch plant or a turnkey operation which originates entirely in another place and culture. The capital-goods industry – machines which produce other products – is a key element in the diffusion of technology among firms and among sectors (Enos and Park 1988; Fransman 1986b; Sabel *et al.* 1987).

A standard of competitiveness, *technological mastery*, is a goal for firms and nations alike. Adherence to this target affects, in turn, all the important decisions involved in technology development (Mytelka 1985). Such mastery involves a series of steps:

1. Production engineering, or the operation of existing plants;
2. Project execution, or the establishment of new production capacity;
3. Capital-goods manufacture, embodying technological knowledge in physical equipment and facilities; and
4. R&D, the specialized activity to generate new technological knowledge (Dahlman and Westphal 1982).

But, at a more practical level, the mastery of imported machinery is the first step of technology learning (Mytelka 1985).

Further learning involves explicit effort and investment in the acquisition of technological capacity in such a setting, although still perhaps outside the scope of conventional R&D (Bell 1984). The most elementary form of learning is *learning by operating*, a variant of learning by doing or by using. On the whole, the enhancements to operating capacities which result from this learning process are rather small. The second form of learning is *learning by changing*: improving upon equipment and techniques subsequent to gaining experience with them. When the 'black box' of technology is opened by investments in successive projects, technical changes can be quite major as principles are acquired and confidence in manipulating technology is gained.

A third form of learning involves monitoring and recording the performance of a technology. *System performance feedback* can generate understanding about why certain things work and others do not. It is clear that this information is neither automatic nor costless, not a function of time or production volume. Instead, it depends on the allocation of resources to generate the flow of data. Bell's fourth type, *learning through training*, retains an explicit element of dependence on external sources of technology. Personnel can be trained to operate machinery and to produce items – the *how* of technology – clearly without necessarily learning *why* – what is behind the technology. That this is a relatively high level of technological learning is illustrated by Bell's example of a set of firms in Thailand where very little training beyond basic operation of the plants took place over a nine-year period, and virtually no technical change had taken place either. Contrast this with the steel and petrochemical industries in Korea, where large numbers of engineers received several months' training at other sites. They subsequently made significant improvements in technology and productivity (Bell 1984: 196–7; Enos and Park 1988: 125–30).

Greater control over the knowledge carried about in people is represented by the fifth form of learning, *learning by hiring*. This form allows firms to create technological capacity, not simply to accumulate it. These may be newly trained professionals from local institutions, such as those hired by the Ducilo rayon plant in Argentina, where periods of technical change coincided with peaks in hiring technicians and engineering staff (Bell 1984: 197–8). Learning by hiring may be just as important as intra-firm accumulation of experience (Enos and Park 1988).

Finally, *learning by searching* assumes that an organization has the capability to investigate various sources of information, to absorb 'disembodied' knowledge and information about several types of technology, and to choose the most suitable one. This form of learning requires an explicit allocation of resources for non-production tasks, usually R&D. All of these costly means of learning serve to alter the capacity for technological change within a firm directly and, it can be argued, an R&D capability encompasses most, if not all, of these means (Cohen and Levinthal 1989).

Specialization into technological 'searching' activities is likely to take place within an organization as individuals specialize in devising better ways of performing tasks, better products, and better ways of producing existing products (Stiglitz 1987). These improvements likewise entail learning how to learn these better methods. Finally, it is important to recognize that learning itself has a cumulative character to it. What has been learned previously, by people and by organizations, has a strong influence on what is chosen (or in some cases, even possible) to be learned subsequently (David 1975). Thus, technological trajectories tend to be followed until there is a relatively abrupt shift to some new paradigm (Dosi 1988; Perez 1983). This sort of shift has been taking place in the clothing industry, which is beginning to be more automated.

The new technologies, however, require a capability with microelectronics that neither the equipment suppliers nor users in the industry have had. Many firms, led by the Japanese firms Juki and Mitsubishi, are now forced to increase their R&D effort markedly (Hoffman and Rush 1988: 169–72).

Likewise, knowledge can be forgotten: 'When one does not perform a task, one forgets the best way of doing it, and it takes a while to relearn the best way of doing it' (Stiglitz 1987: 126). The issue is related to concern at the national level about firms in an industry 'forgetting' how to produce products. American observers decry the tendency by American firms to subcontract production to firms in Asia and elsewhere (*Business Week* 1983; Cohen and Zysman 1987; Reich 1984). As less manufacturing takes place within a firm (or a nation), its employees gradually but steadily lose the capability to make improvements in production, and this proficiency shifts to the actual producers of the products. The producers learn by doing, by modifying machinery, and by incorporating enhancements from other producers, suppliers, and customers. In short, industrial competition now revolves around manufacturing processes and production skills (Cohen and Zysman 1987: 119–29; Rosenbloom and Abernathy 1982). Efficient production, by resulting in markedly lower costs, can drive competing firms out of business even if the latter are innovative in product design. The world-wide success of Japanese firms in semiconductors and consumer electronics – televisions, video recorders, and audio systems – has resulted from determined investment in production processes. Simplified designs using a small number of integrated components could be assembled by automated equipment. These new designs, in turn, sparked more sophisticated products (Arnold 1985; Porter 1983b; Rosenbloom and Abernathy 1982). Offshore production by firms attempting to produce using cheap labour likewise detaches a firm's design and marketing people from the manufacturing end.

The pace at which the monopoly profits from an innovation are eaten away depends on two things (Cohen and Zysman 1987: 126). First, it depends on the speed with which the product itself can be imitated. Second, it depends crucially on the speed with which a competitive manufacturing position can be built. It follows that an innovator's manufacturing skills determine his ability to continue to design and develop product innovations. In part, this stems from the loss of profits which would be ploughed back into R&D. In part, also, the experience associated with repeated production leads to the diverse array of information sources and knowledge which lead to ideas for further innovation. Equipment suppliers, input suppliers, and customers are among the possible sources of new ideas (Pavitt 1984; von Hippel 1988). Thomson (1989) adds *learning by selling* to the ways in which technical knowledge can expand. Especially for equipment and machinery producers, the experience of industrial customers and users gives rise to new improvements and inventions as alternatives to problems are sought. The inventiveness thus stimulates new sales, closing the circle on a circular, cumulative process of sales, learning, and invention.

Stiglitz (1987) and others stress the 'localized' nature of learning – localized not only within the firm, but localized to a specific technology or narrow range of production processes. Basic or scientific knowledge is applicable to many branches of industry, while technical knowledge may apply only to a single product and production process. Technical knowledge is more subject to obsolescence, as the motor car made obsolete much of the accumulated knowledge on efficient horseshoeing. Technical knowledge is so specific to particular processes, and improvements in one technique may leave others unaffected, just as each successive improvement in the capital-intensive technology of motor cars has changed horse-and-buggy technology little (Stiglitz 1987: 129).

The synergies as well as the direct linkages among firms will vary according to economic sector and the ways in which innovative activities proceed in various sectors. For example, Pavitt (1984) categorizes industries into four types, according to their dominant source of technological innovation: supplier-dominated industries; specialized suppliers; scale-intensive sectors; and science-based sectors. These correspond to the ways in which, and where, knowledge is accumulated and utilized in innovation. The linkages among sectors inherent in Pavitt's approach (Fig. 4.12) suggest, in turn, the presence of what Bienaymé (1986) calls an 'innovation multiplier' which affects and benefits user and supplier sectors as well (Cohen and Zysman 1987; DeBresson 1989; Pavitt, Robson, and Townsend 1989; Robson, Townsend, and Pavitt 1988; Scherer 1982). A rich network of linkages between any sector and its related and supporting industries is a key element in Porter's (1990) model of national competitive advantage. To a large extent, this is a rediscovery of the original concept of growth poles elaborated by Perroux (1955), Lasuen (1969, 1973), and others (Ch. 3).

Knowledge is an important part of total cost in industries where technologically advanced production methods, advanced management practices, and large inputs of R&D are significant factors. The stock of intellectual capital – intangible assets embedded in employees through experience and education – needed by firms in these industries to produce their knowledge requirements can represent a large portion of their total assets. In the pharmaceutical industry it is estimated that the stocks of intellectual capital account for about 30 per cent of total assets (MacCharles 1987: 28).

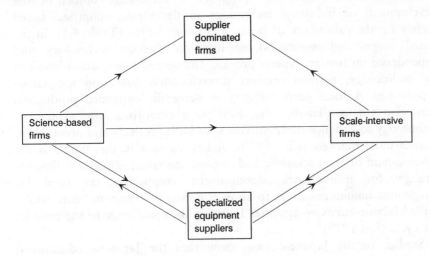

Figure 4.12 Linkages among firms as sources of innovation. *Source:* Pavitt (1984: 364 (Fig. 1))

In order for a firm to succeed in innovation, it must have not only an R&D capability, it must have production engineering skills and a network of information about competitors and other sources of technology and product ideas. This multifaceted view of innovation contrasts with most prevailing approaches, including one of the most common models for relating innovation to corporate activities, the product cycle model.

In addition to internal learning and specialization, localized learning can also be thought of in the context of a firm's external environment (Harrington 1987). Organizations (including firms and national institutions) need to gather information from a wide array of sources in order to optimize their activities. Reliance on agglomerated local labour markets of experienced workers helps in this regard, as do localized clusters of suppliers and customers, who furnish information that can be useful for technological innovation as well as for more routine endeavours. When a region or locality experiences synergy in innovation, in entrepreneurship, and in information, its firms tend to be technically progressive and to adopt best-practice techniques (Sweeney 1987: 201–7).

Learning and economic development

In the context of an uphill climb from dependence on external or foreign technology, Kim (1980) has proposed a three-stage model of the development of industrial technology in developing countries, based largely on the experience of Japan and South Korea (Table 4.5). In the initial stage, *implementation* of imported foreign technology and dependence on foreign experts prevail. The second stage, *assimilation* of the technology, permits product diversification based on indigenous capabilities. A local parts industry to serve the expanded production may also emerge. Finally, the third stage comprises *improvement* of technology to enhance competitiveness of both products and processes in international markets (Kim 1980). A key element in this model is the development of local scientific and engineering talent. Thus, a 'followers' strategy for technological development' emphasizes the need for indigenous human resources to allow a nation or region to 'shift' to the skilled labour-intensive activities found in the early stage of the product life cycle (Sen 1979).

Studies of the Japanese case show that the Japanese educational system was producing a stream of technically competent workers in the 1920s, in marked contrast to the situation in India at the same time, where illiteracy was still the norm (Otsuka, Ranis, and Saxonhouse 1988: 84–6). Japanese firms were able to imitate innovations from advanced countries, and then to 'accelerate the international product life cycle' by shifting routine production to nearby Asian countries (Chiang 1989: 344–5). Taiwan, on the other hand, has major bottlenecks of personnel in key engineering and management jobs, to some extent reinforced by the satellite production for Japanese firms (Chiang 1989). The speed with which these production shifts can occur is seen in the case of South Korea, which quickly acquired production of video cassette recorders (VCRs) shortly after their introduction by Japanese firms (Morgan and Sayer 1988: 74). Japanese firms then moved on to R&D and design of newer products (Chiang 1989).

Thus, 'a real catching-up process can only be achieved through acquiring the capacity for participating in the generation and improvement of technologies as opposed to the simple "use" of them' (Perez and Soete 1988: 459). The Taiwanese computer industry appears to have followed this model. In the mid-1980s, it was known primarily for its manufacture of low-cost 'clones' of personal computers. By the end of the 1980s, Taiwanese producers had moved into newer technology not far behind the industry state of the art (Yang 1989).

Table 4.5 Characteristics of Kim's three stages of industrial technology development

	Stage 1	Stage 2	Stage 3
New firms established through	Transfer of foreign technology	Mobility of local technical and entrepreneurial personnel	
Technical task emphasis	*Implementation* of imported technology	*Assimilation* for product diversification	*Improvement* for enhancing competitiveness
Critical human resources	Foreign experts	Local technical personnel trained at supplier firms	Local scientific and engineering personnel
Production technology	Inefficient		Relatively efficient
Predominant source of technological change	Transfer of 'packaged' foreign technology		Indigenous efforts
Predominant form of international transfer of technology	'Packaged'		'Unpackaged'
Predominant sources of external influence on technological change	Supplier and government		Customer, competitor
Market	Local (low competition)		Local and overseas (high competition)
Emphasis on research, development, engineering	Engineering (E)	Development and engineering (D&E)	Research, development, and engineering (RD&E)
Supply sources of components and parts	Mostly foreign		Mostly local
Important government policies	Import substitution and control of foreign investment		Export promotion
Role of local applied R&D institutions	Consultative	Adaptive development	Research and development (R&D)

Source: Kim (1980: 270 (Table 2)).

In areas other than military goods, developing countries have made great inroads as exporters of technology, mainly but not exclusively to other Third World countries. This has occurred across a spectrum of technological capabilities, albeit most frequently in routine production (Lall 1980). In each case, a great deal of local learning took place.

In the Third World, scientific and technological activities are also less well developed with regard to their linkage and integration with production activities (Fransman and King 1984; Silveira 1985). The separation in developing countries between R&D and production highlights the fact that neither fully accounts for technological capability (Bhalla and Fluitman 1985; Segal 1987; Sharif 1986). R&D is essential to the improvement and innovativeness of technologies over the longer run, and it makes it possible to obtain or adopt technologies which are nearer to the international state of the art. Production skills allow for continual innovation and significant inputs to the design of new products and processes.

The same separation of production from R&D and other forms of technological learning confronts the Soviet Union. Its foremost R&D capabilities are in the military and space sphere, where it has made notable accomplishments. However, transfers between the military and civilian spheres are few, and change is often actively resisted (Winiecki 1987). Soviet-type economies, according to Winiecki, absorb technology so slowly and so late that a 'followers' strategy' does not work, and virtually all products are mature. Chapter 7 returns to the issue of innovation in the Soviet system.

Firms must be able, once they have created or acquired technology, to nourish and sustain it in production. Especially for process technology, learning by using and by doing are essential components of the technological activities of firms. Firms have three ways of adding to their stock of knowledge: through learning by doing and using, through R&D (or learning by studying), and through imitating others (diffusion between firms) (Dutton and Thomas 1985). These types of knowledge and skills, once acquired and learned by an organization – Nelson and Winter maintain that 'individuals are complex organizations too' – then reside as organizational memory or routines of people and organizations (Nelson and Winter 1982). Technological capability, although often measured by R&D effort or R&D personnel, includes a wider array of technological assets, and draws upon a wider variety of information sources, including customers and suppliers as well as in-house capabilities. Local concentrations of R&D provide an important, but not a comprehensive, indicator of regional technological capability.

Best practice and regional competitiveness

Technological capability in firms or in regions is not fixed or permanent, since technology and the abilities of competitors are constantly changing (Dosi 1984; Nelson and Winter 1982). It relies on firms, on their activities, and how close they are to the state of the art or best practice – itself a constantly moving target – at any point in time. Firms and countries which are not near the current 'technology frontier' both in science and in production find it increasingly difficult to keep up with changes in other places (Cohen and Zysman 1987; Katz 1982a; Spence and Hazard 1988).

The technological environment or climate of a region or nation depends in part on the legal and institutional structure within which firms must operate. Perhaps the most important role for government is the provision of a legal system which defines property rights, contracts, and bankruptcy (Crook 1989). Among the effects of such a legal system are rules by which companies know with some certainty the degree to which they can appropriate, or profit from, an innovation. Teece (1986b) provides a valuable typology of the 'regimes of appropriability', those environmental factors, other than firm and market structure, that govern an innovator's ability to capture the profits generated by an innovation. The most important dimensions of such a regime are the nature of the technology and the strength of legal mechanisms of protection. These mechanisms frequently work to the disadvantage of less developed countries in particular (Giaoutzi 1985). Patents, copyrights, and trade secrets (such as recipes and chemical formulas) are legal means of keeping a technology out of competitors' hands, but these do not guarantee that imitation cannot take place.

Most common is the evolutionary transformation of products and processes over time. In the early, 'preparadigmatic' stage of a technology's development, an innovation can become worthless quickly as new designs leapfrog it. As standardization of design takes place, an industry paradigm emerges, and the firm which gains most will tend to be the one which has a full array of *complementary assets* in order to exploit not only the innovative design but also to manufacture it competitively and bring it successfully to market (Delapierre 1988; Dosi 1984; Teece 1986a, b). Indeed, one kind of production may be physically impossible without another. Strength in electronic components facilitates the development and production of robots and computers (Morgan and Sayer 1988: 21). These assets include competitive manufacturing, distribution, service, a successful trade name, and complementary technologies (Chaudhuri 1986; Cohen and Zysman

1987; Itami 1989; Porter 1987; Teece 1986b: 289). A firm's technological capability, then, includes all of its complementary assets, and these allow a firm to profit from 'bunches' of related technologies when they appear (Delapierre 1988).

Traditional neoclassical trade theory refers to regimes of 'tight' appropriability and zero transactions cost, where it does not matter whether an innovating firm has an in-house manufacturing capability, domestic or foreign. It can simply engage in arms-length contracting (such as licensing and coproduction) for the sale of the output of the activity (R&D) in which it has a comparative advantage. In a regime of 'weak' appropriability, manufacturing may be necessary if an innovator is to appropriate the rents from an innovation. This is especially true where the requisite manufacturing assets are specialized to the innovation. If an innovator's manufacturing costs are higher than those of imitators, for example, the bulk of the profits will go to the imitators. Stobaugh (1985) illustrates with the example of petrochemical firms the importance of these activities beyond R&D.

Complementary assets are a concern in the USA at present, where 'hollow corporations' have lost manufacturing capability relative to those in Japan and elsewhere in Asia (*Business Week* 1986; Cohen and Zysman 1987). Employees in foreign plants of transnational corporations acquire skills and know-how not maintained elsewhere in the firm. Malaysian experts from Intel's Penang factory had to be called in to help set up the chip assembly line of the firm's new factory in Arizona in 1983, because none of its US employees had that expertise any longer (Dreyfack and Port 1986). In essence, this situation is a result of separating R&D from the firm's complementary assets (Teece 1988: 277).

The spectrum of complementary assets encompasses a range of capabilities which support and sustain the development and enhancement of technology. R&D capability is but one part of the range. The capabilities across the entire production process need not be present within a single firm. In fact, the strength of industrial districts lies in the inter-firm sharing of ideas and capabilities. The interconnectedness of sectors is a key aspect of the diffusion of product innovations from supplier sectors to user sectors (Pavitt 1984; Robson, Townsend, and Pavitt 1988; Scherer 1982).

For a region, links to outside information networks and a high level of technical skill define a level of best practice. Most of all, technological capability relies on people in technological roles within organizations (Håkansson 1989). A concern for regional innovative capability became a focal point in regional research a few years ago, but it was concerned almost equally with large and small firms: the overall technological level of the region was the focus (CURDS 1979; Ewers and Wettmann 1980; Thwaites 1982).

Not all firms are alike in their knowledge about, or capability to operate at, industry best practice. Small firms, in particular, vary widely in their orientation and capabilities (Taylor and Thrift 1983). Orientation towards national and international markets is one important distinction among types of small firms. Brusco (1986) suggests three types: the traditional artisan common to many rural areas; dependent subcontractors who produce parts and components to be included in products of larger firms which sell nationally and internationally; and small firms in industrial districts. The technological sophistication of the latter group of firms must be quite great, often for a large number of clients, which indicates that their level of quality and technical standards are high. In industrial districts, an additional type of firm is found: producers of machinery necessary for the production of the district's specialized commodity. In industrial districts, 'the local communities have accumulated managerial, technical and commercial competence and capacity. . . . This knowledge is not characteristic of one particular group but . . . is spread to all social strata' (Brusco 1986: 198). Case studies of the ceramic tile industry (Russo 1985) and of German textile-machinery firms (Sabel *et al.* 1987) illustrate that interaction between machinery firms and their users generates a continual flow of information concerning improvements and modifications (Fransman 1986b). In other words, information based on 'learning by using' flows back and forth between industries. Lorenzoni and Ornati (1988) suggest that firms in such 'constellations' of linkages are more willing to seek information from outside sources, such as consultants, universities, and other firms. Chapter 6 addresses in greater detail the workings of industrial districts.

Conclusion

Current thinking about economic change, including technological change, is that it is an evolutionary process (Dosi 1988; Nelson and Winter 1982). In the course of this process, firms which have better routines for search, production, marketing, and management generally 'will prosper and grow relative to those firms whose capabilities and behavior are less well suited to the current situation' (Nelson 1987: 21).

Much of technical change is cumulative within firms (Pavitt 1986a) and, as this chapter has attempted to demonstrate, within regions. First, firms accustomed to organized R&D on old technologies may be more likely to learn how to attract and use scientists and engineers in the new technologies. Second, any new development often depends on upstream or downstream improvements in more conventional technologies. Third, there is evidence that effective assimilation of electronics-based technology depends on the more general competence in management (Cohen and Zysman 1987). Technological accumulation along known technological 'trajectories' is much more likely than is leapfrogging of new countries and regions.

Despite considerable evidence that a linear process rarely occurs, and certainly not without complex interactions among activities, R&D remains an important means of searching for new technical knowledge and is a measure of the input to technological change at the national level, although perhaps not at the level of the firm or the region. However, R&D is costly and poses barriers to entry for new and small firms, and this has regional dimensions. Important informal processes, not captured by R&D statistics, take place through the monitoring of information and technological capabilities, through publications, technical associations, watch-and-learn processes, and personnel mobility. Learning by doing and learning by using provide major opportunities for technological improvements and represent 'externalities, internalized within each firm' (Dosi 1988: 1125; Dutton and Thomas 1985). Taken together, these mechanisms broaden considerably the scope of information-gathering activities required of firms. As informal activities, they are difficult to measure, if not to identify, and they are to a large extent embodied in people and in organizations.

The extent to which innovation depends on people, their accumulated knowledge and capabilities gained through experience, and the information and contact networks on which they draw, is a characteristic often overlooked in prior research (Dosi 1988; Doz 1989; Malecki 1989). Much of this knowledge is *tacit* rather than public and frequently consists of the know-how needed to link distinct bodies of technology. Kash (1989: 49) refers to this as 'fingertip knowledge'. A regional setting or environment likewise has its fixed skills and capabilities, and its trajectories, like those of firms, are defined by the prevailing technological paradigm, as embodied in its population of people and of firms. Even in the absence of inter-firm linkages, a 'collective asset' of groups of firms and industries within countries or regions represents a set of technological externalities (Dosi 1988: 1146).

These are the synergies referred to less explicitly by Andersson (1985) and Stöhr (1986a) in the context of creative regions and territorial innovation complexes, to which Chapter 7 will return. Economists have begun to realize that 'experiences and skills embodied in people and organizations, capabilities, and 'memories' overflowing from one economic activity to another tend to organize *context conditions* that are (a) country-specific, region-specific, or even company-specific and (b) as such determine different incentives/stimuli/constraints to innovation, for any set of strictly economic signals' (Dosi 1988: 1146). Thus, policy options also will differ from place to place.

The advantages in the innovation process appear to lie strongly on the side of large firms, which have the resources to afford both R&D and informal learning and to absorb the risks inherent in innovation (Ettlie and Rubenstein 1987; Hoffman and Rush 1988; Stobaugh 1985). On the other hand, small firms possess a natural flexibility and informality which can enhance innovation (Malecki 1977; Rothwell and Zegveld 1982). The chapters to follow address these two possibilities. Chapters 5 and 6 examine the structure and activities of large firms in their innovation and production activities, respectively. Flexible production, a hallmark of small firms and increasingly of large ones as well, is also a theme in Chapter 6.

Chapter 5

Innovation in the firm: high technology in the corporate context

As Chapter 4 has shown, technological capability has often been closely related to capability in R&D. While this relationship is accurate to some degree, it leaves aside a variety of ways in which new knowledge is acquired outside of formal R&D. The process of organized R&D remains a necessity for firms, regions, and nations that aspire to have control over their technological destinies. For nations and regions, indigenous R&D is essential if local values are to be incorporated in the products and process undertaken there.

The activity of innovation within and among organizations is the focal point of this chapter. 'The technical enterprise', as Fusfeld (1986) calls it, is propelled by industrial R&D. The importance of science and of uncertainty and unpredictability in the R&D process have made technology, broadly defined, a key element of the competitiveness of firms, regions and nations (Guile and Brooks 1987; Horwitch 1986). The direction and the timing of technological innovation, which remain controversial topics, provide a long-term perspective on technological change. The discussion leads to the recent preoccupation with high technology, a set of industries which has been at the centre of policies to rejuvenate old industries and regions and to set the pace for new sectors and places. There is considerable confusion, however, about what 'high tech' is, where is it going, and where will it go in the future. The chapter concludes with an account of the technological strategies of firms, an increasingly international enterprise.

The growth of corporate innovation

Before 1900, there was little, if any, organized research anywhere and individual inventors dominated the course of technological progress

(Jewkes, Sawers, and Stillerman 1969). The principal exception was the in-house R&D by German chemical firms, and the early efforts by a few major US corporations (Molina 1989: 183–95). By the 1930s, however, industrial research had become a major economic activity. In a 1928 survey of 600 American manufacturing firms summarized by Noble (1977: 111), 52 per cent of the companies engaged in research, 29 per cent supported co-operative research efforts, and 7 per cent had testing laboratories. As Hall and Preston (1988) suggest, the electrical and chemical industries depended on science from the outset of their development in the late 1800s. In addition, even prior to 1940, those firms which had the greatest technological capability, represented by in-house R&D, were also among the major contractual users of independent research organizations, such as Battelle Memorial Institute and Arthur D. Little (Mowery 1983a). 'Firms without in-house research facilities were handicapped in their ability to pursue R&D and innovation' (Mowery 1983a: 369).

Corporate R&D at that time, as now, was dominated by a small number of very large firms. In 1938, just 13 companies employed one-third of all American research workers; 45 employed one-half (Noble 1977: 120). 'The organization of industrial research gave rise to a new field of expertise – research management. If science was to be effectively controlled, scientists had to be effectively controlled; the means to such control was the fostering of cooperation among researchers second only to a spirit of loyalty to the corporation' (Noble 1977: 119). Describing research at General Motors in the 1930s, Leslie (1980: 499) also notes that 'the morale of researchers was crucial', particularly for the long-range research cultivated by the firm at the time. The 'professionalization of industrial R&D', and the distinct labour force on which innovation relied, demanded a different approach to management (Freeman 1982: 10–14; Mandel 1975: 248–73). Universities, even 60 years ago, were considered 'knowledge factories' and were called upon to meet the science requirements of modern industry in two ways: as suppliers of trained research personnel, and as suppliers of applied research (Noble 1977: 158).

The 'age of big science' – the 1950s and 1960s – saw a huge growth in corporate R&D, and saw both the distinction and the division of labour between university and industrial research organization diminish. R&D became an 'institutionalized and integrated sector of corporate activity' by the 1960s (Kay 1979: 69). This was particularly the case in the electrical industry, where the importance of physics became clear as electronics technology emerged (Graham 1985). During and after the Second World War, government funding of research by industrial firms had become especially important in industries that served military priorities: aircraft, electrical goods, and instruments (Graham 1985; Hall *et al.* 1987). Large firms continue to dominate the performance of industrial R&D, and thereby control many of the causes and effects of technological change.

Management of technology

As R&D across a number of fields assumed greater importance, research management itself became a growth business by the 1980s. Issues of staffing, structure, and strategy are the principal dimensions believed to influence successful innovation (Betz 1987; Roberts 1987). Firms require of their technical and professional staff a set of 'critical behavioral roles', in addition to technical skills (Roberts and Fusfeld 1981). Many texts and evaluations are concerned with managing people in whom creativity and contact with outside information are key to success (Allen 1977; Badawy 1988; Kono 1988; Roberts 1987; Westwood and Sekine 1988).

Corporate structure or organization is a constraint on the place of R&D both within the structure of the firm and in corporate strategy, and is regarded as a critical part of technology management (Dumbleton 1986; Galbraith and Nathanson 1978; Klimstra and Potts 1988; Roberts 1988; Souder 1983; Twiss 1980). Structure also constrains strategy, and companies have adapted to changing competitive conditions with new organizational forms (Graham 1986b; Kagono *et al.* 1985; Port 1989). A currently fashionable corporate organizational pattern includes the formation of internal venture groups to work on emerging technologies and to foster greater creativity and entrepreneurship in research. These venture groups attempt to counter the highly structured company hierarchies which have evolved during the past several decades (Burgelman and Sayles 1986; Graham 1986b; Mitchell 1989).

Strategy formulation, and attempts to anticipate technological developments and actions of competitors, incorporate R&D efforts as a long-term, as well as a short-term, investment (Ansoff 1987; Friar and Horwitch 1986; Kantrow 1980; Maidique and Patch 1988). The reason for a concern with tight management procedures and links to corporate strategy stems in part from the fact that R&D expenditures, in general, are a drain on company profits (Collier, Mong, and Conlin 1984; Hitt, Ireland, and Goryunov 1988; Morbey 1989). In the institutional framework of R&D, 'not only development but also basic research has essentially become an *economic* activity. Investment in R&D is an investment like any other, although it might include higher risks and higher potential pay-offs' (Kleinknecht 1987: 118, *emphasis in original*). Mitchell and Hamilton (1988) suggest that standard corporate concern with return on investment (ROI) is inappropriate in the case of R&D, which is directed towards knowledge building and reducing uncertainty. The short-term focus has led many companies to cut long-term basic research dramatically (Freundlich 1989). Instead of evaluating R&D expenditures in the same way as other investments, R&D programmes are more similar to stock options, which provide an opportunity to make a profitable investment at a later date (Ellis 1988; Mitchell and Hamilton 1988; Steele 1988). Indeed, much of corporate innovative effort is 'learning by trying', guided by a consistent corporate strategy (Rosenbloom and Cusumano 1987).

Because of its institutionalized nature, R&D expenditure by firms tends strongly towards some stable target, such as a fixed percentage of sales or allocations by other 'comparable' firms, rather than to fluctuate annually (Kay 1979: 72–7). The importance of consistency is seen in Hitt, Ireland, and Goryunov's (1988) study, in which R&D intensity was positively related to corporate performance only among firms which adhered to a single business or narrow spectrum of core businesses. Parisi (1989) reports on a longitudinal study of US companies, for whom R&D per employee was highly correlated with corporate profit margins, although not with return on assets. Among small firms, those with a full-time commitment to internal R&D consistently outperformed firms for which R&D was only a part-time activity. At the same time, it is not uncommon for small firms in both the USA and the UK to fail to account for, or to estimate accurately, either the cost of R&D or its contribution to firm performance (Oakey, Rothwell, and Cooper 1988: 118–34).

Science push or market pull?

A long-running controversy has ensued over whether scientific discoveries 'push' innovation in both timing and direction in a linear fashion or, alternatively, market demand 'pulls' innovations from company laboratories. Schmookler's (1966) work suggested that market needs and demands 'pulled' innovations out into the open (Langrish *et al.* 1972). Mowery and Rosenberg (1979: 139) have disputed such a conclusion: 'The uncritical appeal to market demand as the governing influence in the innovation process simply does not yield useful insights into the complexities of that process.'

Nelson (1987: 65–71) shows that links to science are important in accounting for process R&D, whereas the appropriability of an innovation is most important in the case of product R&D. For both types of innovation, the contributions of users – the market – are at least as important as the industry's own R&D. In the extreme 'customer-active' case, the idea for a new product, and occasionally the product itself, is created by the customer, who then finds a manufacturer able to produce it. The more conventional approach, the manufacturer-active model, also demands input and feedback from users and customers, particularly in the early stages of a product's life (Baker 1983: 47–49; von Hippel 1979).

'Science push' (or 'technology push'), by contrast, suggests that innovations must await technological progress, perhaps even major innovations.

'Everyone knows' what the customer wants, but progress in technology is required before the desired product can be realized. In my work in the computer, plastics and semiconductor industries, I have often been told that new product needs were often not a problem: 'everyone knows' that the customer wants more calculations per second and per dollar in the computer business; 'everyone knows' that the customer wants plastics that degrade less quickly in sunlight; and 'everyone knows' that the semi-conductor customer wants more memory capacity on a single 'chip' of silicon. Under such circumstances a new customer request is not required to trigger a new product – only an advance in technology (von Hippel 1979: 106–107).

Mogee (1980a: 181) summarizes current thinking: 'Basic scientific research seems to underlie technological change in complex and indirect, but important, ways.' The traditional dichotomy of technology or market demand may be instead roundabout. Voss (1984) suggests that users who are familiar with technology stimulate technical advance, whereas suppliers tend to seek to broaden their set of customers or markets for their limited range of technologies. User-active innovations, then, are stimulated by technology push. A major 'technology push', where basic science acts as a 'trigger' for a cluster of basic innovations, has been a factor in the case of biotechnology as it was with electricity and chemicals and, more recently, with plastics and polymers. These arguments seem to imply an exogenous explanation of technology, by which the economy receives strong development impulses from erratic forward leaps in natural sciences. An additional role may be played by suppliers of research equipment and materials, who are major determinants of what can be done and how science can advance (Mowery and Rosenberg 1979).

There is an alternative explanation for science–push, related to the purposeful effort undertaken by industry to create new products and processes. The evolution and rapid growth of industrial R&D laboratories were to some extent patterned after that of Thomas Edison at Menlo Park, and the success of the large corporations which capitalized on them. As Hall and Preston (1988) show, these were predominantly developed in Germany and in the USA during the late 1800s in electrical applications. The industrial laboratories of the large electrical manufacturing and chemical corporations provided industry with its first experience in large-scale organized scientific research. These labs typically supported fundamental, or basic, research as well as applied research (Noble 1977: 121).

From a more mundane perspective, Rosenberg (1982: 141–59) suggests that technological knowledge based on cumulative experience and learning often results from dissatisfaction with current technologies. The unreliability of the vacuum tube, and of the early transistors which replaced them, led to basic research programmes in solid-state physics, which in turn sparked microelectronics discoveries. Market demand is then 'created' for products for which no need was previously felt (Shanklin and Ryans 1984).

It is unrealistic to expect that either science push or market pull will hold in all cases; both technological opportunity and market demand must exist simultaneously (Mowery and Rosenberg 1979). Coombs, Saviotti, and Walsh (1987: 103) conclude: 'If any generalisation can be made, technology-push tends to be relatively more important in the early stages in the development of the industry while demand-pull tends to increase in relative importance in the mature stages of the product cycle.' Even this conclusion varies across sub sectors as well as across the life cycles of industries, as demonstrated in the chemical industry (Walsh 1984).

Technology: a long-wave view

New technology is not a recent phenomenon, nor is excitement over the promise of high-technology industries. The observation of 50-year cycles

Table 5.1 The four Kondratieff waves.

	First 1787–1845	Second 1846–95	Third 1896–1947	Fourth 1948–2000(?)
Key innovations	Power loom; puddling	Bessemer steel; steamship	Alternating current; electric light; automobile	Transistor; computer; CIT
Key industries	Cotton; iron	Steel; machine tools; ships	Cars; electrical engineering; chemicals	Electronics; computers; communications; aerospace; producer services
Industrial organization	Small factories; *laissez-faire*	Large factories; capital concentration; joint stock company	Giant factories; 'Fordism': cartels; finance capital	Mixture of large 'Fordist' and small factories (subcontract); multinationals
Labour	Machine minders	Craft labour	De-skilling	Bipolar
Geography	Migration to towns (coalfields, ports)	Growth of towns on coalfields	Age of conurbations	Suburbanization; de-urbanization; new industrial regions

Table 5.1 continued

	First 1787–1845	Second 1846–95	Third 1896–1947	Fourth 1948–2000
International	UK, workshop of world	Germany, American competition; capital export	USA, German leadership; colonization	American hegemony; Japanese challenge; rise of NICs; new international division of labour
Historical	European wars; early railways	Opening of North America; global transport and communications	World wars; early mass consumption; Great Depression	Cold war; space race; 'global village'; mass consumption
Role of state	Minimal army/police	Early imperialism	Advanced imperialism; science and education	Welfare state; warfare state; organized R&D

Source: Hall and Preston (1988: 21 (Table 2.2)).

of boom and bust, identified by Kondratieff in 1919 and later by Schumpeter and Kuznets, saw a resurgence in the late 1970s and 1980s. To summarize this work, four alternating cycles of boom and bust have characterized capitalist development since the late 1700s, when the innovations that marked the onset of the Industrial Revolution were put into action. Entire industries, created or transformed as a result of key innovations, have allowed capitalism to maintain its vitality and 'creative destruction' as new sectors – and regions and nations – became the seedbeds of each upswing of innovation-based growth (Table 5.1). Mensch (1979), in particular, helped to spark the resurgence of interest in these cycles or long waves, and placed emphasis on the swarms or bunches of innovations which have preceded each Kondratieff upswing. Especially important, according to Mensch, are basic innovations, which establish new branches of industry, and radical improvement innovations, which rejuvenate existing branches. Recent *basic* innovations include xerography and satellite remote sensing (Steele 1989: 268). Within those new or renewed sectors, the pioneering innovations typically are followed by a series of improvement innovations, corresponding to the life cycles of industrial goods, such as successive generations of computers, as discussed in Chapter 4 (Mensch 1979: 52–4). As Mandel (1980: 25) notes, 'a real technological revolution means, at least in its first phase, large differences in production costs between those firms that already apply the revolutionary technique and those that do not or do so only marginally'.

Radical innovations which constitute new technological revolutions or 'techno-economic paradigms' may well be disruptive for existing firms (Freeman and Perez 1988; Kleinknecht 1987: 202–3). This will show up in the competition among firms and industries over time. 'Entrepreneurs who have invested in the old technologies are of course not interested in the emergence of . . . substitution competition. And they might have ways and means to oppose it' (Kleinknecht 1987: 120). This is much the same as Cooper and Schendel (1976) and Soukup and Cooper (1983) have found empirically as established firms respond to technological threats, such as fountain pens following the emergence of ballpoint pens and transistors replacing vacuum tubes. Firms respond to new technologies – or the anticipation of new technologies – in their allocations on R&D and among sectors and types of technologies (Kleinknecht 1987: 123).

Ray (1980: 21) makes the important point that

> from the point of view of its impact on the economy, it is not the basic innovation but its *diffusion* across industry or the economy,

and the *speed* of this diffusion, that matters. Only the widely-based rapid diffusion of some major innovations can be assumed to play any part in triggering off the Kondratiev – or any other – long-term upswing (*emphasis in original*).

Innovation ceases to be a leading influence when improvement innovations are replaced with increasing frequency by *pseudo-innovations*, minor improvements and variations found particularly in consumer goods (Haustein and Maier 1980; Mensch 1979: 58). This syndrome is especially common among large firms and older sectors. 'New' food products which are merely variations of existing successful brands and 'new' annual models of motor cars with superficial exterior changes have often fallen into this category. The new flurry of micro-wave foods, which must be significantly different in composition and packaging from conventional products, are perhaps a step up from the 'pseudo-innovation' category. Recent automotive innovations have also gone beyond their former tentative character, now going far beyond the annual cosmetic changes common during the 1960s (Clark 1983).

Hall and Preston (1988: 25–6) summarize the role of technology in the context of Kondratieff long waves:

> Clusters of key interrelated technologies, developing through backward and forward linkages, are the real triggers of long waves. . . . We believe that they tend to come forward when the returns from existing investments are declining. . . . If the transition from one Kondratieff to the next requires not merely clusters of hardware innovations but transformations of the entire socio-economic framework, then in a sense the whole process is endogenous; the underlying mechanism is indeed the laws of motion of capital, represented by a falling rate of profit, which eventually must trigger not merely a set of technological innovations but also changes in the economic, social and political superstructure.

The clusters or swarms of innovations in each Kondratieff wave spawned the high-technology industries of their day, from the telegraph and telephone and the electrical innovations (batteries, generators, motors, and electrical transmission) in the 1800s, to radio, television, radar, electronics, and computers in the twentieth century. These are now merging, in Hall and Preston's view, into a 'convergent information technology' (CIT), integrating computers and telecommunications, which may be the carrier into the fifth Kondratieff. This wave – if the cycles adhere to their 50-year pattern – should begin its upswing about the year 2000 (Hall and Preston 1988). Information technology (IT) in manufacturing, with its emphasis on reprogrammability, offers the potential to extend capital-intensive techniques to areas of industry that have been notorious for their slow rates of technical change (Coombs, Saviotti, and Walsh 1987; Hoffman and Rush 1988).

Although the consensus view of long waves centres on the position of technology, other causes may be related to them as well. Wars also follow up-and-down cycles and, when they do not come as needed for economic renewal, can be substituted by military spending to stimulate an economy (Mager 1987). Wholesale prices and inflation also follow 50- to 60-year cycles, as do prices of key resources whose supply is strained. These resource constraints thus induce R&D on alternative resources (Volland 1987). The upsurge of coal and then of petroleum and natural gas in the third and fourth Kondratieff waves is cited as evidence which reinforces this hypothesis. Even after examining such factors, 'what becomes apparent is that the dynamics of an upswing in cycles are the result of increased spending induced by massive investment in new opportunities for profit', akin to Markusen's (1985) 'super-profits' (Mager 1987: 208; Mandel 1980: 60; Rosenberg 1982: 156). This view, based on the concept of industry life cycles, 'assumes that the steep part of the allegedly S-shaped lifecycle of leading industries will coincide with the long-wave upswing, and that the exhaustion of improvement potentials will be most clearly visible in the long-wave [downturn]' (Kleinknecht 1987: 122). Mandel (1980: 119) believes that the theory of long waves of capitalist development integrates capitalism's capacity for flexible adaptation to new and radical challenges, such as changed social and institutional environments into the overall history of the system.

Some elements of a historical assessment show striking similarity between the past and the present. In the late 1800s, the new electrical industries exhibited from their beginnings a tendency towards concentration of capital and the formation of national monopolies based on patent rights. This era also saw the birth of radio and the advent of entertainment broadcasting, a socio-organizational innovation which moved the forces of change away from the inventor-entrepreneur to the public and private networks and government regulatory bodies (Hall and Preston 1988). Technological revolutions have also demanded changes in types of machine systems and their own specific forms of organization of the labour process (Kaplinsky 1984a; Mandel 1980: 42–3; Marshall 1987).

Infrastructure investment both by the state and by the private sector has continued to be significant, first in railways, telegraph, and electrification, and currently through investment in satellites, space technology, and improved communications technology. Communications is becoming the basis of international competition (Cohen and Zysman 1987). Standardization and co-ordination are essential for infrastructure, however, and delays in this regard can thwart competitiveness. Such infrastructure systems have demanded very large-scale organizations. Several technological clusters or chains, such as the telephone and television, span across Kondratieff cycles, and these remain core technologies of the convergent IT.

These large-scale changes in technological regime, not in a single technology or industry, are the crux of the technological view of long waves (Perez 1983; Soete 1986). Radical innovations have important effects on neighbouring sectors and on the build-up of a supporting infrastructure. For instance, the rise of an motor car industry entailed not only the sale of cars, but also investments in road construction and traffic regulation, and in service and repair networks (Kleinknecht 1987). Likewise, the idea of frozen food originated in 1912, well before the infrastructure for transportation and storage were developed (Steele 1989: 62). The development of clusters of innovations depends not only on a radical breakthrough in technology, but also on the nature of the linkages among firms and industries (DeBresson 1989).

The geography of long waves

The location of success in new technologies can lead to dramatic decline in national influence. British decline stemmed largely from an inability to compete with the giant business organizations that had emerged in Germany and in the USA by 1900. Innovation no longer depended on individual inventors but on systematic laboratory research, an educated workforce, and a knowledgeable management who could combine technology and markets in complex combinations. The German and American firms established licensing and patent channels as well as production facilities in the UK, marking the technological dependence of the UK in new technological developments. As Jewkes, Sawers, and Stillerman (1969: 181) put it: 'The British have been peculiarly unenterprising in trying out new ways of organising research.' The largest part of the UK's failure may have stemmed from its inability to make the necessary sociopolitical adaptations to new technologies (Hall and Preston 1988). After the Second World War, despite a series of important inventions, the UK again faltered in industrial R&D, this time *vis-à-vis* the USA and Japan, and established a state bias towards military R&D without commercial applications, removing the nation from the first rank of technological powers (Hall and Preston 1988; Jewkes, Sawers, and Stillerman 1969).

The geographical locus of the core technologies thus appears to shift from one wave to the next. The high-technology industries of the 1800s were primarily developed in Germany, particularly Berlin, and in the UK, especially London. In the late 1800s, Thomas Edison's inventions were capitalized on by large US firms, such as General Electric, RCA, and Westinghouse. New international dominance and the success of Japan at production and marketing new technologies, although perhaps not at inventing them, is used to validate the role of the broader organizational and socio-institutional innovations, and of co-ordination by the state, in national prosperity. The decentralization of new IT to Asia, especially the newly industrializing countries (NICs) – Hong Kong, Singapore, South Korea, and Taiwan – also raises questions about the geography of the fifth Kondratieff at the global scale.

Some writers believe 'that regional economies, like national economies and the world economy, go through long waves of economic activity' (Booth 1987: 447). A strict 50-year interval is less frequently mentioned in the pattern of regional growth and decline. Like their international counterparts, however, regional long waves hinge in large part on 'waves of innovation that result in the creation of high technology industries. These industries in turn go through a life-cycle of development characterized by initial rapid growth and eventual retardation of growth as their markets become saturated' (Booth 1987: 448). These new industries tend to emerge in regions not already dominated by major industries, or in regions which have been in decline for some time.

Two dimensions of the regional long-wave view are important. First, the idea that innovation is relatively more abundant in new regions owes its logic to the idea that old industries and large, established firms thwart the development of new ideas and the finance of new firms. Mature enterprises (and regions) lose their entrepreneurial energy, and innovations and new sectors emerge in other places. This 'spatial succession' or 'regional rotation' has been a recurring observation (Goodman 1979; Markusen 1985: 42–50; Watkins 1980). Older regions, with infrastructure and a social and political structure geared to support the dominant employers, are less able to respond to entrepreneurs and to the needs of new industries (Hall 1982; Norton and Rees 1979; Scott 1988c; Watkins 1980).

Second, the growth of an industry in a region puts upward pressure on labour and land costs which drives economic activity to other, lower-cost regions. Particularly for mature products late in their product cycles, 'filtering down' the urban hierarchy is thought to take place, a dispersion from large urban regions where innovations unfold, based on an agglomeration of local skills and linked activities, to smaller regions where firms utilize unskilled labour as tasks are routinized (Markusen 1985; Thompson 1965, 1968). The prolonged decline of a regional economy due to the demise of major industries makes a labour market willing to accept work under less attractive conditions. This has been the case in Wales and the northern region of the UK, and much of the 'manufacturing belt' of the north-eastern USA (Flynn 1986, 1988; Goodman 1979; Morgan and Sayer 1985).

These are the conventional explanations based on product cycles and industry cycles. Non-routine activities, especially R&D and innovation, are the core of this logic. However, even activities which are non-routine can be standardized, as Hall *et al.* (1983) have shown in the case of the American computer software industry. In an effort to keep labour costs down, the development of computer software has been divided into relatively routine tasks, which programmers with fewer qualifications can perform, and critical systems integration tasks which are assigned to a smaller number of 'supertechs'.

Not all authors are convinced that the long-wave view is adequate to account for innovative activity, particularly in predictable 50-year intervals. Technology is a social outcome of the diverse decisions of large firms in their pursuit of profits, but it is not co-ordinated in such a way to predetermine cycles (Morgan and Sayer 1988). Further, it may not always be the case that new industries will not emerge in yesterday's regions (Marshall 1987), although the evidence from previous waves leads in that direction.

High-technology industries: problems of definition and significance

The discussion of long waves of technological development indicates that the periodic rise in prominence of certain industrial sectors has been a characteristic of innovative technology. From textiles and iron to motor cars, aircraft, and electronics, some families of interrelated 'high technology' industries have played significant roles in economic growth in their particular eras.

A basic obstacle relates to the very diversity of 'high technology'. Everyone knows what it is, but no two definitions are alike. The common understanding of the term encompasses a range from state-of-the-art basic research through the (rather routine) production activities of the chemical and electronics sectors. Thus, it is fair to say that a high degree of heterogeneity is implied by the popularly recognized label of high technology. A single definition is difficult, both because of multifaceted perceptions and expectations and because, in practice, available data prevent classification that can be meaningful at the level of the establishment as well as for entire industries. The prevailing approaches, reviewed by Markusen, Hall, and Glasmeier (1986: 10–23), McArthur (1990), and Thompson (1988b), utilize indicators available only at the industry level. Once industries are categorized as high technology, all establishments of all firms in such industries are de facto considered to be high technology. Industries, however narrowly they are defined, exhibit a wide range of technologies and behaviours (Dosi 1988; Walker 1985: 236). This diversity is ignored in virtually all classification schemes.

Two indicators are most commonly used to define high-tech industries: (1) R&D intensity, or the percentage of sales expended on R&D and (2) technical workers (scientists, engineers, and, often, technicians) as a percentage of the workforce. Thompson (1988b) refers to these as 'R&D/SE&T-based definitions', which ascribe an output (innovativeness) based on inputs (R&D or technical labour). R&D intensity is intended to capture the rapid rate of change in products and technologies and the importance of technological endeavours to an industry or firm, as indicated by the relative amount invested in the generation of new products and processes.

Employing R&D or technical employment criteria separately results in different lists of sectors – sometimes broader, sometimes narrower. For example, using a criterion of an average expenditure on R&D exceeding 3 per cent of sales, Maidique and Hayes (1984) found only five US industries to be defined as high technology. These are: chemicals and pharmaceuticals, machinery (especially computers and office machines), electrical equipment and communications, professional and scientific instruments, and aircraft and missiles. The OECD (1986a) classifies only six industries as high R&D intensity: aerospace, office machines and computers, electronics and components, drugs, instruments, and electrical machinery (Table 5.2). Given the growth of services, it is advisable to include within high tech service industries in which R&D intensity is also high, such as computer programming, data processing, and other services; R&D laboratories; management consulting; and commercial testing laboratories (Browne 1983; Malecki 1984a). It is still uncommon, however, to see these or similar services classified as high technology.

Table 5.2 Industries classified according to R&D intensity (R&D expenditure/output)

Industries	Intensity, 1980
High intensity	
Aerospace	22.7
Office machines, computers	17.5
Electronics and components	10.4
Drugs	8.7
Instruments	4.8
Electric machinery	4.4
Medium intensity	
Motor cars	2.7
Chemicals	2.3
Other manufacturing industries	1.8
Non-electrical machinery	1.6
Rubber, plastics	1.2
Non-ferrous metals	1.0
Low intensity	
Stone, clay, glass	0.9
Food, beverages, tobacco	0.8
Shipbuilding	0.6
Petroleum refineries	0.6
Ferrous metals	0.6
Fabricated metal products	0.4
Paper, printing	0.3
Wood, cork, furniture	0.3
Textiles, footwear, leather	0.2

Source: OECD (1986a: 59 (Table 2.11)).

The second standard definition of high technology, technical (or technical and professional) occupations as a percentage of the labour force, is intended to measure the technical inputs into production in addition to R&D. Using the criterion that employment of scientists, engineers, and technicians must exceed the national average for manufacturing industries, Markusen, Hall and Glasmeier (1986), for example, present a set of 29 three-digit sectors and 100 four-digit sectors. Such fine-tuning on an industrial level implies a precision ill-suited to a definition which lumps labour-intensive assembly plants together with new-product development.

Using both R&D intensity and technical employment indicators, the US Department of Labor has standardized to some extent the definition of high-technology industry (Burgan 1985; Riche, Hecker, and Burgan 1983). Despite the appeal of standardized definitions, they are often difficult to apply to places other than where they were developed. Hall *et al.* (1987), in a UK study parallel to that in the USA by Markusen, Hall, and Glasmeier (1986), were unable to replicate very closely the US definitions of high technology based on occupations because of discrepancies in industry classifications and in data. The OECD standard employs R&D, strictly delineated according to its 'Frascati manual', to define high technology industries (OECD 1981c, 1984, 1986a).

One also finds high technology defined in terms of industry growth rates, or on the basis of a more intuitive, but arbitrary, identification of science-based, emerging products and processes based on non-routine, state of the art knowledge (Botkin, Dimancescu, and Stata 1982). As Thompson (1989a: 136) puts it, 'if there is any "essence" to high technology, it is surely the "newness" and "difference" brought to products and processes through the application of scientific research'. Pre-market activities, largely incorporated in R&D spending that represents a financial drain rather than a profit, are key to a high tech identity. The quantitative (R&D/SE&T-based) definitions are strongly influenced by the way in which they are measured. The motor car industry, in which huge amounts of R&D and technical workers generate highly sophisticated products, is not usually defined as high tech because its annual sales are so large that R&D spending as a percentage of sales is below average. Likewise, firms in the capital-intensive petroleum industry have relatively small numbers of employees, so they are identified as high tech only when SE&T employment indicators are used.

Japanese thinking on high technology is somewhat different (Imai 1988: 206). High technology is a technology that is R&D intensive and 'system oriented', involving 'a package of technologies rather than individual technologies'. This definition is notable for its exclusion of R&D-intensive technologies that are isolated and independent, and its focus on 'those high technologies that are strongly system-oriented as well as forming the basis of the new economic infrastructure', including microelectronics, biotechnology, and new materials. The Japanese definition, then, appears to be oriented towards core technologies associated with potential long-wave upswings. This line of reasoning is also behind McArthur's (1990) diffusion-based approach. In this view, both *newly emerging* technologies and *widely diffusing* technologies are considered high tech. As Table 5.3 indicates, widely diffusing technologies are those which are already well along the way to being part of a new infrastructure. Newly emerging technologies are less widespread at the moment. This perspective retains the importance of technological flows among sectors (Ch. 4).

Table 5.3 Types of high technology: widely diffusing and emerging technologies

Widely diffusing technologies	Newly emerging technologies
Incorporation of microprocessors into products	Biotechnologies
Robotics	Alternative energy technologies (photovoltaic cells, solar, wind, coal gasification and liquefaction)
Computers in process control	
Electronic office machinery	New materials technologies (ceramics, superconductors)
Information technology and telecommunications equipment	
Software	
Composite materials technologies	

Source: McArthur (1990: 818–19).

Any sectoral definition of high technology has a serious shortcoming for the study of economic activity and its differential locational processes and impacts: namely, an identification as high tech is not disaggregated to the level of the individual plant, office, laboratory, or other business establishment. Secondary data sources group establishments and firms into industrial sectors according to final product or service, and this aggregation combines the R&D, new product production, mass volume production, and other activities related to that output. Thus, it is inevitable that new-product production and experimental development taking place at a manufacturing plant, perhaps utilizing the expertise of many technical people, will be considered high tech only if the entire sector or firm is so classified. The converse of this situation occurs when a sector is proclaimed to be high tech on the basis of aggregate indicators, despite the fact that most (or even all) of its output is quite standardized, produced in large volume, and mainly employing low-wage workers for routine assembly. Electronics and computers, two 'core' sectors of high technology, largely follow this pattern. Moreover, large multi-sector corporations operate across a wide spectrum of industry and product types, although the standard practice world-wide is to allocate firms to a single sector. In part as a result of this diversity, the R&D activities within an industry group vary widely (K Hughes 1988).

In addition, innovation and technological change occurs as well where no formal R&D is conducted, as Chapter 4 has stressed, through modifications, learning by doing, and other informal means of technological learning. These efforts are ignored in R&D figures and represent the bias evident in enthusiasm over high technology (Malecki 1984a; Morgan and Sayer 1988: 22, 37–40). Thus, concern over the state of R&D in the Malaysian electronics industry and its dependence on R&D conducted by the parent companies in Europe, Japan, and the USA (Fong 1986) may well be misplaced. Recall from Chapter 4 the fact that Malaysian experts were needed to set up an American semiconductor factory, demonstrating a high level of expertise outside formal R&D (Dreyfack and Port 1986). Fransman (1984) makes a similar observation with respect to the Singapore machinery industry, which has had a high level of technological capability despite a virtually total absence of R&D.

The technological intensity of an industry may be thought of as a composite of three distinct components:

1. The technological intensity of labour, relating to the amount of expertise and skill;
2. The technological intensity of capital, embodied in sophisticated machinery; and
3. The technological intensity of the product, indicated by the degree of industry investment in new products and processes (Bar-El and Felsenstein 1989).

This multidimensional approach to defining high technology results in the dispersal of sophisticated capital equipment, but little decentralization of R&D and skilled labour inputs to peripheral areas.

The product life cycle and its variants (Ch. 4) can help to clarify the meaning of high technology in a regional context. The non-routine work associated with R&D, prototype manufacturing, and administrative functions of a firm are found in different locations – a spatial division of labour – from routine manufacturing, goods handling, and back-office clerical activity (Glasmeier 1986). Non-routine and innovative activities common at the beginning of a product's 'life' require technical workers. Volume production largely entails the de-skilling of work tasks and/or automation of production, and tends to seek out favourable local labour conditions. Even the product cycle's critics (e.g. Taylor 1986) do not deny the relationships among labour skill requirements, production characteristics, and location types.

Thus, emphasis in high technology can be placed on innovative and non-routine activities, including R&D, but also on the application of technologies as they are adopted across a range of user sectors, including services. On the one hand, as Ch. 6 will further show, one of the principal locational constraints which firms face, both in their own R&D and in applications of technology from elsewhere, is a technical labour force. In addition, within a region, the magnitude of regional economic benefits or multiplier effects from inter-firm linkages may vary directly with the amount of non-routine activity, such as R&D, conducted at individual establishments (Hagey and Malecki 1986; O'hUallachain 1984b). Thus, it is more important to know what a firm is actually doing at a location than to know simply to which industrial sector it has been allocated. The point is often acknowledged, but the constraints imposed by data shortcomings typically prevent a suitable analysis at the establishment level unless extensive survey research is conducted.

In their strategy and behaviour, successful high technology firms exhibit a paradoxical combination of continuity and chaos (Maidique and Hayes 1984). Continuity is revealed in firms' adherence to a relatively narrow spectrum of products and technologies, reinforcing the ideas discussed in Chapter 4 concerning technological trajectories and cumulative learning within firms. At the same time, successful firms are adaptable and able to change fairly rapidly as new technologies present themselves, but only within a cohesive organization that relies more on communication than on structure. Marketing or selling high-technology products is also different than for other products, whose demand is known or readily estimated. Products may need to be tailored to specific customers, or markets may have to be created for new products whose characteristics and advantages are unfamiliar to customers (Hlavacek and Ames 1986; Shanklin and Ryans 1984). This 'supply-side marketing' reinforces once again the science-push flow of technological innovation.

Other elements of company strategy and organization may be somewhat distinct in high-technology firms from those in other industries. Co-ordination of marketing, research, and engineering must be tighter for products whose window of opportunity is small and whose total market is uncertain but may well be small. In such a market niche, quality and dependability rate more highly than price (Porter 1983a; Riggs 1983). In actual practice, the concern with high technology exaggerates its significance. As Jean-Jacques Dudy, European director of science and technology for IBM, says: 'In modern industry there's no such thing as high-tech and low-tech. You need all-tech to be competitive' (Peterson and Maremont 1989: 34).

The military dimension of high technology

The attributes of high technology – high amounts of R&D spending and a significant proportion of scientists and engineers in the workforce – suggest that military expenditure is closely intertwined with high technology (Browne 1988; Tirman 1984). Modern military spending incorporates a heavy reliance on expensive and sophisticated weaponry (Gansler 1980; Malecki 1984b). The driving force of military technology is one of technology push, production for a market that tends to buy the newest technology available, regardless of price. This relative insulation from conventional market pressures has also led to 'the establishment of a class of technologically advanced firms which are organized in such a way that they are quite unlikely to produce commercially successful innovations' (Horwitz 1979: 284). Kaldor (1980, 1981) in the UK context calls this the 'mummification' of defence manufacturers, channelling 'technical innovation along a dead end'. Thus, commercial spin-offs from military production are infrequent in comparison to civilian production. This remained largely true in both American and British military technology in the 1980s (Dickson 1983; Office of Technology Assessment 1989). The problem of constricted technology affects military–industrial enterprises in any national setting (Ball and Leitenberg 1983; Todd 1988).

Evangelista (1988: 4) estimates that, together, the USA and the Soviet Union account for 80 per cent of the world's financing of military R&D. He further estimates that nearly 50 000 scientists and engineers in the American workforce are employed in defence research and production. This R&D activity shows no immediate sign of abating, despite a *rapprochement* between the two countries, which some believe to be intended on the Soviet side as a means of freeing up resources from military expenditure for consumer-goods production (Galuszka and Brady 1989; Goldman 1987).

A wide group of industrial sectors profit from large defence budgets (Table 5.4). Aerospace and shipbuilding, in particular, rely heavily on military spending, as do large subsectors of the electronics and communications sectors. The American economic boom, generated by the Korean War during the 1950s and continued during the cold war of the 1960s, fuelled the development of important technologies, notably microelectronics. The growth of Silicon Valley south of San Francisco, California, was to a large degree a result of large budgets for research and procurement of experimental electronic devices (Saxenian 1983a; Wilson, Ashton, and Egan 1980). The 'turnaround' of the New England region, centred on Boston, likewise owed much to the expansion of military-oriented firms which produced missiles, computers, and communication equipment (Barff and Knight 1988; Ferguson and Ladd 1988). California has remained dominant in military spending in the USA across a spectrum of industrial sectors, but to a particularly marked degree in R&D contracts and high-tech industries (Malecki and Stark 1988). Thus, the 'defence perimeter' and sunbelt military spending may exaggerate the actual concentration of R&D and high-technology manufacturing, which continue to be directed towards the North-east and California (Rees, Weinstein, and Gross 1988).

Table 5.4 US industrial sectors most reliant on defence, 1979

SIC	Industrial sector	Value of defence shipments (in millions of dollars)
3662	Radio and TV communication equipment	$10 249.5
3721	Aircraft	7 932.5
3761	Guided missiles and space vehicles	7 778.6
3724	Aircraft engines and engine parts	4 649.3
3731	Shipbuilding and repair	2 850.7
3728	Aircraft equipment, nec	2 344.6
2911	Petroleum refining	2 183.2
2819	Industrial inorganic chemicals, nec	2 056.8
3764	Space propulsion units and parts	1 416.6
3573	Electronic computing machinery	1 268.0

Source: US Bureau of the Census (1981).

SIC = standard industrial classification;
nec = not elsewhere classified

To a large degree, however, the technological sophistication of military goods of warfare which accompanied the cold war era had some unanticipated impacts. Table 5.5 shows how US military production changed from an orientation on ships and low-tech products (tanks, vehicles, ammunition), which dominated during the Second World War and into the 1950s, to a product mix focused on missiles, electronics, and communication equipment (*Business Week* 1980). Electronic devices became especially important when made into 'avionics' as part of military aircraft. In addition, several projects in advanced manufacturing technology are subsumed under the MANTECH (manufacturing technology) programme of the US Department of Defense (Noble 1984). Other major R&D efforts are funded by the Pentagon's Defense Advanced Research Projects Agency (DARPA). These include the Strategic Computing Plan and, most prominently, the Strategic Defence Initiative (SDI), also known as Star Wars (Molina 1989: 69–89).

Table 5.5 Changes in the composition of military hardware in the USA (%)

Type of military goods	Fiscal year				
	1942–44	1953	1962	1980	1986
Aircraft	27.3	31.5	25.7	31.9	35.2
Missiles	0.0	0.5	33.7	19.6	21.5
Ships	26.2	6.8	7.4	13.1	9.6
Electronics and communication equipment	6.6	11.2	16.6	22.4	21.0
Tank-automotive, weapons ammunition and other	39.9	50.0	16.5	12.9	12.7

Sources: Bolton (1966: 123); US Department of Defence (1981, 1986).

Regional consequences as well result from the phenomenon of military spending. In thorough analyses of American and British high-technology industry, Markusen, Hall, and Glasmeier (1986) and Hall *et al.* (1987) found very high correspondence between industrial sectors directed towards military production and the growth of high technology regions in the two nations. Markusen (1986a) suggests that, although the USA has had no industrial policy, its *de facto* industrial policy for many years was based on military spending. A 'defence perimeter' along both US coasts has benefited much more than have interior sections of the country (Markusen 1986b, 1988; Markusen and Bloch 1985). Indeed, O'hUallachain (1987) suggests that US employment growth between 1977 and 1984 was primarily in those states and sectors oriented towards defence production, corroborating Markusen, Hall, and Glasmeier's (1986) findings. Subcontracting linkages, often thought to be a mechanism that disperses defence spending geographically, instead acts to concentrate such expenses in a small number of states (Malecki 1982, 1984b).

Likewise, in the UK, the south-west region, centred on the Bristol to London axis and the M4 corridor, has seen a steadily increasing concentration of defence contractors and government research establishments (GREs) since the Second World War (Boddy and Lovering 1986; Buswell, Easterbrook, and Morphet 1985; Hall *et al.* 1987; Wells 1987). Per capita defence spending is over 2.3 times the UK average in the south-west region, making it in some ways the British counterpart of California (Lovering 1988). However, the differences between the two regions are important; fewer spin-offs have characterized the British South-east and South-west, despite the similarity in high technology in the two countries (Breheny 1988).

The aggregate reality of British high tech is less auspicious than in America: substantial job losses were experienced in key high-tech sectors from 1971 to 1983. Only a few sectors were exceptions to this trend: computers and radio, radar and electronic capital goods – both of which owe their fortunes to the Ministry of Defence. On the whole, British high tech is a much more narrow, and less dynamic, phenomenon than in the USA. However, the prosperity present is geographically concentrated in the area surrounding London, which is virtually the only area of the UK to show high-tech job gains (Hall *et al.* 1987).

The agglomeration of military production in large urban areas also characterizes the situation in other countries. Canadian aerospace firms, which depend upon US military contracts, are geographically concentrated in central Canada, Ontario, and Quebec (Todd and Simpson 1985). German military production has its core area in Bavaria, especially in the electronics-related sectors, despite the presence of significant production centres in northern Germany (Kunzmann 1988). French aerospace production, much of which is military, has decentralized somewhat outside Paris, to Toulouse, Marseille, and Bordeaux, although the highest skilled activities such as R&D remain for the most part near Paris (Pottier 1987). These examples demonstrate the geographical concentration, rather than the dispersion, of technology, even in the face of strong government influence. On the whole, much of what is considered 'high technology' is military in character, masked by the nature of industrial and occupational categories. Their potential for benefit to the wider regional, national, and international economy remains limited.

Information technology

Information technology is a term commonly used to describe the combined utilization of electronics, telecommunications, software and decentralized computer workstations, and the integration of information media (voice, text, data, and image) (Frisk 1988). As IT has developed from several distinct technologies during the past few decades, it has had fundamental impacts on the way economic activity takes place. Manufacturing or production, once a very distinct operation from research, development or design, is now integrated into computer-integrated manufacturing and other configurations for flexible production (Ch. 6). More importantly, this integration has transformed the way in which companies are organized geographically. Computer networks can be organized in many different ways, corresponding to a company's overall structure, behaviour, and strategy, and according to the demands of the industry.

Information technology as an infrastructure-related set of technologies has allowed firms in services, such as banking, tourism, and consulting, to offer their services to their global customers in the locations where the latter do business (Moss 1987; Nusbaumer 1987b). Thus, the network of the computer service firm I P Sharp (Hepworth 1987: 170) looks very similar to that suggested by Friedmann's (1986a) 'world cities'. Moss (1986) and Warf (1989) argue that the new telecommunications infrastructure will favour large urban areas, especially those which serve as headquarters and financial capitals, and particularly the three major financial centres, London, New York, and Tokyo. However, Moss (1988) shows that the new urban hierarchy being shaped by telecommunications is favouring cities which are now also critical to corporate operations, including Hong Kong, Singapore, and Seoul. Howells (1988) observes that in Europe 'information rich' and 'information poor' regions are developing. Indeed, although centres and peripheries are being redefined, with few exceptions one can expect cumulative causation forces and agglomeration economies to account best for the future pattern of regional comparative advantage (Gillespie and Williams 1988). New communications services are unlikely to be introduced uniformly across space, and 'are increasingly likely to favour investments in the existing concentrations of economic activity and areas of current economic advantage' (Goddard and Gillespie 1988: 144). The evidence thus far supports this conclusion. Langdale (1983) has shown how innovations in telecommunications in the USA were introduced to connect the major metropolitan areas and, only years later, the networks became complete enough to serve many small places. Hepworth (1986) has shown vividly that the computer networks of several Canadian computers are structured to centralize control in Toronto. The concentration of data flows in Europe shows that these are heavily overrepresented in capital cities there as well (Gillespie *et al.* 1989).

On an international scale, certainly, the ability of a country to adopt new information technologies requires that it has an array of knowledge and skills and a number of prior technologies in place. Consequently, the nature of IT, as an example of a complex technological system, is that it appears to be widening the gap between leaders and followers (Antonelli 1986a). To some degree, this appears to be a result of government regulation of information technologies, which has slowed their diffusion in developing countries (Katz 1988).

Harvey (1988) contends that although telecommunications technologies have generated 'time–space compression', *control* over space and over far-flung corporate networks has become a competitive objective (Bakis 1987). Communications technology allows firms to overcome geographical restrictions, take advantage of time compression, and restructure business relationships by bypassing intermediaries while linking with desired organizations (Hammer and Mangurian 1987). Information can also be seen as a resource, to be managed or mismanaged (Bar *et al.* 1989). Antonelli (1988b) writes of 'the network firm' which takes advantage of IT or telematics for procurement, manufacturing, and marketing functions, in addition to control activities, such as accounting, forecasting, and planning. Generally, IT on a global, standardized network, serves to reduce co-ordination costs both within and among firms (Antonelli 1988a). IT is a key ingredient in systems of flexible production, which may utilize facsimile (fax) machines to transmit instantly orders for the next shipment of parts or components to arrive in the next few hours. Chapter 6 carries the implications of this technology further.

For developing countries, IT is seen more as a threat than an opportunity. The breadth of institutional changes demanded by major technological systems such as IT is much more difficult to effect in the Third World. Manufacturing technologies render labour costs less significant, and thus reduce the comparative advantage of poor countries for the manufacture of products such as clothing and electronics. The new technologies also threaten to make developing countries more technically dependent on developed countries (King 1982; Su 1988). Within these countries, should they be connected to a global network, the new technology is likely to remain concentrated within the large, primate city which serves as access point to the world economy. Little widespread diffusion can be expected (Clapp and Richardson 1984).

Technological strategies of firms

Technology is a principal means by which firms compete. It can be 'created' given sufficient effort and expenditure, evidenced at the national scale in the 'big science' space and energy exploits of advanced economies, in contrast to persistent problems of housing and waste disposal. At the same time, some problems appear to await scientific breakthroughs before they can be solved through technological effort, such as cures for cancer or AIDS. Technology begins, as Chapter 4 has shown, with basic research findings that might or might not result in marketable products. Such research is long term and thus easily avoided by short-sighted management who do not see a near-term pay-off (Rosenbloom and Abernathy 1982). The success of Japanese companies in capturing the VCR market over competitors in Europe and the USA is attributed to this short-sightedness. In addition, complementary assets, such as competitive manufacturing, suppliers, distribution, service, a proven trade name, and complementary technologies, are of course involved, but a firm cannot survive – or will not grow – unless it undertakes periodic, if not constant, technological change.

R&D is an activity with uncertain outcomes, whether due to technical, market, or policy matters (Freeman 1982: 148–55; Mogee 1980a). Some products end in failure on technical grounds; others are technical successes but fail to find a market; still others fail because of manufacturing problems. Overall, marketing factors, more than technical ones, have been found to differentiate between successful and unsuccessful new-product innovation (Johne and Snelson 1988; Rothwell *et al.* 1974). Clearly, 'pseudo-innovations' and other minor changes involve the lowest uncertainty of any kind, and will tend to be preferred by managers (Kay 1979: 78–83). Substantial research by Robert Cooper (1984a, b) suggests that a firm can reduce uncertainty even in new markets and new technologies if new products are closely related to existing products and if the firm's existing R&D, engineering, and production capabilities, and other 'complementary assets' are drawn upon (Cooper and Kleinschmidt 1987). In this way, both a producer and its customers 'learn' about the new technologies (Maidique and Zirger 1985).

Markets are more difficult to predict in the case of truly innovative products (Shanklin and Ryans 1984). Mitchell and Hamilton's (1988) comparison of R&D investments to options in financial markets is a response to this uncertainty. Finally, as emphasized in Chapter 4, communication and information flow are critical elements in innovative firms, whether with suppliers, users, or competitors (Håkansson 1987, 1989; Håkansson and Laage-Hellman 1984; von Hippel 1988).

R&D as an ongoing company activity may be especially vital for firms in stagnant industries, those whose product life cycles are longer and less numerous, such as household appliances and commodity chemicals (Clark, Freeman, and Hanssens 1984; Linn 1984). In these sectors, R&D is often for product improvements to increase quality – again keeping the technological frontier in sight. The innovation process in these industries is perhaps more important than in rapidly growing industries, because innovations are less frequent and more pivotal (Hammermesh and Silk 1979).

There are a number of ways in which firms utilize technology as a basis of competition (Freeman 1982; Porter 1980, 1985; van der Meer and Calori 1989). The path or trajectory a firm chooses will be based to a large degree on its accumulated knowledge (Ch. 4). A firm with an *offensive* strategy maintains a world-class research capability, usually in a central research laboratory, that is able to keep abreast of virtually any development that a competitor might unveil. This strategy obviously can be very costly, in terms of overheads on research laboratories and equipment as well as large staffs of well-paid researchers, but many large firms that are leaders in their industries follow this strategy to some degree. If a company cannot keep a state-of-the-art R&D capability in all fields, research leadership in a few key technologies can carry over into other areas. R&D personnel prefer to work for employers who prize and reward research, thus making personnel recruitment and retention easier (Steele 1975).

Offensive R&D is aimed at being first to market with innovations (Maidique and Patch 1988; Spital 1983; von Hippel 1983). Especially in research-intensive sectors, such as pharmaceuticals and semiconductors, being first to innovate is a common strategy. Although it is usually a profitable and successful strategy, technological 'pioneers' also bear significant costs and can 'burn out' and be overtaken by later rivals (Henderson 1984; Olleros 1986; Porter 1985: 189-91). There are several causes of this burn out, which differs from the usual shake-out of small firms that occurs somewhat later as a product or industry matures.

First, 'pioneer externalities' benefit later entrants as technologies and markets turn, to some degree, into public goods which later entrants can access for a fraction of the pioneer's development cost. Second, market uncertainty, particularly a long payback period common to radical breakthroughs, can prompt a pioneer to leave the market to others. Xerographic copying technology, solar energy, and robotics are examples of markets which have been – or still are, in the case of solar energy – slow to develop. The market may develop slowly for other reasons, such as the initial lack of infrastructure or complementary technologies, for example software for robotics and airports for airplanes. The importance of infrastructure and of complementary innovations has been stressed by Brown (1981) and by McIntyre (1988). Finally, technological uncertainty may prevent rapid adoption of first-generation technology. Users' expectations that improved versions are just around the corner, as well as major incompatibilities as second-generation technologies emerge, can thwart the success of pioneers (Olleros 1986; Porter 1985; Rosenberg 1976; Rosenbloom and Cusumano 1987).

> At this fluid and uncertain stage, pioneers (somewhat like the early users on the other side of the market) will have to face up to a high probability not only of choosing the wrong initial technology, but also of being entirely leapfrogged by a later group of entrants riding a far superior and totally different technological approach (Olleros 1986: 15–16).

Because of these uncertainties only a relatively small number of firms will be willing to follow an offensive strategy, and even those which do may well retreat somewhat to consolidate their former successes (Freeman 1982: 176). Solutions for a pioneer firm include subcontracting to keep investment costs low, and joint ventures with established marketers to speed up sales, and therefore returns on R&D investment, as have been common in the biotechnology industry. Licensing can also facilitate the commercialization of a firm's technology over that of rivals (Olleros 1986).

Firms with a *defensive* strategy are able to react rapidly when a competitor firm unveils a new product, introduces a lower-cost version of a relatively new product, or may be known to be on the verge of new discoveries in some field. Firms that tend towards defensive R&D, or 'technological followership', also must maintain a sizeable research and engineering presence, and may be pioneers in lowest-cost product design and manufacture (Porter 1985: 176–93). Such technological threats often emerge from other sectors, where it is more difficult to monitor them than from known competitors (Cooper and Schendel 1976; Foster 1986).

A third approach to technological strategy is simply *imitation* of the technological moves of more innovative firms. An imitative strategy will usually entail a somewhat greater degree of engineering expertise, oriented towards low-cost manufacturing processes, and relatively little actual scientific research capability, at least in comparison with firms in either the offensive or defensive categories. Such engineering efforts tend to be decentralized and located at a number of production sites. Japanese R&D, particularly development, continues to be more closely 'coupled' with manufacturing than is the norm in most firms elsewhere (Hull, Hage, and Azumi 1985; Westney and Sakakibara 1987).

Freeman also suggests that there are *dependent* firms, which are wholly reliant on others, such as suppliers, to initiate technical changes. A dependent firm will typically conduct no R&D at all and have no capability in product design. *Traditional* firms, which are usually suppliers to larger organizations, change only in response to specification from the outside. They may be especially strong in craft skills. Finally, *opportunist* or *niche* firms may be quite innovative, but serve a market with few, if any, competitors, but opportunities for differentiation in settings where economies of scale or experience provide little advantage (Charles, Monk, and Sciberras 1989: 157–8; Cooper, Willard, and Woo 1986). By serving only small markets, niche producers can easily switch from product to product. Niche firms are often remarkably innovative and provide the foundation of many regional economies (Doeringer, Terkla, and Topakian 1987: 82–96; Sabel *et al.* 1987).

A variant of the niche strategy that is used to enter new markets in order to perfect new technology is the *'thin markets' strategy* (Abernathy 1980; Lifton and Lifton 1989). It makes use of 'learning by selling' and feedback from lead users (Thomson 1989; von Hippel 1989). Best applied in limited, specialized markets, this strategy involves convincing a small number of firms to purchase the new technology, even though its cost typically is higher than presently used products. Without these thin markets, some R&D would never even be attempted, and the essential experience of users would not be taken into account (Abernathy 1980; Lifton and Lifton 1989).

The technological capabilities of firms largely determine their relationship with other firms and the strategies which are open to them (Taylor and Thrift 1982a; Thorelli 1986). Arnold (1985: 119–24) has applied Freeman's strategy types to the technological strategies of firms in the television industry, and found Japanese firms such as Sony to be the most consistently offensive. Finally, in addition to the technological aspects, strategies are influenced by the firm's complementary assets and the nature of its external linkages. Firms which rely primarily on suppliers are in a different position from firms which can maintain 'science-based' strategies (Pavitt 1984; Pavitt, Robson, and Townsend 1989). These conditions serve to define the technological trajectories of firms (Nelson and Winter 1982; Pavitt 1986b).

Corporate organization of R&D activities

In addition to the strategic importance of R&D, its place of in the firm's organization is consistently emphasized by management researchers, and this has direct geographic implications. Commonly, the principal organizational need for R&D is proximity to the firm's administrative headquarters, to facilitate communication among the various functional sections of the firm, especially between R&D and marketing (Burgelman and Sayles 1986; Carroad and Carroad 1982; Souder 1977) and between R&D and manufacturing (Hughes Aircraft 1978). If an organization fails to keep its innovation-related functions together, 'intraorganizational dislocations' can create formidable barriers to innovation (More 1985). Much of the lack of communication and co-ordination between marketing and R&D stems from the different priorities and styles of the people working at the two tasks (Graham 1986a; Gupta and Wilemon 1988). Likewise, the interface between R&D and production becomes more significant as R&D is 'coupled' to a greater degree with manufacturing operations (Charles, Monk, and Sciberras 1989: 139–42; Rubenstein 1989: 344–62).

The issue of centralization versus decentralization, commonly addressed in work on R&D management, is also one of the few issues which directly relates to regional technological capability, through its impact on R&D location (Malecki 1980a; Rubenstein 1989). Because the corporate R&D function typically falls directly under top management – either at the corporate or the divisional level – it has often been remote from the other functions of the company. Beyond a largely organizational concern, however, there are few clues to the geographic structure of a firm's R&D effort (Howells 1984; Malecki 1980a; Rubenstein 1964, 1989). Centralized R&D within a firm reflects a centrally run and tightly organized structure. Important corporate functions, such as R&D, are retained close at hand, virtually always at or very near the firm's administrative headquarters. Only rarely does such a firm have its central R&D lab located away from corporate headquarters.

A firm with centralized R&D (as well as other firms' central labs) will tend to have such research efforts close at hand, usually at or very near the firm's headquarters. A central location allows, for example, economies of scale in facilities and equipment, minimal barriers to interaction, the potential for interdisciplinary efforts, and the ability to utilize large numbers of people on major projects (Gibbs *et al.* 1985; Graham 1986b; Thomas 1983). A central laboratory with a broad charter will be able to attract the best people because it will closely resemble the academic environment with which the firm competes for researchers. R&D personnel prefer to work for employers who prize and reward research, thus making recruitment easier in those firms and facilities, (Graham 1985; Steele 1975; Thwaites and Alderman 1990).

The more common corporate administrative structure is a decentralized organization, organized on the basis of product line divisions or subsidiaries (Rubenstein 1964, 1989; Malecki 1980a). Long-range and basic research remains centralized in one (or very few) central labs, whereas product development is decentralized to a number of manufacturing plants where it can be more closely linked or 'coupled' with manufacturing and marketing (Maidique and Hayes 1984; Rubenstein 1989; Souder 1983; Tushman 1979; Westwood 1984). These plants are usually dedicated to the production of relatively new products rather than to standardized production of established products (Schmenner 1982). There are other advantages to dispersed R&D locations. Costs may have become excessive in current locations (Saxenian 1983b), and new sites allow firms to tap other sources of technical talent (Kanter 1984). Even in a seemingly decentralized structure, however, firms continue to concentrate their more innovative research at central laboratories and their less technical efforts, especially process R&D, at some – but not all – branch plants (Dalborg 1974; Kelly 1986; Thwaites and Alderman 1990).

When R&D is decentralized in this manner, communication with other parts of a division is facilitated (Fuller 1983). However, geographically dispersed R&D tends to decrease the amount of communication possible among researchers (Tomlin 1981). There are other advantages to decentralization, such as encouraging entrepreneurship in large corporations, which can be facilitated by separate locations for R&D (Anderson 1969; Leavitt 1968; Quinn 1979). Firms must strike a fine balance between promoting entrepreneurial behaviour and creativity or integrating R&D with other corporate functions.

The issue of organization is a controversial one. Japanese firms tend to favour decentralized R&D in order to keep it close to the manufacturing process (Hull, Hage, and Azumi 1985; Westney and Sakakibara 1986). At the same time, most Japanese firms also maintain large central research laboratories in which basic and applied research are concentrated (Kono 1984). In the USA, it is more common for a fairly large proportion of R&D to be centralized, a tendency that hurt American firms in video recording technology as they were overwhelmed by the success of Japanese video cassette systems. This experience showed vividly the costs of a gap between research and production (Graham 1986a). Increasingly, firms outside Japan are upgrading engineering expertise, oriented towards both flexible and low-cost manufacturing processes, at production sites. Locationally, such process engineering tends to be decentralized and found at a number of production sites (Ahlbrandt and Blair 1986; Dumbleton 1986). As with production plants and other parts of a corporation, R&D structures are largely the outcome of the historical development of the enterprise, including mergers and acquisitions (Thwaites and Alderman 1990).

Factors external to the firm and its organization are also important. These include the advantages of proximity to other firms which can minimize the cost of, and maximize the opportunities for, acquiring information, whether from rivals, suppliers, customers, universities, or other sources of information (Andersson 1985; Dalborg 1974; Rogers and Larsen 1984; Oakey 1984; von Hippel 1988). When a firm has decided on geographical dispersion of R&D, it tends to be constrained to areas where workers can be attracted and retained. Increasingly, location is seen as part of the 'creative climate' which companies are striving to create for their R&D operations (Martin 1984).

The contact networks of technical workers are a principal source of technical knowledge (Allen 1977; Håkansson 1987, 1989). Firms rely on the heterogeneous contacts and networks of individuals in order to acquire information from other organizations. The information 'gatekeeper' role is a well-known one, and one useful throughout the life cycle of products (Badawy 1988; de Meyer 1985; Roberts and Fusfeld 1981). Only recently has the diversity of contacts of technical workers been studied (Hamfelt and Lindberg 1987; Rubenstein and Ginn 1985). Increasingly, it is seen to be vital for firms to recruit personnel with certain contacts and personal networks in place, to encourage contacts to be formed and maintained, and to circulate people in order to develop more fully their networks of contacts to the advantage of the firm (Ettlie 1980; Håkansson 1987; Shapero 1985). The common contact networks of rival firms even result in informal trading of know-how between competitors (von Hippel 1987). However, large firms are best positioned to exploit information sources. Oakey, Rothwell, and Cooper (1988) found that external information contacts were relatively rare among small firms, illustrating yet another disadvantage faced by small firms *vis-à-vis* their larger counterparts.

Global competition and strategic alliances

Despite the pace of technological effort expended by large firms, the diversity of technology and markets has led firms to a 'make or buy' decision concerning technology (Rubenstein 1989: 266–87). Uncertainty about the cost, performance, and timing of internally developed innovations, and especially the narrow time frame within which innovations must be taken advantage of, has prompted combinations among competing corporations, especially among very large companies (Ballance and Sinclair 1983: 158). These *strategic alliances*, partnerships, or coalitions combine the strengths of two (or more) competitors, often based in different countries, against other (usually) global firms. Such external relationships have become a significant form of competition among transnational and other large companies (Chesnais 1988; Cooke 1988; Friar and Horwitch 1986; Mowery 1989; Perrino and Tipping 1989).

'Global strategies' are believed to be essential in order to compete (Hamel and Prahalad 1985; Hout, Porter, and Rudden 1982; Porter 1980, 1985; Spence and Hazard 1988). Japanese firms have led the way in demonstrating the competitiveness provided by co-operation with suppliers and an emphasis on organization and manufacturing skills (Chesnais 1988: 69–78). Such a *network strategy* in a growing 'postmodern' corporation, thrives on the basis of its linkages with other firms, whether as sources of technology or other complementary assets (Fig. 5.1). Product differentiation and the rise of technology as a basis of competition have rendered it impossible for any firm to have technological capability in all areas of interest, as 'classic' firms attempted. More important, in-house R&D is seen to be less significant than production capabilities and other complementary assets (Pisano, Russo, and Teece 1988). Linkages among competitors, as well as between dominant firms and suppliers, became probably the predominant form of corporate strategy during the 1980s (Chesnais 1988; Fusfeld 1986; Friar and Horwitch 1986; Spekman 1988). The 'triad' – Europe, Japan, and the USA – contains the dominant markets and producers in the international economy today, and firms in all three regions are central to the web of alliances being formed in semiconductors and other high-technology sectors (Chesnais 1988; Ohmae 1985).

Examples abound. Japanese semiconductor firms have allied with competitors both within and outside Japan (Hobday 1989). Swedish companies have also widely adopted the use of a network strategy for

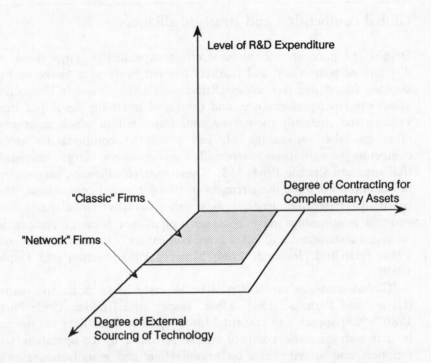

Figure 5.1 The corporate strategies of 'network firms'. *Source:* Pisano, Russo, and Teece (1988: 26 (Fig. 2–1))

R&D (Håkansson and Laage-Hellman 1984). The development of international collaborative ventures is currently common among US firms in industries as diverse as steel, motor vehicles, robotics, and biotechnology (Mowery 1988a). In the case of an emerging technology such as biotechnology, such alliances permit three advantages to be exploited:

1. A *window strategy* allows firms to identify and monitor leading-edge technologies developed elsewhere. Linkages will usually take the form of research grants and contracts to outside organizations, including universities and research institutes, or equity investments in new firms;

2. An *options strategy* is more selective, designating a small number of market or technical areas in which to participate. R&D contracts, licences, equity investments, and joint ventures are common;

3. A *positioning strategy* reflects a commitment to a technology for commercial exploitation. R&D contracts, licences, and joint

ventures aimed at production are the types of alliance seen with this strategy (Hamilton 1986: 111–12; Olleros and Macdonald 1988; Pisano, Shan, and Teece 1988).

Relationships between industry and universities in particular have grown dramatically, prompting critics to deride the pacts as making knowledge a commodity for sale (Dickson 1984; Peters 1989). These links often demand changes in policies on both sides of the relationship (Geisler and Rubenstein 1989).

Especially in Europe, lagging behind both the USA and Japan, national firms are constrained to relatively small domestic markets. This will change with the initiation of a 'single market' in the European Community (EC), set to begin in 1992. The EC has been instrumental in organizing and funding co-operative R&D programmes among firms located throughout Europe to increase European technological capability in the broad area of IT (Nueño and Oosterveld 1986).

These programmes, such as ESPRIT (European Strategic Programme for Research and Development in Information Technologies), EUREKA (European Research Co-ordinating Agency) and JESSI (Joint European Submicron Silicon), are intended to 'make Europe's productive system in high technology sectors more competitive in the tough confrontation with the USA and Japan' (Pistella 1989: 205; Bylinsky 1986; Peterson 1989; Peterson and Schares 1988). EUREKA, launched in July 1985, is a mechanism for promoting co-operative, transnational R&D not only among firms, but also between firms and universities, government agencies and research organizations (Onida and Malerba 1989).

Consequently, a flurry of co-ordinated R&D and technology programmes in information technologies and in biotechnology are taking place (Nueño and Oosterveld 1986; Sharp and Shearman 1987). Under EC sponsorship, ESPRIT, EUREKA, R&D in Advanced Communication Technologies for Europe (RACE), and Basic Research in Industrial Technologies (BRITE), among others, all attempt to bring a competitiveness to Europe via European producers – 'to allow European manufacturers to be part of the core rather than part of the periphery' (Davis 1988: 85; Frisk 1988). Policies for technological competitiveness will be returned to in Chapter 7.

The formation of a network of organizations is a common practice among Japanese firms as a means of reducing risk and uncertainty. These networks may be formed between competitors, between manufacturers and suppliers, within channels of distribution, with spin-off enterprises, and within corporate groups (Kagono *et al.* 1985: 77–83). The flexibility afforded by external sources of R&D outweighs for many firms a loss of control (Riedle 1989: 219). However, network strategies are pursued differently by different types of firms, and illustrate that control is still possible. Japanese firms prefer joint ventures in which they are dominant, taking advantage of the technology or market presence of smaller firms (Tyebjee 1988).

Co-operative industrial research is not new. In the form of research associations (RAs) it has been in existence in the UK since 1917. RAs account only for 2.2 per cent of all R&D, but the proportion is much higher in some industries, such as clothing (29%), and timber and furniture (52%). These sectors, however, generally do little R&D overall. High-tech industries, such as aerospace, tend not to have co-operative R&D programmes (Johnson 1973, 1975: 190–208). Co-operative R&D has also taken place for several decades in the USA, but also generally in low-research sectors. Perhaps because of this sort of bias, it is also noted that co-operative R&D may be a 'second-rate venture' or low priority for the companies involved, or at least of their researchers. Link and Bauer (1989: 95–6) found that, in two large chemical firms, the best scientists were perceived to be kept for in-house research, rather than for joint venture research projects.

Informal types of co-operative research are prominent in contemporary Japan. Cutler (1989: 22) describes close ties between Japanese university professors and their former students, who may be working for a number of competing firms. These people meet in professional societies and other meetings and share a great deal of information about current research.

Joint ventures, aimed at the development or marketing of individual products, have been a strategy for decades as a means of sharing risks. They have proliferated during the 1980s, both as combinations of large firms and as a connection between a large and a smaller enterprise, usually for the former to tap into the R&D and technology of the latter. Wolff (1989) points out that technological alliances between a large and a small firm may be unsuccessful, because small firms are often not sufficiently commercial in their orientation or are inexperienced in a larger market-place.

Small firms are targeted as important potential beneficiaries of such a co-operative R&D programme, in light of the fact that 90 per cent of small firms are less likely to develop innovative products in-house (Riedle 1989). In addition to general risk-sharing, a major motivation for co-operative R&D is the ability to monitor technological frontiers, get advice on technological trajectories, prepare for technological jumps and changes in the competitive environment (Onida and Malerba 1989: 162). However, large firms play a preponderant role in European programmes to date, which focus on technology push rather than on barriers to innovation, a fact that may well limit their success (Piatier 1984; Woods 1987).

For other firms, the arms-length alliances are less attractive than are conventional means of acquiring technology, via merger and acquisition of other companies. Even here, successful mergers and acquisitions occur primarily when the technological environment of the acquired firm is known and understood; better yet is a 'patient partnership' instead of an expectation of a quick fix (Chakrabarti and Souder 1987). However, there is some evidence that corporate mergers decrease R&D funding, especially for basic research (National Science Foundation 1988).

Finally, it is worth mentioning the results of Link and Bauer's (1989: 96–7) study of co-operative research among US firms, which shows the strength of geographic relationships. In 15 of 18 cases studied, 'when research partners were in a close proximity, the ease with which the projects were conducted and completed improved'.

Licensing strategies and technological capability

The licensing, or acquisition and sale of technology, provides an alternative to internal R&D for the purchasing firm, and provides a return on prior R&D investments and a source of revenue for further R&D by the firm selling its technology. From the point of view of a purchasing firm, licensing results in faster commercial development and market entry or enhanced market share than costly internal R&D would permit. It can also provide specialized knowledge and skills, and even stimulate a licensee's technological capability by making it more aware of the advantages of outside developments (McDonald and Leahey 1985). The advantages of licensed technology depend heavily on how current the technology is, and whether the licensee is permitted to retain the rights to any improvements made.

Licensing of production to other companies is a strategic option for an innovating firm. In order to gain access to a foreign market where direct foreign investment is not feasible or attractive, for example, a firm might choose to 'license' a domestic firm to be the sole authorized producer of a product within that country (Kumar 1987). (Much licensing has nothing to do with technology, but only represents use of a trade mark and results in immediate market recognition, such as clothing with Mickey Mouse or other characters. Whether such a creative symbol or image is 'technology' is open to question, but it does fall within the complementary assets important to technology.) In production under licence, a local vendor negotiates a licence for a specified period from the original vendor which permits the licensee, for a fee, to produce products identical to the licensee's original design. Alterations from an original design may be allowed, such as packaging oriented to the local market. The terms of technology licences vary but may give the licensee access to the licensor's research on the vintage of the technology licensed, or allow personnel to be trained by the licensor. The export markets in which the licensee may operate and the extent of design changes permitted may, however, be restricted under the terms of the licence (S Thomas 1988: 327–8).

Restrictive agreements often prohibit the domestic firm from some activities, such as exporting, improving upon the product, or setting prices (Odle 1979; Prasad 1981). It is these restrictions which inhibit technological capability and, as a result, multiplier effects, such as linkages and spin-offs of new firms, are most remote (Giaoutzi 1985). Tied-in purchases of inputs from the licensor, rather than local purchases, were required in all 126 licence agreements examined by Adikibi (1988) in Nigeria. Nearly as frequent (118 of 126) were requirements that all R&D be supplied by the licensor. In 81 cases, no R&D was to be performed locally at all, and in 55 agreements, 'grant-back' conditions ensure that any innovation belongs to the foreign firm.

Even in cases where restrictions were not prominent, MNC subsidiaries have had a relatively minimal impact on technological development, because of low levels of local linkage and the lack of R&D and design capability. In effect, a two-tier, dualistic structure of MNC subsidiaries at the higher, more complex end, and domestic companies at the lower end, has been the norm except in cases where local content or linkages have been required (Marton 1986: 282). Indeed, government policies that address the issue of technological dependence, by requiring local R&D and local purchases, have been most effective (Marton 1986).

Development of an independent design capability, the ability to design and produce a product, is an indicator of decreased technological dependence.

> Generally the costs of building a manufacturing capability are direct, requiring factories, production lines, and training for operatives, whereas the costs of building a design capability are primarily opportunity costs, requiring the recruitment and training of talented design staff who might otherwise be engaged on

alternative challenging design projects. Whilst the process of acquiring a design capability is reasonably certain, there can be no guarantee that a strategy of acquiring a design capability will be successful. If the personnel are unable to develop a sufficient understanding of the technology (notably the design intent), it will not succeed (S Thomas 1988: 328).

Korean firms, for example, which have been very successful in a number of high-tech sectors, have yet to develop independent design skills, according to observers. Korean semiconductor producers have depended entirely on foreign design tools (automatic design packages) for designing new products, and conflicts over patents and restrictions on exports have resulted. Such dependence also constrains linkages to design houses in other countries (Byun and Ahn 1989: 652).

In recent years, however, there has been an increase in the use of licensing for the purpose of the sale and transfer of technology. One stimulus for this change is the capability of firms in an increasing number of countries to acquire technology in this way and national unwillingness to permit unrestricted or unnecessary foreign direct investment. Japan, Western European and Eastern European countries have been successful in using licensing, and the more advanced less developed countries prefer licensing. It has become a widely available option because of the increased competition among suppliers of technology and the resulting need to sell existing technology to be able to finance future R&D programmes (Adam, Ong, and Pearson 1988). Licensed technology is not necessarily state of the art. In the study by Mansfield *et al.* (1982: 48), the technology transferred via licences and joint ventures tended to be older (13 versus 10 years old) than that transferred to subsidiaries.

The firms most likely to have licensing out as part of their strategic planning are large firms and those firms which have the highest R&D intensities (the ratio of R&D expenditure to annual sales) (Adam, Ong, and Pearson 1988; Bertin and Wyatt 1988; McDonald and Leahey 1985; Robinson 1988; Vickery 1988). The short life cycles in R&D-intensive industries seem to be the motivation for firms in those sectors. The price which a technology licence can command depends on its competitiveness and its stage of development (Udell and Potter 1989). The degree of patent protection or other strong appropriability, however, is costly to maintain, especially where product cycles are short.

Most licences are granted to subsidiaries and affiliates and to other multinational firms. The smallest share of licences goes to local firms in developing countries. Thus, the benefit from licensing may be largely within the set of already rather prosperous firms, which have the information networks about who has what and who needs it (Bertin and Wyatt 1988: 69–71). Licensing is but one of the fields in 'the technological battle between multinationals' (Bertin and Wyatt 1988: 94).

For firms seeking to acquire technology, licensing may be a route to enhancing technological capabilities. Reid and Reid (1988) found that among firms in New Brunswick, several characteristics were associated with those that had acquired a licence for product or process. They were younger, larger, and, most importantly, had a greater overall technological capacity than firms without licences. Most of these licences during the mid-1980s consisted of product and process manufacturing licences and patents. Information on the availability of such licences was most often from trade shows, followed by national and international contacts within the industry (Reid and Reid 1988). By contrast, the older licensing arrangements typically consist of trade marks, brand names, and technical and engineering services.

These findings suggest that there is a pattern of licensing which begins with firms' first exploiting the marketing capability of proprietary know-how and then expanding to using the technology capability that arrangements provide for developing manufacturing potential. For licensee firms, there are clear advantages. 'These firms, as a group, show greater increases in employees and sales, product introductions and internally developed patents. They also are likely to be more diversified and have higher levels of formally trained employees than manufacturing firms as a whole' (Reid and Reid 1988: 407).

The ability of a company to absorb and improve upon licensed technology depends greatly on its ability to understand and control embedded technology as well as embodied technology. In effect, the buyer, or user, of transferred technology needs technical expertise nearly equal to that of the supplier in order to absorb the technology. This knowledge includes contract administration and patent management, which are generally considered to be managerial, rather than technical, skills (Rubenstein 1980; Vakil and Brahmananda 1987).

Conclusion

This chapter has reviewed the major streams of research on innovation and technology in firms and in the economy. Large-scale phenomena, such as long waves and the definition of high technology, affect the way technology is understood by those outside firms. Within firms, issues of strategy, organization, and external relationships are the means of competing in a setting of rapid technological and political change. R&D is necessary for competitiveness but, once again, is not sufficient to bring about mastery over all the facets of competition.

The range of activities that affect and are influenced by technology have many geographical implications, which are the focus of the following chapter. The distinction between countries, regions, and places with high levels of technological capability, and those with little such capability, is largely an outcome of the location decisions of firms. Location is among the competitive options which have been transformed as technology has continued to advance. The location of specific labour skills, and their suitability for routine and non-routine technologies, is a major element in the spatial transformation. No less important is the technology of production and its infrastructures, which allows a wide set of options to flexible firms.

Chapter 6

The location of economic activities: flexibility and agglomeration

The empirical 'discoveries' of the past 10 years or so have shown that there is a great deal of complexity in the processes that fall under the label of economic development. The 'Third World' is not monolithic, nor are 'branch plant regions'. It is difficult for regions and nations to hold on to accumulated advantages, and it may be necessary to create conditions under which technological capability can take root. Perhaps most important, the actions of global firms and their relationships with technology, with labour, with small firms, and with other large firms set the conditions for national, regional, and local prosperity and development.

To a large degree, theory has failed to keep pace with these new relationships. Increasingly, *flexible*, as opposed to rigid and stable, links allow firms to adjust their activities to uncertainty and to rapidly shifting conditions brought about by competition and technology, as well as by government regulation. The dilemma has been over how this empirical diversity might best be addressed. Walker (1985: 227) proposes that there are three fundamental, structuring aspects of the capitalist economy and capitalist growth: capital–capital (competitive) relations, capital–labour (class) relations, and technology. Of these, capital–labour relations are the most amenable to manipulation; competitors and technological change are much less predictable or controllable (Sayer 1985).

The issue of flexibility is key to discussion of how the location of economic activity takes place. Conventional industrial location theory has failed to be very useful in understanding or predicting the location of plant closures, of high-technology industry, or even of manufacturing plants for which the theory was developed. Moreover, a massive transition seems to have been taking place since about 1970 which has altered the 'rules of the game' for both corporations and for localities. The emergence of flexible production and inter-firm arrangements has vastly changed the way in which economic activity takes place as well as *where* it takes place. The tendency for economic activities to agglomerate in large urban regions is a persistent geographic outcome.

As firms increase their use of flexible methods of production, of work, and of inter-firm relationships, new interactions with governments also appear. Flexibility, its heterogeneity, and its implications for location theory, will be the focus of this chapter. Its significance for policy will be examined in Chapter 7.

Industrial location and the role of labour

The location of economic activity was once a relatively simple phenomenon to explain. Manufacturing was done by small, single-location companies, utilizing simple inputs to produce a single product. Unfortunately, the body of location theory developed to account for this type of firm is unable to illuminate or anticipate the actions of global companies and their networks of upstream and downstream linkages (Fredriksson and Lindmark 1979).

Location theory

Traditional, 'classical' location theory (Isard 1956; Smith 1981) had its origins in the work of Weber and the maturation of neoclassical economics. Weberian theory illustrates the idea that an optimal location for a firm can be derived, concentrating on the influence of distance and minimization of transportation costs related to the location of markets and raw materials. Firms are assumed to be simple, producing a single product, serving only nearby markets. Raw materials appear as the only critical inputs; labour, information, and other inputs are ubiquitous, available everywhere. In this setting, the cost-minimizing (profit-maximizing) location is easily determined (Beckmann and Thisse 1986; Smith 1981). Linear programming models assist in the calculations for large problems, such as the location of a brewery which distributes its heavy product to a large number of distribution centres (Schmenner 1982). Non-linear and integer programming models and multiple objective decision models extend the general idea that an optimal decision can be attained concerning a location or other objective

(Beckmann and Thisse 1986; Nijkamp and Rietveld 1986). The characteristic of location theory which is prized by its admirers and derided by its critics is summed up in the description of Beckmann and Thisse (1986: 87): 'location theory has been developed through the incorporation of spatial variables – localized resources and distances – into microeconomic theory'.

Weberian location, by taking the 'siting decision' as the starting-point, ignores the interplay between the isolated decision to locate a company facility and larger and more intricate investment decisions, only one of which is to invest in a new production facility (Clark, Gertler, and Whiteman 1986; Walker and Storper 1981). 'The primary issues for capitalists are *whether* to invest and *what* to invest in – what products, which processes, or which firms' (Walker and Storper 1981: 483). Seen in this context, a location decision cannot be separated from the larger competitive and profit-seeking context within which capitalist firms operate (Massey 1979a; Walker 1988). Some of these decisions involve new investments (Schmenner 1982), whereas others largely revolve around disinvestments, usually described by the euphemism 'restructuring' (Beauregard 1989; Bluestone and Harrison 1982; Massey and Meegan 1982).

Among the more valuable conclusions of classical location theory is the principle of agglomeration – that firms benefit through lower costs of production by operating in close proximity to other firms. These *agglomeration economies* may be classified as *localization economies* when they concern firms in the same or linked industries, or as *urbanization economies* when they concern a general cost saving from location in an urban area. The critical question is again whether to 'make or buy' the goods and services needed – to produce them internally or to purchase the inputs from firms through market linkages (Scott 1988c). Firms can attempt to internalize these linkages by being vertically integrated, which has been a primary mode of growth of large firms. Alternatively, firms can vertically disintegrate and purchase necessary inputs in external markets. Vertical disintegration is inherently more flexible than a system of internal transactions which must be monitored and managed (Casson 1987). Since the transmission of economic growth throughout an economy takes place mainly through inter-firm linkages, the location of these linked firms is central to understanding the empirical outcome of development processes. The failure of growth centre policies (Ch. 3), for example, was primarily a result of insufficient local linkages in assisted regions and of entrepreneurship, the entrance of new firms into the economy in a particular location.

As it stands, vertical disintegration, or the fragmentation of production into a division of labour and production tasks undertaken by many different firms, is likely under certain circumstances (Scott 1988c: 25–7). In particular, as product differentiation increases and production runs become smaller, firms shift the uncertainties of volume on to subcontractors and input suppliers. Second, specialized subcontractors can achieve internal economies of scale by doing similar work for many different customers, to the advantage of all. Third, segmented labour markets, with clearly defined primary and secondary segments tend to be characterized by disintegration of low-skill tasks to low-wage workers, most of which are in small firms, both in advanced countries and in Third World countries. International subcontracting has grown remarkably on a global scale, and is one of the major aspects of the international capitalist system (Adám 1975; Dicken 1986: 183–90; Germidis 1980). Fourth, geographical agglomeration can induce vertical disintegration, since transactions are easily found and conducted in metropolitan areas and industrial districts (Amin 1989b; Scott 1988c). In sum, whether there are a few large firms or many small ones depends on the products and services produced and the nature of production. In turn, a setting in which there are many small firms producing unstandardized products will engender a markedly different regional economy from one in which standardized products are the output of large plants (Scott 1983b). Innovative, rather than mature, products and 'information-intensive production systems' are more likely to be managed either internally or through contractual agreements (Hepworth 1989; Willinger and Zuscovitch 1988: 249). Indeed, the tendency towards internalization and control is at odds with vertical generalizations about disintegration (Porter 1986; Thrift and Taylor 1989).

In reality, production systems are much more complex than the above suggests. Subcontracting, in particular, is a middle ground between internalization and open-market transactions. Like joint ventures and other collaborative arrangements, subcontracting is 'neither purely firm-like nor purely market-like' (Casson 1987: 48). It also leaves the subcontractor free to form similar arrangements with other firms. The flexibility, if we may call it that, of subcontracting relations is why it is seen both in clustered settings and at great distances (Walker 1988: 389–91).

Technology, flexibility, and plant closures

Depending greatly upon the industrial sector, production restructuring or reorganization may take the form of rationalization, intensification, or investment and technical change. The first two of these largely entail employment losses, whether through plant closings in the case of rationalization, or capital intensification. 'Technical change', as Massey and Meegan (1982) use the phrase, encompasses the array of shifts from one product line to another which involves opening new plants and closing old ones. It may involve a situation of 'jobless growth' as output rises while employment is stagnant or declining.

Disinvestment and outright closure are major phenomena in contemporary economies. For example, plant closures were the largest cause of job loss in the USA from 1975 to 1982 (Howland 1988a), but they were less important in the UK during the early 1970s relative to widespread contraction of employment at plants which remained in operation (Watts 1987).

Watts and Stafford (1986) place plant closures within the broader context of corporate decisions. Cessation closures result from the decision to abandon a product or product line, a decision made where geography or specific locations are not really part of the analysis. Default closures are related to the decision to expand one (or more) existing plant in circumstances of slow growth and the less efficient plant(s) are closed by default. Finally, selective closures, which account for the majority of plant closures, involve the selection of, for example, the least profitable plant among several. Casson (1987: 211–22) considers the closure decision to be the converse of that for greenfield investment. In fact, the causes of closures are much more complex, as firms restructure and automate production (Peck and Townsend 1984, 1987).

Schmenner's (1982) detailed survey findings on US firms substantiate the point that most company closure decisions involve a combination of the elimination of old product lines and the retirement of old, inflexible technology. However, Howland (1988a, b) found that closures cannot be predicted by lists of variables which measure profitability of a location; instead, it seems that non-spatial attributes, such as plant age, status, and size were able to 'predict' closure. In Ireland, O'Farrell and Crouchley (1983, 1987) also found it difficult to predict closures with regional or local variables, such as town size. Ownership and the presence of a female labour force were better predictors of plant closure.

Closure, like the location decision, is influenced by the state of technology, especially best-practice process technology, and profitable product technologies. As a result, production investments are much more transitory than in the past. A plant's 'life' may be relatively short in any location, particularly if it is tied to the production of a single product or product line. Only more 'flexible' plants can weather the turmoil of uncertainty. Those most likely to be considered flexible are plants affiliated with headquarters and R&D functions, and employment declines are less likely in those locations (Bassett 1986; Schmenner 1982).

Large corporations and the spatial division of labour

In contrast to the simple, single-product firms of classical location theory, the internal complexity of firms, especially of the large organizations which dominate global, national, and local economic activity, has spawned a massive literature on large, particularly multinational, corporations (Bade 1983; Casson 1987; Dicken 1986; Doz 1986; Dunning 1988; Hamilton and Linge 1981; Taylor and Thrift 1982b, 1986; Watts 1980).

One stream of this effort stresses the organization and growth of firms, their relationships with the external environment, and the ability of firms – especially large ones – 'to manipulate or modify their environmental context' (Hayter and Watts 1983: 161). Enterprise organization – at the corporate or group level, among subsidiaries within firms, and of the individual plants, offices, and other establishments of an enterprise – and the environment within which firms operate determine whether and how the goal of firm growth takes place (McDermott and Taylor 1982). As Chandler (1962) suggested, organization and strategies for growth interact in the pursuit of profit and growth, with effects on location (Håkanson 1979).

Large, especially transnational, firms have wide information networks and treat the location decision in a much more 'rational', informed manner than do small firms (Oster 1979; Schmenner 1982). Indeed, they are able to manipulate their environment – locations, labour, and subcontractors, and even their competitors via joint ventures – in a highly effective manner.

Small firms, competing against larger competitors in the segmented economy, typically must do so on several fronts simultaneously, but largely confined within the power networks of large firms (Fredriksson and Lindmark 1979; Taylor and Thrift 1982a, 1983). Using the core–periphery analogy to describe classes of firms, Danson (1982) suggests that labour market segmentation is a result of 'centre' or 'core' firms being responsible for most of the best jobs (Baron and Bielby 1980). Skilled jobs related to R&D, innovation, marketing, and finance tend to be located in the dominant urban regions. Similarly, Clarke (1985: 234), in investigating this continuum, found that plants within a firm (in his case, the British chemical firm ICI) are increasingly either central or peripheral. Central plants are technologically advanced and have more highly skilled labour, whereas peripheral plants are likely to have only older, tried and tested technologies with very little on-site innovation. In between, semi-peripheral plants tend to have a mixture of old and new generations of equipment, but not the most recent forms. These plants are becoming either more central or more peripheral over time.

Labour and location

Labour, a fundamental factor of production in economic theory, is much more heterogeneous than conventional economic theory recognizes (Clark, Gertler, and Whiteman 1986; Malecki and Varaiya 1986). Within advanced economies as well between advanced and Third World societies, distinct – dual or segmented – labour markets are identifiable (Berger and Piore 1980; Wilkinson 1981). These have been reinforced spatially by branch plant economies and the increasing concentration of control functions in large centres. In the spatial division of labour which largely persists, rural and other previously unindustrialized regions tend to be home to workers in the secondary segment, in contrast to the primary segment of workers (Cooke 1983; Savey 1983). The diversity of worker functions and power results in a mosaic of labour costs, skills, and characteristics from which firms can pick and choose those most suited to profitable activities (Cooke 1983; Harvey 1988; Peck 1989). Territorial clusters developed around a particular set of knowledge and skills, or around local customs regarding women's work, attract the portions of an industry suited to those locations (Storper and Walker 1989: 144–7).

Labour is far from homogeneous in quality or quantity among locations, and companies require different labour skills for different work tasks (Fig. 6.1). As one generalization, technical, engineering, and scientific – *non-routine* – skills are critical to R&D and new products. By contrast, especially for assembly-line mass production, little skill is needed for routine production tasks (Shapero 1985). Such segmentation occurs at least between *primary workers,* whose work is more often non-routine and critical, and more difficult to control. Professional and craft workers have traditionally fallen into this segment, although many of their skills are being replaced by computer technology. This is most true of craft workers, such as machinists, but also of the more routine work of architects and computer programmers.

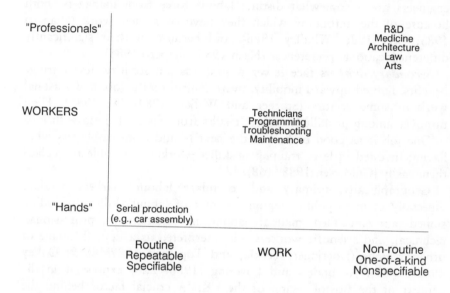

Figure 6.1 Labour skills required of workers for different kinds of work.
Source: Shapero (1985: xiii (Fig.I–1))

Professional and technical workers are a small but important part of the spatial division of labour in each firm, and they constitute a significant segment of the labour market. Technical workers are important to – indeed they serve to define the presence of – high-technology industries (Ch. 5). They are recognized as part of a rather separate labour market, made up of managerial, R&D, and technical workers (Bailyn 1985). In the conventional segmentation of the labour market, these workers clearly hold independent primary jobs (Harrison and Sum 1979; Noyelle 1982; Storper and Walker 1984). High-technology industry is in fact prominent in its internal contrast between technical labour and less skilled labour (secondary jobs) utilized for standardized production (Batten 1985; Hall and Markusen 1985).

The growth of these two ends of the skill spectrum has generated a 'vanishing middle' – the subordinate primary jobs – which have continued to shrink (Appelbaum 1984; Harrison 1982, 1984; Harrison and Bluestone 1988).

Not even technical jobs are secure. Baldry (1988: 55–71), Hall *et al.* (1983), Perrolle (1986), and Smith (1986) have vividly illustrated how technical work such as computer programming and other professional tasks are being automated and de-skilled to allow less skilled workers to perform them (see also Kaplinsky 1984a). It has been noted that engineers are a somewhat distinct labour force from managers, both because of the autonomy which they have over their work (Shapero 1985; Urry 1986; Whalley 1986), and because of their significantly different locational preferences (Nash 1985; Shapero 1985).

Secondary workers face lower wages, less job security, fewer fringe benefits, limited upward mobility, overt control on the job, and seasonal work in some sectors (Storper and Walker 1984: 28; 1989). High turnover among unskilled workers results from dissatisfaction with jobs – 'One job is as good or bad as the next' – and from employers who, having invested little in training unskilled workers, are able to replace them easily (Gulowsen 1988: 168).

Geographically, primary and secondary labour markets overlap, especially in metropolitan regions. A mix of labour skills – including skilled and unskilled manual labour in addition to professional, technical, and scientific workers – is interpreted as a key advantage of urban regions (Doeringer, Terkla, and Topakian 1987: 88–9; Oakey 1981, 1985). As Boddy and Lovering (1986: 226) express it in the context of the Bristol region of the UK: 'A crucial factor behind the locality's retention and attraction of technologically advanced activity has been the ability of firms to combine highly qualified staff and more basic production workers, as a basis for both specialized R&D activity as well as production.' The tendency for R&D and managerial jobs to agglomerate in regions which are internally dualistic – having large numbers of primary workers as well as low-wage production and service workers – is perhaps the principal contradiction of contemporary high-tech industry (Noyelle 1982; Saxenian 1983b).

A key distinction among the labour markets described thus far is their geographical extent (Bouman and Verhoef 1986). The labour market is local for unskilled, semi-skilled, and clerical personnel. It is nation-wide for managerial and marketing talent, and for key technical specialists may be world-wide. The geographic market provides the comparisons for pay, fringe benefits, and other conditions of employment (Riggs 1983; Storper and Walker 1984).

Because people with different labour skills are found in different places, firms generally take advantage of, and thereby reinforce, the existing spatial distribution of labour types within their internal organization. Employment in rural areas, in particular, is associated with low wages, fewer fringe benefits, and less attractive working conditions than in urban areas (Doeringer 1984; Moriarty 1980; Smith 1981: 325–48). In urban areas, the presence of a larger labour supply – of immigrants, for example – may drive down wage costs in some industries (Sassen-Koob 1989; Scott 1988b). Metropolitan areas, especially larger ones, have the significant advantage for workers of providing a number of job opportunities without requiring a change in residence. At the same time, firms have an incentive to disperse their facilities: costs may become excessive in current agglomerations (Saxenian 1983b), and new locations allow them to tap other sources of technical talent (Kanter 1984).

The 'branch plant economy' is the prototypical outcome of the tendency towards corporate locations which discover, reinforce, and create disparities in labour. The principal consequence for a region or locality of an economy which depends predominately on branch plants is a limited labour force. The workers in such a region will tend to have few skills, since few are needed for the routine tasks of the factory or office. No innovation or decision-making takes place; these are performed elsewhere in the firm's network (Firn 1975; Watts 1981).

The description thus far is rather starkly presented, and overlooks variations along a continuum from plant centrality to peripherality (Clarke 1985). A key constituent of plant centrality is the mandate of the plant. A plant with a *world product mandate* will have the necessary R&D personnel, engineers, and decision-makers to maintain a global state of the art in products and processes (Young, Hood, and Dunlop 1988). The regional development outcomes will also vary. For instance, the degree to which input linkages are local or non-local directly determines the multiplier effect for the region. Hagey and Malecki (1986) have summarized previous research, supported by evidence from firms in four high-technology industries in Florida. Local purchases of inputs were higher among firms which had higher levels of R&D and higher employment of scientists and engineers, controlling for plant size and autonomy. Kipnis and Swyngedouw (1988a) found similar results among firms in Limburg, Belgium.

The labour force structure of corporate operations affects, in turn, the geographical pattern of industrial location. High-technology firms, like all firms, differ in the degree to which R&D is done at a given site; some operations are labour-intensive and do not need to be at central locations. Consequently, R&D employment intensity is greater in metropolitan locations, a pattern which applies to large and small firms alike (Felsenstein and Shacher 1988). This generalization does not preclude the presence of R&D in peripheral areas (Kipnis and Swyngedouw 1988b; Malecki 1980a).

The resulting spatial division of labour within firms corresponds to the fact that some skills are in short supply, such as those needed in R&D, whereas other tasks are readily performed by workers in a wide variety of locations (Clark 1981; Massey 1984). Although there are intermediate skills which have traditionally been in somewhat short supply, such as skilled machinists, the strong tendency of management has been to 'de-skill' these jobs – usually by means of automation – in order to reduce firms' dependence on these skills. Such de-skilling has taken place in a number of industries, including clothing, where computer-aided design (CAD) and manufacturing have largely eliminated the need for skilled graders (who produce patterns in various sizes), markers (who fit the patterns on fabric to minimize waste), and fabric cutters (whose accuracy and uniformity are essential) (Hoffman and Rush 1988). These skills have been subsumed in automated systems operated by a much smaller number of operators who can complete the required work tasks in radically shorter time.

Labour costs vary markedly from place to place, both at the international scale and within nations. Consequently, labour is the central factor among the possible influences on the location of economic activity (Storper and Walker 1989). Labour tends to be place-bound, and each place has its own distinctive industries, types of jobs, and communities which form around the local mix of jobs (Pudup 1987; Storper and Walker 1989: 155–7). Geographical differences attract different specializations in production (Massey 1979b).

That said, however, 'the restructuring thesis has had relatively little to say about the role of small and medium sized firms' (McArthur 1989: 201). The bulk of attention has been devoted to large corporations which, through restructuring, have provided opportunities for existing small firms and new firms by vacating markets and by subcontracting (McArthur 1989: 202). Small firms are, however, far from a residual factor, poorly integrated into the capitalist economy. The evolution of regional economies involves, perhaps more than anything else, the development of localized labour skills and the agglomeration of linkages among local firms (Breheny and McQuaid 1987b; Hekman and Strong 1981; Scott 1988b). The development of skills and the generation of economic activity which exploits those skills may come about through the actions of large multilocational firms or of small local firms. The interrelationship between these two groups of firms is central to the understanding of the phenomenon of the industrial district and of entrepreneurship in general. In industrial districts, the set of skills tends to be artisanal; the 'new industrial spaces' engendered within high technology, by contrast, are different, based as they are on highly scientific and technical skills (Brusco 1989).

The location decision of the multilocational firm

The typical corporate location decision is a three-step process, in which a firm selects a broad region which best serves the market, labour, and resource needs of the firm. As Schmenner, Huber and Cook (1987) have found in the USA, and Lloyd and Reeve (1982) found in the UK, a dominant pattern in recent years has been an avoidance of unionism, even if this is associated with poorly educated or unskilled workers. Second, a selection is made among communities, especially as a firm attempts to secure the best 'package' of incentives from various communities. The second step is more difficult to account for, but in general taxes are unimportant (Doeringer, Terkla, and Topakian 1987; Mulkey and Dillman 1976; Schmenner, Huber, and Cook 1987). Incentives from governments may include the provision of roads, tax incentives, power facilities, and cash, all of which were recently offered to, and accepted by, Texas Instruments in connection with a plant in Avezzano, Italy (*Wall Street Journal* 1989). The third step is a real estate decision regarding the actual site selection within a local area.

The focus in Weberian analysis on distance as the geographical variable on which location decisions are made greatly distorts the relative importance of manipulable inputs. In partial recognition that distance and transport costs have diminished in importance (almost to the point of irrelevance for many products and firms), the most frequent reaction has been the provision of lengthy lists of 'location factors', variables which 'influence the location decision'. Lists which include dozens of variables are common, serving to identify the possible items which a firm might consider (Blair and Premus 1987; Moriarty 1980; Schmenner 1982: Vaughan 1977: 42–3; Watts 1987: 168–74).

The length and complexity of some lists of location factors have led to a composite variable, *business climate*, which 'is a rough metric of a location's expected ability to maintain a productive environment over the foreseeable future' (Schmenner 1982: 53). Neoclassical price competition between firms 'seems less the crucial variable than how adaptive firms, workers and communities are in terms of the changing economic environment' (Clark 1986a: 418). Clark, Gertler, and Whiteman (1986: 35) emphasize that local economies can 'be differentiated along the basic dimension of economic uncertainty. . . . Local economies form a means for absorbing within the workforce the fluctuations and uncertainties which industrial economies generate for the productive process.' In general, labour market segmentation is such that unskilled labour – which tends to be local, less mobile, relatively abundant, and found in many locations – is less able to create a labour relation in which the cost of uncertainty falls on the firm rather than on workers (in the form of unemployment). Thus, as manufacturing tasks have become de-skilled, workers with less skill, and with little or no industrial experience, can be hired. This has been identified with the demise of the labour union movement in several countries.

Labour is a central element in the notion of 'business climate'. In the USA, an annual report of 'state manufacturing climates' has been published since the late 1970s, amid great publicity (Grant Thornton Inc 1989). The concept of business climate has led to a wide variety of incentives for industry in general, as part of an attempt to improve a state's ranking from year to year. A reduction in state-regulated labour costs, such as unemployment benefits and workmen's compensation, or a halt in the growth of state expenditures can markedly improve a state's ranking *vis-à-vis* its competitors – i.e., other states. Even with a broadening of the concern with variable costs to include more than taxation, it is still impossible to state that a favourable business climate leads to industrial growth (Erickson 1987). Skoro (1988: 151–2) is more blunt: 'business climate indexes . . . are useless as predictors' and are actually political statements.

If anything, when low costs and low wages are the basis for competition, the infrastructure, services, and public facilities at the regional and local level are less able to meet the needs of a population and of industry (Power 1988: 162–8). The trade-off between taxes and public services in the USA has been the focus of several recent studies, by the Corporation for Enterprise Development (1988), Ameritrust and SRI (1986), and the Committee for Economic Development (1986). Cobb (1982, 1984) forcibly argues that the 'selling of the south' in the USA has been based virtually entirely on low wages and taxes. He asserts that the relatively low importance of labour factors on many surveys is attributable to a lack of willingness on the part of business executives to admit openly that labour costs and pliability are critical. 'Business climate' then is a euphemism for the labour factors (Harrison 1984; Malecki 1986a). Schmenner (1982: 150–1) presents similar findings: low wage rates and non-union sites are relatively unimportant, whereas 'favourable labour climate' is by a wide margin the most important factor in a location choice.

Indeed, in the widely cited growth of industry in the US South – the 'sunbelt' – little development has taken place in the competition to rank highly in business climate listings. The mix of industries common to the region in the mid-1980s were both low tech and high tech, but are for the most part low-skill, low-wage routine production operations (Falk and Lyson 1988; Lyson 1989; Malecki 1985a; Park and Wheeler 1983; Rosenfeld, Bergman and Rubin 1985; Rosenfeld and Bergman 1989). Overall, there may be little about which to be pleased in US sunbelt growth (Ullmann 1988). This generalization has its exceptions, however, because individual sectors are in different stages of product life cycle – or degrees of innovativeness – and each has its own pattern of labour preference (Johnson 1988).

Sloan (1981: 12–13) describes the business climate of Greenville, South Carolina, which 'has the enviable reputation of having the best business climate in the nation and, indeed, in the Western world':

> Greenville's climate is not simply a matter of its wages, its level of unionization, or its taxes. . . . A captain of industry in Greenville is guaranteed that he won't have to put up with the headaches that make the life of an executive such an unending trial in other parts of the country – troublesome reporters, egghead intellectuals, antibusiness politicians. Greenville offers a different kind of life to harassed executives – respectful reporters, complimentary academics, worshipful politicians.

Similarly, Morgan and Sayer (1985) identify 'labour skill' as a comparable euphemism. 'Skill' more often means the behavioural characteristics of labour: 'qualities as "good company employees", in terms of attendance, flexibility, responsibility, discipline, identification with the company and, crucially, work rate and quality' (Morgan and Sayer 1985: 390). The practice of choosing locations and workers on the basis of such behavioural traits is most frequently associated with Japanese firms (Hampton 1988; Henry 1987; Morgan and Sayer 1985), but it is by no means confined to them. These behavioural traits are also referred to as worker attitude, a subtle quality which outweighs the conventional concept of skill (Doeringer, Terkla, and Topakian 1987: 90).

Rural areas and other 'greenfield' sites are favoured for new production because people can be employed who do not think of themselves as belonging to an industrial workforce (Berger and Piore

1980; Clark, Gertler, and Whiteman 1986). The establishment of non-union 'parallel production facilities' by aircraft engine makers as a means of weakening the ability of unions to strike at main plants is another tactic (Bluestone, Jordan, and Sullivan 1981). In the American apparel and shoe industries, almost all companies 'have located their US production plants in low-wage areas with a ready supply of female labour – isolated rural areas, many in mountain regions like the Appalachians and Ozarks; Mexican border communities; and large cities with substantial poor and/or immigrant populations' (Schmenner 1982: 121). Similar 'regional labour reserves' are common throughout the UK and Europe (Townsend 1986; Hudson 1983).

Manual assembly operations, which are standard in large-volume products, including clothing, electronics, and watches, are located deliberately in low-wage areas, especially in Asia (Bluestone and Harrison 1982: 164-88; Dicken 1986; Grunwald and Flamm 1985; Steed 1981). An added advantage of Asian locations is that working hours of 48 hours per week are common (Suarez-Villa 1984: 100).

A favoured location of production for many products headed for Western markets has been the Special Economic Zones in the People's Republic of China, where wages for young women are among the lowest in the world (Table 6.1). Few linkages or wider industrial benefits have materialized for China (Wang and Bradbury 1986). The entire array of products which are subject to this division of labour is remarkable. The Chinese factories are preferred for toys, electronic goods, garments, and artificial flowers (Lee and Brady 1988; Lee, Engardio, and Dunkin 1989). China is not the only country to receive investment because of low-wage labour. Factories in Haiti, the poorest country in the Western hemisphere, produce textiles, baseballs, equipment for electric parts, television receivers, office machines, gloves, luggage and handbags, tape recorders and players, and toys and dolls, largely for the American market. The product mix produced in Mexico is similar, with the significant addition of motor vehicle parts (Grunwald and Flamm 1985).

This consequent geography of international production is concentrated in export processing zones and free trade zones throughout Asia, thus perpetuating the dualism common in earlier colonial industrialization (Marton 1986: 41-4; Salita and Juanico 1983; Singh and Choo 1981). At the same time, Japanese investment in Asia, in search of low-cost labour, is largely responsible for the continued success of the NICs (Kraar 1989; Rodan 1989: 198-200; Yang *et al.* 1989).

Women are particularly preferred for many such jobs, in part because they tend to work for lower wages and have less inclination to join unions than do men, and because their 'behavioural characteristics' are suitable (Morgan and Sayer 1985; Standing 1989). In Japan, 5 million women workers are classified as part-timers, despite working up to 48 hours per week. Female pay is low in Japan, and fringe benefits are few, especially in service occupations. In fact, women's pay as a percentage of men's fell from 56 to 52 per cent from 1978 to 1985 (Helm, Takahashi, and Arnold 1985).

Table 6.1 Average monthly earnings in selected Asian countries, 1988 (in US dollars)

Country	Average earnings
South Korea*	633
Taiwan*	598
Singapore*	547
Hong Kong*	544
Thailand	80
Philippines	75
Malaysia	55
Indonesia	55
China**	40

* Includes bonuses and overtime
** Excludes state subsidies data: BW
Source: Lee, Engardio, and Dunkin (1989: 78).

Perhaps even more important are the decisions of multinational electronics firms, which prefer female labour for their assembly operations, both in Asia (Fuentes and Ehrenreich 1987; Lim 1980) and in Europe (Morgan and Sayer 1988). These Asian plants, especially the labour-intensive assembly activities, overwhelmingly employ female workers, whose 'passive, "cooperative" personalities' are indispensable (Lim 1980: 118). The shift to flexible production is likely to increase world-wide the types of jobs in which women predominate (Baldry 1988; Standing 1989). At the same time, the docility of women in Asia and elsewhere may well have been exaggerated (Lutz 1988).

The primary activity of a company facility, then, greatly affects its location. One of the few surveys that systematically compares a set of location factors across various types of corporate facilities is found in Browning (1980: 56–9). The survey indicates that manufacturing plants consider a large set of factors to be important, led by availability of labour, availability of energy/fuel, and highway transportation. Distribution centres respond mainly to dimensions of market accessibility. Three other types of facilities, which largely employ primary workers, are also given separate consideration: regional divisional offices, corporate headquarters, and R&D facilities. For the first two of these, the three chief location factors are the same: air transportation facilities, highway transportation, and availability of executive/professional talent, which was distinguished from 'availability of labour'. R&D facilities had the same leading location factors, but availability of professional talent was considered the most important factor.

Although problems of aggregation remain, in this instance across sectors, this location survey addresses directly the spatial division of labour by demonstrating that different location factors influence the various functions of firms. More importantly, it distinguishes between 'labour' and 'executive and professional talent' as distinct types of labour (Ady 1986; Browning 1980; Lund 1986; Time Inc 1989: 19–21). If a functional distinction within firms is not made, rather confusing survey results can occur, where both production labour costs and availability of technical talent may be found to be important, when they are actually influential to contrasting activities in different locations (Premus 1982, 1986). Concerning transportation facilities, Dunning and Norman (1987: 625) note that 'the *availability* and *convenience* of travel facilities appears to be rather more important than the cost of these facilities' (emphasis in original).

The location of non-production activities

When compared to the body of theory on the location of manufacturing, non-production activities of firms are less well supported by theoretical ideas, despite the fact that non-manufacturing facilities comprise one-third of all new corporate facilities (Ady 1986). Non-production activities vary in their reliance on face-to-face contact and information, which are essential for non-routine administrative and R&D activities. By contrast, more routine office jobs have, to a large degree, shown the same dispersal tendencies as manufacturing plants (Goddard 1975; Daniels and Holly 1983).

In the context of technology, the concern with non-routine activities is especially crucial for R&D, much of which is centred around the generation of novel products and processes and variations of existing products and processes. The non-routine nature of the work of R&D strongly influences the types of workers who will be employed (Shapero 1985; von Glinow 1988).

The location theory literature framework of cost minimization is effective for examining the location of non-routine activities, such as R&D, company headquarters, and producer services. Many of the influences on location resemble traditional efforts at cost – especially transportation cost – minimization, despite arguments about the footloose nature of such activities. Czamanski (1981) usefully distinguishes between three components of 'friction of distance': (1) transport of goods; (2) movement of persons; and (3) transfer of ideas. For the similar case of producer services, Coffey and Polèse (1987a: 605) suggest that each producer-service establishment is subjected to three locational pulls: towards the *market* (minimizing costs for communications, travel, and courier service); towards *specialized labour pools* (minimizing recruitment and retention costs); and towards *large diversified service centres at the top of the urban hierarchy* (minimizing input costs).

The prominence of information in modern economic activities obliges firms to attempt to minimize the cost of contacts and of acquiring information by locating in agglomerations of other, especially similar, firms (Johansson 1987; Oakey and Cooper 1989; Pascal and McCall 1980). In the context of non-routine activities, an additional objective and advantage of agglomeration is to maximize opportunities for acquiring information, rather than only to minimize the costs of obtaining it (Rogers and Larsen 1984; Oakey 1984).

An R&D location decision is a particularly infrequent one for a firm, compared to the more common decisions for offices, plants, and other facilities. Most firms have relatively few R&D establishments – far fewer, at any rate, than their number of other facilities. Large US firms, for example, average around eight labs per firm (Malecki 1979: 311). But even for firms such as General Electric (US) with over 100 R&D facilities, and General Motors with dozens, one of the necessary characteristics of an R&D operation is stability – both for the firm's organizational structure and for the scientists and engineers who work in it. Despite some recent consolidation following mergers and acquisitions, the number of company sites where R&D is conducted is growing, as firms follow the Japanese model for R&D (Hull, Hage, and Azumi 1985). In these 'technical branch plants' (Markusen, Hall, and Glasmeier 1986: 77–8), the attraction and retention of technical workers must enter the location decision, although it is generally not significant for plants where only production takes place (Browning 1980: 58). Because of the need for both engineers and production workers, high technology manufacturing in the USA has decentralized very little to rural and small-town locations (Barkley 1988). In the semiconductor industry, Scott and Angel (1987) have shown that R&D and the more sophisticated manufacturing operations remain strongly concentrated in the Silicon Valley area. Increasingly, firms are upgrading engineering expertise, oriented towards both flexible and low-cost manufacturing processes, at production sites. At peripheral production sites, however, few linkages with local firms may develop (Gripaios *et al.* 1989). Linkages are more likely to develop in regions where R&D is conducted, and R&D gravitates towards urban regions (Hagey and Malecki 1986).

The location decision for R&D thus incorporates both organizational and labour market demands. The organizational pull keeps a majority of R&D labs at or near the firm's headquarters (Lund 1986; Malecki 1979). The joint location of R&D and headquarters is further demonstrated by the similarity of the location factors stressed by firms for the two facilities (Browning 1980: 58; Molle, Beumer, and Boeckhout 1989). For both, availability of executive and professional talent and air accessibility are the most important factors; these are clearly among the typical advantages of large urban regions. Similarly, a marked clustering of British pharmaceutical R&D appears to be largely attributable to the location of firms' headquarters; most of the outlier laboratories are associated with production plants (Howells 1984). The aggregate geographical effect of the tie to headquarters is to constrain the location of most R&D to regions where firms' headquarters tend to cluster, especially large urban regions. The advantages of large urban regions – quality of life, labour pools, and transportation accessibility – are considerable in their attraction to critical corporate activities such as administration and R&D. The ready movement of technical workers to other sites – in order to visit other researchers and to interact with manufacturing facilities as production begins – all require rapid (and therefore usually air) transportation. Firms see a site with acceptable air service (frequent flights to many destinations) as a way of minimizing the time workers need to be away from their base of operation.

The attraction and retention of managerial and technical personnel are vital for white-collar facilities. 'Locations that have a demonstrable supply of these workers will have an advantage; other locations must prove that the local quality of life factors are such that these workers can be readily attracted to the area' (Ady 1986: 80). Among the local factors indicated by 3000 research engineers in a 1982 survey were: housing cost and availability, climate (warm, coastal, dry, mountain), quality of primary and secondary schools, recreational opportunities, job opportunities for spouse, community attitudes, cultural opportunities, and taxes and municipal services (Ady 1986: 81).

Survey findings such as these should not be taken to mean that worker preferences *determine* the location of R&D or of high technology generally (Storper and Scott 1989: 37). Rather, they exemplify that the common interests of mobile professionals and of their employers are satisfied best in urban agglomerations. Urban regions provide a large number of alternative employers, which attracts professional workers, whose skills are readily transferred to other employers (Angel 1989). The importance of skilled production workers is also critical in the small-volume production of non-routine products that are common in all high-tech sectors and that predominate in several, such as aerospace and instruments (Bluestone, Jordan, and Sullivan 1981; Oakey 1981). A new array of checklists of location factors specific to high-tech firms has appeared in response to the allure of high-tech growth. These emphasize the importance of technical skills or of technical labour, and the relative unimportance of material or physical inputs (Castells 1985b, 1988; Cox 1985; Galbraith 1985; Henry 1984; Hicks 1986; Jarboe 1986; Malizia 1985; Premus 1982; Rees and Stafford 1984; Rosenberg 1985).

The critical point is that R&D workers, like other professionals in short supply, are mobile in the sense that their relative scarcity gives them labour market mobility, but they are in effect geographically somewhat immobile in that they are willing to live only in distinct types of places (Buswell 1983). What sorts of places? The 'quality of life' factor cited on most lists of location factors for high-tech industry largely represents urban commercial amenities – 'the potential for sophisticated leisure and consumption' – and other correlates of city size sought by this well-paid, mobile segment of the population (Castells 1988: 60; Malecki 1987b; Thwaites 1982: 379).

Entrepreneurial possibilities are also a consideration of research workers (Malecki 1987b). R&D workers belong to the social group likely to have the knowledge, contacts, and information conducive to spin off from other firms. Since the spin-off process is relatively common in several sectors, places acquire reputations among high-tech workers on the basis of their entrepreneurial climates. The reputation or image of a place weighs heavily in job location decisions of such workers (Macdonald 1986; Oakey 1983). New high-tech areas can create a high-tech image by marketing efforts in order to attract corporate R&D and its workers. A high quality of life and the reputation or perception that a place is 'high tech' can weigh more heavily than other factors for scientists and engineers, outweighing the need to locate near major universities (Begg and Cameron 1988; Keeble and Kelly 1986).

The local labour market and the opportunities for job mobility and entrepreneurship do not deter firms from locating in major urban agglomerations. An R&D facility, for example, located away from headquarters, in an attractive university locale, is thought to encourage entrepreneurial energy and innovation within large corporations (Quinn 1979; Kierulff 1979). Separate location for R&D in known entrepreneurial areas raises the risk of losing prized employees through job-hopping or spin-off, but it also increases the likelihood of finding workers who are leaving other firms, thus providing information and knowledge which are otherwise difficult to obtain (Angel 1989; Ettlie 1980; Oakey and Cooper 1989; Rogers and Larsen 1984). Firms in high-tech sectors appear to be willing to accept this risk, since high-tech firms are significantly more likely to locate in existing agglomerations of technical workers (Armington, Harris, and Odle 1984; Harris 1986). It benefits both employers and professionals to locate in the established agglomerations of an industry, where such information interchange is most likely (McDermott and Taylor 1982; Oakey 1984). The need for agglomeration economies or advantages is neatly summarized in the context of corporate priorities by Robert Noyce, president of Intel Corporation: 'Our industry tends to cluster geographically. Why? Because it is to take advantage of the infrastructure of talent pools, support services, venture capital, and suppliers' (Noyce 1982: 14). He terms this clustering 'unconscious' and 'involuntary' co-operation 'which quickly diffuses knowledge . . . throughout the industry' through the mobility of technical personnel.

Finally, universities, an almost universally cited 'factor' accounting for the location of R&D and high technology, must be considered an overstated ingredient. The two 'products' which universities produce – scientific and technical personnel and research findings – are equally available for purchase at a distance. The experience of Stanford University in Silicon Valley, of MIT in the Boston (Route 128) area, and of the 'Cambridge phenomenon' in the UK all point to the fact that the growth of these areas as seedbeds of innovation and high technology owe their success more to the large urban regions in or near which they are located than to any close or direct relationship between firms and universities (Gillespie *et al.* 1987; Howells 1986; Keeble 1989; Malecki 1986b, 1987b). If there is any effect, it appears to be related to the research activities of clusters of premier universities located near one another and where other R&D programmes are prominent (Antonelli 1986b; Malecki 1980b; 1985b). In such settings, the 'external R&D' of firms – that obtained from universities, research institutes, and from other firms – is much higher (Davelaar and Nijkamp 1989a; Lorenzoni and Ornati 1988). As a source of engineering labour, universities are a minor influence on firm location. Experienced workers are recruited in major agglomerations; firms recruit entry-level engineers quite widely outside any local area (Angel 1989: 109).

The recounting above of the 'location factors' involved in R&D and high technology is not a simplistic connection between location preferences of engineers and scientists for attractive locations and high-tech boom areas. Preferences matter, but are constrained by the locations of employers. To increase the likelihood of finding alternative employers – and of firms finding employees – large agglomerations are the best locations for both firms and technical workers. At the same time, not all urban agglomerations associated with high-technology activity will evolve into hubs of self-sustaining growth. Technical branch plants, in particular, seem to have little effect on more extensive growth in their locales (Glasmeier 1988a). Some firms, especially outside Silicon Valley, have generated a loyalty among their employees which thwarts raiding and spin-offs (Levin 1982).

Urban agglomerations are particularly attractive to educated people with job mobility (Angel 1989). Employers generally have difficulty attracting professional and technical workers to locations judged remote (Cooke 1985b; Glasmeier 1986; Pacione 1982). Von Glinow (1988: 48) found that '"being able to work on exciting projects" in a "nice geographical location" was repeatedly cited as attractive to potential employees'. Metropolitan areas, especially larger ones, and prominent 'high-tech' locations have the significant advantage of providing a number of job opportunities without requiring a change in residence. The attraction of large urban areas is likely to be even stronger for dual-career couples, which have become an important component of the labour force (von Glinow 1988: 50). Dual-career married couples reinforce the effect of agglomeration: large urban areas are more attractive when initial employment is sought, and they are more appealing locations at subsequent times by increasing the likelihood of changing jobs without residential change (Malecki 1987b; Gibbs *et al.* 1985).

The success of corporate R&D programmes depends to an important degree on the ability of firms to attract and retain highly qualified technical and professional workers. The location of a firm's R&D facilities largely determines the degree to which workers will be initially attracted to, and be likely to remain with, the firm at that location. Firms recognize their employees' geographic preferences, as seen in the agglomeration of R&D in successful locations, in major metropolitan regions, and in attractive new areas. Scientists and engineers, as a scarce resource, have a large influence on corporate location decisions regarding R&D facilities. Firms' R&D facilities, especially when clustered near those of other firms, constitute an attractive pull factor. Thus, each location decision in R&D constrains and at the same time reinforces the other (Harding 1989; Lund 1986; Malecki 1987b).

The study by Lund (1986) supports the conclusion that there are two very distinct tendencies in firms' location decisions for R&D. Roughly half of 158 US firms which had added R&D facilities since 1975 saw it essential to be located near corporate headquarters or manufacturing plants. This was also the case for a set of pharmaceutical firms studied in the UK by Howells (1984). The second set of firms was more footloose, seeking out supplies of scientific personnel directly or, more indirectly, by locating near universities or research centres or locations with high quality of life.

The geographical concentration of R&D

In an international context, Vernon's 'product cycle' hypothesis suggests that new products will tend to be developed where multinational enterprises are able to centralize R&D, notably where scientists and engineers are found (Vernon 1966, 1974, 1979). This hypothesis has led to formal models which show that product development tends to remain concentrated in countries with large pools of skilled labour (Nelson and Norman 1977). For the similar case of regional headquarters, Dunning and Norman (1983) found that, in Western Europe, accessibility by air and availability of skilled labour are most important in the location decision of multinational firms. Less attention has been paid to the R&D location decision within individual countries. The concerns of firms favour large urban regions, since air connections as well as technologically advanced telecommunications are available earliest in large regions (Molle, Beumer, and Boeckhout 1989).

At the international level, the choices of large firms go beyond locations where scientific and engineering personnel can be found. Doz (1986) depicts the dilemma facing large multinational firms: they must, on the one hand, strive to co-ordinate and control their far-flung operations and, on the other hand, they face considerable pressure to decentralize activities – including R&D – to their various host countries. The dilemma is not an easy one. As a result, firms may perform R&D outside a handful of advanced countries, but the amounts tend to be small, the focus tends to be on minor adaptations, and it is often separated from local R&D or from commercial applications outside the local market (Mansfield *et al.* 1982; Prahalad and Doz 1981; Teece

1981). Nearly all R&D by major multinationals is kept within the advanced economies (in Europe, Japan, and the USA) in each of which skilled professionals and 'pockets of innovation' in specific technologies are found (Etemad and Séguin Dulude 1987; Mansfield, Teece, and Romeo 1979; Perrino and Tipping 1989; Ronstadt 1984). In addition, for custom production of some products, such as application-specific integrated circuits (ASIC), Japanese semiconductor firms have opened chip design centres in Europe and the USA (Hobday 1989). Countries with educated, English-speaking workforces, such as Israel and Malaysia, have become sites for significant amounts of R&D and engineering work (Rossant, Reed, and Griffiths 1989; Salih, Young, and Rasiah 1988).

Multinationals also engender a spatial division of labour within the countries in which they operate, developed and underdeveloped alike. Among multinationals in the UK, for example, Hall *et al.* (1987) found that proximity to Heathrow airport was a major consideration. Most of these firms also allocate their various functions to different locations in the UK; R&D and administration are kept within Berkshire in the South-east, whereas production is found at peripheral Welsh or Scottish sites. For similar reasons, in Canada, US-based multinationals cluster in the Toronto region (Britton 1985). Molle, Beumer, and Boeckhout (1989) rank London first among major European cities for both headquarters and R&D functions on the basis of a set of location factors. They further conclude that the relative attractiveness of regions for the location of such functions 'will not give way easily to deconcentration' (Molle, Beumer, and Boeckhout 1989: 171).

The locational concentration of R&D in every country favours established regions (especially national capitals) and reinforces the regional economic growth of those areas where universities, industrial R&D, and national government R&D facilities and contracts are plentiful (Breheny and McQuaid 1987a; Buswell, Easterbrook, and Morphet 1985; Malecki 1980b). In the UK, the South-east dominates in the location of high-technology manufacturing and services, with lesser concentrations in the South-west and East Anglia – all south of the North–South 'divide' (Begg and Cameron 1988). The tendency for high tech to bypass areas with 'traditional' manufacturing is widely considered to be in part a circumvention of unions and inflexible workers in mature labour markets (Castells 1985b; Markusen, Hall, and Glasmeier 1986; Morgan and Sayer 1988; Scott and Storper 1987).

Although this may overstate the case somewhat, it is clear that suburban zones of major metropolitan regions, where the demands of professional workers for suburban locations and of firms for access to transportation and communication, reinforce the development of 'silicon valleys' within large urban regions (Antonelli 1987; Britton 1987: 174–8; Camagni and Rabellotti 1988; Howells and Charles 1988; Scott 1986, 1988c: 60–104; Keeble 1988). As Begg and Cameron (1988) note, for reasons related to the locational preferences of professional workers – a critical group of workers and among the few growing segments in sectors such as electronics (Morgan and Sayer 1988: 131) – a proper economic, social, and physical 'image' is important. Thus, in the UK, new town locations and a lack of traditional manufacturing are seen in the location pattern of British high-tech industry (Begg and Cameron 1988; Keeble and Kelly 1986). The ability of these workers to influence corporate location is evident from the work of Macgregor *et al.* (1986) on the Newbury district of Berkshire in the UK South-east. Their detailed survey results of firms there showed conclusively the importance of an existing agglomeration of managerial, professional, and technical staff already employed by other firms, from which new employees are recruited. In short, labour market and urban agglomeration advantages are by far the most important considerations for high-technology firms (Breheny and McQuaid 1987b; Oakey and Cooper 1989).

In France, high-technology industry is spreading throughout the country, although 'Paris's domination is in no way under challenge' (Pottier 1987: 220). The 'sunbelt' region of France, Provence–Côte d'Azur on the Mediterranean coast, and including the cities of Nice and Marseille, is a growing region of modern industry (Dyckman and Swyngedouw 1988). Likewise, production of high-technology products, such as aerospace and robotics, are distributed throughout France, but R&D remains concentrated in the Paris region (Pottier 1987).

The situation in Japan is different. Both R&D and the production of high-technology products are concentrated in the Tokyo region (Nishioka and Takeuchi 1987). The allocation of corporate functions and the resultant division of labour reinforce this pattern: 'Though today high technology plants are distributed in every part of Japan, they are generally under the very strong influence of the Tokyo area because of their far weaker R&D functions and much lower level of technology' (Nishioka and Takeuchi 1987: 286). The geographical dualism of the Japanese industrial system has a distinct core–periphery pattern to it. Perhaps only Osaka competes in any way with Tokyo but, as a large urban region, it is also a cultural, commercial, and industrial agglomeration (Morita and Hiraoka 1988).

Morgan and Sayer (1988: 236–8) also present several examples of the importance of high-status locations and, indeed, a 'herd instinct' involved in the tendency towards agglomeration of high tech in the south of England. The latter is especially observable among US multinational firms in the South-east (Hall *et al.* 1987: 177) and in Scotland's 'Silicon Glen' in the Glasgow region (Haug 1986). However, even in the slow but steady rise in R&D activity by US multinationals in Scotland, it may well be that Scotland is primarily a 'hardware' rather than a 'software' location for these firms (Hargrave 1985; J Henderson 1989: 134–8; McDermott and Taylor 1982: 186–8). The R&D functions in Scotland tend to be of a product development nature and are part of the necessary infrastructure of customer service (Henderson and Scott 1987: 68).

Post-Weberian location theory

To address corporate decisions only within the context of 'choosing locations' is to ignore the competitive and profit-seeking context which is behind location (as well as product, technology, and other) decisions. As Massey (1979a: 70) puts it, it is in the 'a-spatial – rather than in the simply spatial – sphere that the primary causal links exist between the wider economy and the individual company'. The basis for Massey's point was the foundation of her book, *Spatial Divisions of Labour*, which meshes company location practices into the realm of capitalist firm behaviour, if incompletely (Walker 1988). At nearly the same time that the notion of spatial division of labour, as an explicit policy of large enterprises, was becoming widely accepted, a shift in empirical reality was noted, away from large firms and 'Fordist' mass production, and towards a more 'flexible' mode of accumulation centred around networks of small firms (Aydalot 1983; Lipietz 1986; Piore and Sabel 1984; Sabel 1982).

The behaviour of large, usually multinational, firms was the background for greater comprehension of what has come to be known as 'Fordist' production or the Fordist regime of accumulation. This is characterized by a mass production culture, involving mass markets for standardized products, and a social contract among firms, workers, and governments. As Bluestone and Harrison (1982) describe the 'Pax Americana', the interplay between corporate strategy, government policy, and labour generated some stability that has since ended, such as the concentration of power within large firms, a tangible effect in the evolution of global firms. The 'economic crisis' of the 1970s exposed the weaknesses of giant firms. Not only were the costs of vertical integration high, but producers in Japan and the NICs began to penetrate the American and European markets of the major multinationals. Consequently, 'flexible' production methods have been integrated into the organization of production in large firms.

Agglomeration revisited

The role of agglomeration economies has been revived as a widely held influence on regional growth dynamics. As Scott (1988b, c) has shown, inter-firm linkages, which can occur readily in large urban agglomerations, but only with difficulty over long distances in other geographical settings, are the principal mechanism by which the process of regional growth takes place. Aydalot (1988: 40–1) suggests that the only high-technology complexes which have developed and grown are those where local linkages – with customers, suppliers, subcontractors, or with scientific and technical knowledge – have flourished. Without local linkages, the regional development benefits are relatively small (Glasmeier 1988a; Hagey and Malecki 1986).

The importance of local linkages arises outside of high-technology industries as well. The importance of an agglomeration of customers – rather than of raw materials – in the recent transformation of US steel production is a major conclusion of Markusen (1986c) on the US steel industry. She describes the growth of steel-using industrial complexes, especially in Chicago and in Detroit, replacing Pittsburgh, Youngstown, and other traditional, resource-oriented locations. 'Just-in-time' production in the Detroit area, for example, reinforces the location of steel producers in that region.

Despite the severity of restructuring on industrial cities, the types of jobs are more abundant and varied, and wages are typically higher, in large urban regions. Even in the context of massive sectoral shifts and an increasingly international economic system, the attraction of urban regions seems largely unabated for office, service, and decision-making functions. Peripheral sites provide some production and other routine jobs, but non-routine jobs and large numbers of older jobs continue to cluster in metropolitan regions. There is also a spatial element in economic change, as Table 6.2 indicates, especially the tendency towards reconcentration of economic activity in urban regions, and policy will be forced to respond to the new local and regional disparities that result (Albrechts and Swyngedouw 1989). Cities allow interactions between buyers and sellers, between workers and employers, between allies and among competitors in a most flexible fashion, as Walker (1988: 396–8) has vividly described.

Table 6.2 Contrasts between Fordism and flexibility

Fordism	Flexibility
The production process	
Based on economies of scale	Based on economies of scope
Mass production of homogeneous products	Small batch production
Uniformity and standardization	Flexible and small batch production of a variety of product types (flexible automation)
Large buffer stocks and inventory	No stocks
Testing quality ex-post (rejects and errors detected late)	Quality control part of production process (immediate detection of errors)
Rejects are concealed in buffer stocks	Immediate reject of defective parts
Loss of production time because of long set-up times, defective parts, inventory bottlenecks	Reduction of lost time; diminishing 'the porosity of the working day'
Resource driven	Demand driven
Vertical and (in some cases) horizontal integration	(Quasi-) vertical integration; Subcontracting
Cost reductions through wage control	Learning by doing integrated in long-term planning
Labour	
Single task performed by worker	Multiple tasks
Payment per rate of output (based on job-design criteria)	Personal payment (detailed bonus system)
High degree of job specialization	Elimination of job demarcation
No or little on-the-job training	Long on-the-job training
Vertical labour organization	More horizontal labour organization for core workers
No learning experience	On-the-job learning
Emphasis on diminishing workers' responsibility (disciplining through pacing by assembly line)	Emphasis on workers' co-responsibility (disciplining through co-optation of core workers)
No job security and poor labour conditions for temporary workers; increasing informal activities	High employment security for core workers (life-time employment)

Table 6.2 continued

Fordism	Flexibility
Space	
Functional spatial hierarchy	Spatial clustering and agglomeration
Spatial division of labour	Spatial integration or division of labour
Homogenization of regional labour markets (spatially segmented labour markets)	Labour market diversification (in-place labour market segmentation)
World-wide sourcing of components and subcontractors	Spatial proximity of vertically quasi-integrated firms; formation of regionally specialized *filières*
Organization of consumption space through suburbanization	Organization of consumption space through the urban centre
The state	
Collective bargaining	Division or individualization; local or firm-based negotiations
Socialization of welfare (the welfare state)	Privatization of collective needs and social security. Soup-kitchen state for the underprivileged
Centralization	Decentralization and sharpened interregional or inter-city conflicts
The 'subsidy' state or city	The 'entrepreneurial' state or city
Indirect intervention in markets through income and price policies	Extensive direct state intervention in markets through procurement
National regional policies	'Territorial' regional policies (third-party form)
Firm-financed R&D	State-financed R&D
Industry-led innovation	State-led innovation
Ideology	
Mass consumption of consumer durables: the consumption society	Individualized consumption: 'Yuppie' culture
Modernism	Postmodernism
Totality or structural reform	Specificity or adaptation
Socialization	Individualization

Sources: Based on Albrechts and Swyngedouw (1989: 75–6); Moulaert and Swyngedouw (1989: 336–7 (Table 1)).

Economies of agglomeration are a composite of the distance-minimization efforts of numerous enterprises. Similarly, economies of infrastructure (or network economies) impose spatial organization on a region, permitting some places to obtain advantages which less well-connected places cannot (Perrin 1974: 32). A well-connected place also tends to accumulate infrastructure investments over time, giving it yet further advantages (Clapp and Richardson 1984; Friedmann 1986a).

Of all economic activities, producer services are perhaps most pulled towards urban agglomerations, at least towards those which meet 'certain urban size threshold requirements' which 'will in part be determined by the costs of recruiting (and keeping) specialized labour' (Coffey and Polèse 1987a: 609). The reduction of transaction costs possible in urban agglomerations, which accrue to all firms, as well as the *network economies* (reduction of costs as the number of users increases) important to service firms, result in further strengthening of agglomeration economies (Cappellin 1988; Scott 1988c).

Large firms try to maximize information-intensive contacts by centralizing in large urban centres. Around these cluster service activities for whom large firms are the dominant market. Such linkages benefit firms by stimulating innovation, especially product innovations (Davelaar and Nijkamp 1989a; Macpherson 1988b). Localized clusters or constellations permit consultation, interaction, and rapid response among firms, which is especially important for niche producers outside of high-technology industries (DeBresson 1989; Doeringer, Terkla, and Topakian 1987; Lorenzoni and Ornati 1988).

Hall and Preston (1988: 252) maintain that 'agglomeration economies of a quite traditional type may prove crucial' in the success of new firms a century later, citing as evidence London and its western corridor, the New York–Boston axis, and California's Silicon Valley. However, they refute the idea that only new industrial spaces can be seedbeds of new technology (Scott 1988c, d). The high-tech prosperity of Greater Boston 'shows that an old, indeed very old, region can remain innovative' (Hall and Preston 1988: 261). New regions can emerge, as in California's Orange County, but only where economies of agglomeration for labour, communication, customers, supplies, and services allow new small firms to prosper (Scott 1986, 1988b).

An array of other influences may be more important than urban size alone. Regions dominated by large firms and by branch production plants are unlikely to have the level of information, of R&D, and knowledge of the state of the art needed to spawn new firms, the second· consequence – after linkages – of regional economic development. This is summed up well by Sweeney (1985: 97):

> Areas dominated by large firms tend to have low entrepreneurial vitality. Such firms have internalised their information resources

and networks. . . . Areas dominated by branch plants of large firms also have low entrepreneurial vitality, their networks are with their distant parent company. Sub-contracting, an important source of start-up for new companies, is minimised by the large integrated firm and even more by branch plants. An important means of developing local networks for technology and information transfer is thus lost. Localities which are dominated by large firms are not centres of entrepreneurial vitality, partly because there is a lack of openness in the flow of information from the large to the small.

The development – or lack of development – of local linkages is the fundamental distinction between regions where development can be seen to have taken place and where it has not (Amin 1989c; Miller and Coté 1987; Saxenian 1988). In industrial districts, which need not be large agglomerations, the degree of linkage and interaction among firms makes them another type of 'old' industrial region which succeeds under current conditions.

Industrial districts

Industrial districts are a particular type of agglomeration, characterized by 'a localized "thickening" of interindustrial relationships which is reasonably stable over time' (Becattini 1989: 132). These districts are primarily comprised of small firms, but firms whose markets are national or international, in contrast to traditional artisan firms and dependent subcontractors (Brusco 1986). Inter-firm relations – primarily among the small firms themselves and secondarily with outside customers, agents, and competitors – stimulate innovations that are truly best-practice technology (Rullani and Zanfei 1988a; Takeuchi and Mori 1987). The Benetton Group which markets the products of small, mainly family-owned, knitwear firms in the Third Italy is a prominent, oft-cited example (Rullani and Zanfei 1988a; Scott 1988c). A smaller, but equally significant, example in central Portugal, based on wood and metal furniture, textiles, and pottery, may also be emerging (Lewis and Williams 1987).

Japanese examples are less well known in English, but the examples provided by Friedman (1988: 177–200) and Takeuchi (1987, Takeuchi and Mori 1987) of machinery districts in portions of Tokyo and in the Sakaki district illustrate that innovative, interconnected small firms are the core of such regions. The web of local linkages and subcontracting is the heart of the agglomeration advantages of such industrial districts. Family ties are also important and ease information flow among firms, and co-operation and sharing of new equipment are commonplace in 'Sakaki Inc.' (Friedman 1988: 196–8).

The depiction of a tightly connected web of firms which mutually support one another is often called the NEC (for north-east–central) model, after the region of Italy characterized by small and medium-size firms, also known as the Third Italy (along with the developed North and underdeveloped South). Such production systems are common in Japan and Europe, but less so in North America (Lewis and Williams 1987; Lorenz 1989; Maillat 1990; Rosenfeld 1989–90; Sabel 1989; Sabel *et al.* 1987). The emergence of this 'local production system' in Emilia Romagna, the centre of the Third Italy, has roots in sixteenth-century silk production in Bologna. For centuries, an informal economy in the surrounding region relied on work by women in their homes. It was not until the 1950s, however, that employment shifted out of agriculture into manufacturing, mainly shoes, textiles, and machinery, produced by artisans and craftspeople. The specific skills required for each industry or even for a certain operation are often found only in quite limited geographic areas (Capecchi 1989). The informal social and economic structure persists in the high-technology cottage industry in the Third Italy (Sabel 1982: 220).

This region continues to display its distinctiveness, but its 'organized complexity' and network of linkages have not diffused to the rest of Italy (Amin 1989a, c; Pezzini 1989; Sforzi 1989). The ENI Group maintains petrochemical and machinery manufacturing operations throughout Italy, but the closest relationships, involving collaboration on innovation and sales linkages, are with firms in central Italy (Mariani 1989). The informal nature of the economy and society keeps most linkages local, among other small and medium-size enterprises (SMEs) in the district. Similar to the Japanese Sakaki district, the Third Italy is characterized by sharing of knowledge, machinery, and trust among specialized innovative firms (Sabel 1982: 220–6). Within industrial districts or localities, the *co-operation* among firms is the key element (McArthur 1989: 203–4). The co-operation and trust seen in industrial districts are considered part of the informality and 'preindustrial' character of such places (Sabel 1989: 47–52). Interdependence arises from the intense specialization of firms: 'The moment the firm begins to expand and move beyond its original specialty it finds itself dependent on the help of neighbors with complementary kinds of specialties; and because the neighbors can never exactly anticipate when they too will need assistance, the help is forthcoming' (Sabel 1982: 225). Innovation originates in the specialization as well, as new firms' entrepreneurs start up new businesses making a variation on the original product (Capecchi 1989; Russo 1985).

The most recent phase of development in the Italian industrial districts involves the incorporation of IT. The telecommunications network now permits contact among local firms, suppliers, and buyers, in addition to conditions on the world market (Fornengo 1988; Rullani and Zanfei 1988b).

Wherever such industrial districts are found, informality and reciprocal contracts are the norm (Sabel 1989; cf. Miller and Coté 1987). The 'partnership', or 'mutual dependency that exists between client and subcontractor' is most common among SMEs, who subcontract a number of standard operations on a permanent basis (Lorenz 1989: 124). This interdependence is quite recent in some countries. Lorenz (1989) dates it as beginning in the recession of 1982–3, but it is becoming a common practice (Commission of the European Communities 1989).

Flexibility

More flexible forms of *production*, of *work*, and of *inter-firm relations* are presently forcing changes into the orderly sequence suggested by the product cycle (Ayres and Steger 1985; Goldhar 1986). Flexibility has become part of the prevailing perception about current economic change, especially within firms and regions. Indeed, 'the use of the term "flexibility" has become so common that it cannot simply be dismissed as something not worth attention' (Boyer 1988c: 222). It is useful to distinguish among these three types of flexibility, for they represent truly different forms of organizing production.

Hoffman and Kaplinsky (1988: 331) suggest that there are three 'pillars' to the currently evolving forms of production organization:

1. The adoption of systemic computer-integrated manufacturing technologies;
2. The development of new systematic relationships between plants and firms; and
3. The adoption of a new flexible labour process in which the past tendencies towards the increasing division of tasks, the de-skilling

of work, and the removal of control over production from the worker are reversed.

Taken together, these three forms of flexibility are directed towards the goals of product innovation and quality which are fundamental to the era of 'systemofacture' (Hoffman and Kaplinsky 1988).

Flexible production

At the first level of analysis, production flexibility includes the spectrum of changes which result from automation. Automation based on computers and microelectronics has three qualities which push it beyond the concept of other labour-saving capital investment. First, flexibility is built into the machinery, which is general-purpose rather than product-specific. No set-up is required to switch to a different product. A new product can be built simply by programmed instructions on machines which can produce many different products (Shaiken 1984: 6–7). Second, this flexibility leads to a broader scope of application. The expense of new technology might be greater than of previous generations of machines, but it provides a great deal more flexibility (to use the word in yet another sense) in product variety. This form of flexibility is especially important, because a firm with such a capability is able to handle both routine, volume production and more difficult (and profitable) non-standard orders, which allow it to take the lead on small-volume new-product introduction (Hayes and Wheelwright 1984: 40).

That this is occurring is vividly shown in the case of General Motors Corporation. 'In the early 1970s, GM shipped one million Chevrolet Caprices annually. Today, such best-sellers as the Corsica and Beretta manage annual sales of only 350,000 or so between them, and automakers must cater to regional preferences' (Taylor 1989: 104). Consequently, GM and scores of other firms have been forced to alter their approach to production. Thus, the 'world car' strategy, in which a standardized product which could be made in large, efficient volumes from parts manufactured in giant specialized plants, and sold throughout the world, has not worked as variety has surpassed price as a marketing virtue (Dicken 1986). To cater to consumer tastes, Ford is developing a modular engine that can be assembled three different ways to create four-, six-, and eight-cylinder versions (Taylor 1989: 108). Computer-aided manufacturing (CAM) technology is directed towards responding to diverse customer specifications (Gerwin 1989). Finally, an important goal in itself of programmable technology is improved quality and reliability, a factor more important than variety or price in many markets (Sciberras 1985; Starr and Biloski 1985).

Economies of scope help to explain both the push towards flexible production within firms and the profusion of strategic alliances among firms (Charles, Monk, and Sciberras 1989; Teece 1980). Internal economies of scope accrue to firms which can produce two or more different, related products together more cheaply than in isolation, ideal for situations of product differentiation. External economies of scope result when firms can use their know-how to apply to the activities – particularly innovative activities – of other firms and other markets. Recent alliances between semiconductor and biotechnology firms are an excellent example where advantages are neither traditional nor immediately obvious, but are certainly proliferating (Cooke 1988; Molina 1989: 132–44).

Within firms, the Fordist priority on mass production based on economies of scale has given way to economies of scope, cost savings based on producing a variety of products or services in small – even one of a kind – batches. Ideally, such a factory 'can produce a continuous stream of different product designs at the same cost as an equal stream of identical products' (Goldhar 1986: 27; Goldhar and Jelinek 1983). 'Design for manufacture' simplifies products to minimize the number of parts and to increase the number of common parts in different final products (Charles, Monk, and Sciberras 1989: 135–7). As Hitomi (1989) summarizes, such multi-product, small-size production is characterized as follows:

(a) variety of product items, produced in different volumes and with different due dates;
(b) variety of production processes;
(c) complexity of productive capacity;
(d) uncertainty of outside conditions;
(e) difficulty of production planning and scheduling;
(f) dynamic situation of implementation and control of production.

Figure 6.2 illustrates that automation has different effects depending on where and how it is applied. Any production has three principal spheres of activity: design, information co-ordination, and manufacture (Hoffman and Kaplinsky 1988: 56–61; Kaplinsky 1984a). Automation may take place only within a given activity ('island automation'), such as an individual manufacturing step, like the sewing of buttonholes, or within computerized design, and not affect the others (Fig. 6.2a). A

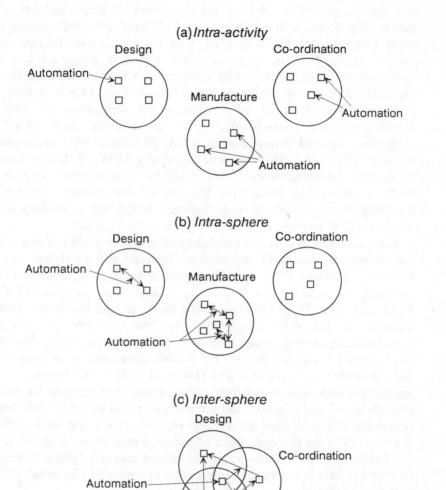

Figure 6.2 Different spheres or scales of automation. *Source:* Kaplinsky (1984a: 27 (Fig. 2.4))

higher level of automation involves the integration of activities within a given sphere of production (e.g. design or manufacture) but where each sphere is still unconnected to the others (Kaplinsky 1984a). This level of automation, a flexible manufacturing system (FMS), may include up to several machining centres and/or robots, and permits the small-batch and custom production needed for economies of scope and product variety (Fig. 6.2b). A British Aerospace FMS produces 2000 variants of small aircraft parts by batches of 5 to 10 units each (Tchijov and Sheinin 1989). FMS is most commonly found in aerospace and in machinery and automotive engine production, where lot sizes are small and demands for performance are high (Warndorf and Merchant 1986).

The ultimate level of automation is the integration of design, manufacture, and co-ordination into an integrated unit, labelled computer-integrated manufacturing (CIM) (Goldhar 1986) or 'system-ofacture' (Fig. 6.2c) (Hoffman and Kaplinsky 1988). R Schoenberger (1986: 99), discussing 'world class manufacturing techniques', notes that 'FMS is all we can handle for the rest of this century. Table 6.3 illustrates some of the contrasts between traditional technology and those in a CIM factory.

Most automation falls short of CIM and instead is largely confined to uncoordinated 'islands of automation' (Bessant and Haywood 1988; Kaplinsky 1984a). CIM is still a goal towards which companies, especially the largest ones, are aiming (Edquist and Jacobsson 1988; Flynn and Cole 1988; Goldhar 1986; Hoffman and Kaplinsky 1988: 146–52; R Schoenberger 1986). As 'the benefits of flexible manufacturing are becoming available in different forms to suit different product/market characteristics', the beneficiaries may have begun to include smaller firms (Bessant and Haywood 1988: 354). However, the automation technologies and their organizational counterparts are easily misunderstood and misused, and are particular obstacles for small firms (Cainarca, Colombo, and Mariotti 1989; Gros-Pietro and Rolfo 1989; Warner 1987). As the experience of previous long waves suggests (Ch. 5), 'flexibility is much more an organisational property than a technical one' and is best managed by large firms (Bessant and Haywood 1988: 359; Child 1987).

Just-in-time (JIT) production

Such production systems require, in turn, a tightly controlled stream of inputs tailored to the needs of production on both a short-term and a long-term basis. On a daily scale, supplies of inputs are best managed at low cost by employing the kanban (just-in-time or JIT) system of inventory management, developed by Toyota during the 1950s. 'In broad terms, the core of the JIT concept is the elimination of waste in all forms – production, materials, labour, time, energy, money and so on' (Arnold and Bernard 1989: 403; Lenz 1989). Instead of the producer (e.g. motor car assembler) maintaining inventories of all the many hundreds of parts needed for the vehicles to be produced, parts suppliers are required to deliver these several times per week – or per day. This sort of system thus demands tight control over suppliers, who are then also required to provide a higher level of quality – fewer defective parts per order with a goal of zero defects. In short, parts producers must operate at a higher frequency ('more often') and in smaller volumes ('less') than was true under traditional production organization (Arnold and Bernard 1989: 415; Fornengo 1988; Hoffman and Kaplinsky 1988; Lenz 1989).

Table 6.3 Traditional technology contrasted to computer-integrated flexible manufacturing

Characteristics of traditional technology	Characteristics of CIM technology
Economies of scale	Economies of scope
Learning curve	Truncated product life cycle
Task specialization	Multimission facilities
Work as a social activity	Unmanned systems
Separable variable costs	Joint costs
Standardization	Variety
Expensive flexibility and variety	Profitable flexibility and variety
Traditional factory characteristics	**CIM factory characteristics**
Centralization	Decentralization
Large plants	Disaggregated capacity
Balanced lines	Flexibility
Smooth flows	Inexpensive surge and turnaround
Standard product design	Many custom products
Low rate of change and high stability	Innovation and responsiveness
Inventory used as a buffer	Production tied to demand
'Focused factory' as an organizing concept	Functional range for repeated reorganization
Job enrichment and enlightenment	Responsibility tied to reward
Batch systems	Flow systems

Source: Based on Goldhar (1986: 29).

The reduction in throughput time, or cycle time, becomes a source of profitability (Schmenner 1988). Focusing on reduction of throughput time results in reduction of inventories, set-up time, and lot sizes. These changes, in turn, induce improved quality, revamped factory layout, stabilized production schedules, and minimization of engineering changes. Mazda, for example, was able to reduce set-up time from 6.5 hours to 13 minutes after it incorporated a new manufacturing system between 1976 and 1980 (Hayes, Wheelwright, and Clark 1988: 187).

Impressive results have been reported as a consequence of implementing JIT into production systems, such as reduction of stocks by 50 per cent, reduction of transportation and stockkeeping costs by 20 per cent, improvement of overall productivity by 25 per cent, and even more impressive figures in the UK are reported (Arnold and Bernard 1989). However, there are a number of intangibles in the measurement of gains from automation which conventional productivity measurements do not take into account. *Time* has become 'a manufacturer's most precious commodity' (Port, King, and Hampton 1988: 104; Schmenner 1988; Stalk and Hout 1990).

JIT production spills over into many other activities of production. It necessitates, for example, integrated data processing for R&D, procurement, production planning, inventory control, and marketing (Arnold and Bernard 1989). Outside the motor car sector, this is illustrated in the systems established by major US retailers, which feed daily sales data to computers at several clothing manufacturers, itemizing the styles, colours, and sizes which must be stocked (Braham 1986; Caminiti 1989). Consequently, flexible technologies appear to promote larger, rather than smaller, firm sizes. Finding customers for the more diverse array of new products makes sales and marketing more important company functions (Starr and Biloski 1985; Teece 1982).

The number of suppliers tends to drop substantially under JIT, since a proliferation of input producers is unwieldy to manage, and single-sourcing becomes commonplace. Suppliers may face increased costs – among them for computer systems and communication links – largely related to a push for high-quality, 'zero defects' production (Arnold and Bernard 1989). The communications needs of flexible production can be very large and costly, but the use of IT (telematics) is a prominent aspect of both industrial districts and of large corporate organizations (Fornengo 1988; Rullani and Zanfei 1988b).

The JIT system leads into requiring greater R&D and manufacturing system improvements at the factories of suppliers. R&D for motor car component firms has grown steadily, as electronics, new materials, and other new technologies have become standard practice (Hoffman and Kaplinsky 1988: 158–69). Bache *et al.* (1987) label the array of relationships between a dominant firm and its suppliers 'supplier development systems'. This reliance on suppliers has the significant effect of reducing the number of suppliers to those few which are reliable in terms of quality, delivery, and response. Manufacturers may even be better off with a single supplier, but must avoid the temptation to abuse the relationship and cut the profitability of that supplier (Burt 1989). Such producer–supplier relations are a cause of the 'dual structure' of Japanese industry which is now being imitated widely (Friedman 1988: 128–61).

The example of clothing

The apparel or clothing industry is one which benefits greatly from faster turnaround in production. For many years now, garment producers have relied on low-wage labour, primarily in Asia, to produce standard, low-cost clothing for industrialized markets in North America and Europe. As a set of standardized products which have large-volume sales through mass merchandisers and chain department stores, clothing manufacture (particularly assembly and sewing, often of pre-cut fabric components) was especially well suited to low-wage locations, such as Asia and countries of the Caribbean. In fact, textile and clothing production were the principal examples of the 'new international division of labour' by Fröbel, Heinrichs, and Kreye (1980).

More recently, however, the clothing industry, like that of motor cars, has become more concerned with fashion, lead times, and inventories. The wholesale cost comparison between a domestically produced dress and an imported one leans by 20 per cent ($50 to $40) in favour of the import, largely because of labour costs running 50 per cent or more below those in the USA, costs counterbalanced to some degree by shipping, duty, and related charges. Increasingly, however, lead times – orders must be placed with subcontractors months in advance – and speed of delivery are important, allowing retailers to charge prices that remain stable without markdowns or discounts for fashionable goods. This trade-off between speed and low costs has long been present in clothing production (Steed 1981), but its significance has grown along with the importance of frequent style changes. Manufacturers who have electronic data processing links to major retailers can fill orders in as little as two days, and generally in less than a month for any item, compared to three to four months from plants in the Caribbean (for the US market) and longer from Asia, which is the major supply region for both the USA and Western Europe. Thus, the 'return of clothing production to Britain' and to the USA is similar to the situation of other industries that have adopted 'flexible' and computerized manufacturing (Gibbs 1987, 1988; Weiner, Foust, and Yang 1988).

Technological change is more difficult to effect in clothing than in the related textiles industry, largely because of the complexity of sizes, fabrics, and styles on ever-smaller production volumes (Dicken 1986; Hoffman and Rush 1988). Where it is being put into place, and even where it is not, low-wage countries are somewhat less appealing locations than previously. Pre-assembly tasks, such as marking and cutting, are those being automated most, shortening the time required from design to assembly. Being able to reduce the lead time provides the jump on competitors – and profitability – which innovation and fashion traditionally have provided. Innovations in sewing have been common primarily only among the largest firms, and along with the pressure to shorter lead times, promise to make even clothing production, long a low-tech entry point for low-wage countries, an industry increasingly found in developed countries and wealthy NICs (Hoffman and Rush 1988). Clearly, trade restrictions have played a role in this predicament, and will continue to affect the competitiveness of poor countries in this and other industries (Dicken 1986: 248–50; Hoffman and Rush 1988: 192–205).

Robots and flexible production

One way in which firms have changed is to adopt automated production equipment, especially robots. To a large degree, it is perceived that unionized workers are unwilling to submit to technological change, because their jobs are indeed threatened (Ayres and Miller 1982; Bamber 1988). Craftsmen and other skilled workers, especially those which use precision industrial machinery in machine shops and factories, are among those most threatened by automation technologies (Blackburn, Coombs, and Green 1985; Harrison 1984; Noble 1984; Shaiken 1984). In clothing as well, small firms are increasingly seen as out of the market for CAD-based systems (Hoffman and Rush 1988; Zeitlin and Totterdill 1989).

Robots are the principal new technology of flexible production. They can be programmed – and in most cases reprogrammed – and require dramatically less labour than do the labour-intensive production operations which they replace. While jobs are replaced by robotics, they tend to be unskilled, repetitive jobs; new jobs created by adoption of robots demand a significant technical background (Edquist and Jacobsson 1988; Hartmann *et al.* 1985; Hunt and Hunt 1983; Salih, Young, and Rasiah 1988). A British study of employment changes in various occupational categories from 1978 to 1984 showed that only professional scientists, engineers, and technologists were increasing, while all others, including managerial, administrative, and clerical staff, draughtsmen, and technical staff, declined (Bessant and Senker 1987). However, the scenario for NICs might not be so bright. Multinational corporations 'will continue to seek cheap labour, the only difference is that the labour they now want has to be skilled' (Salih, Young, and Rasiah 1988: 396).

Table 6.4 Industrial robot populations in selected countries

Year	Japan	USA	UK	FRG	France	Italy	Belgium	Sweden
1974	1 000	1 200	50	130	30	90		85
1975	1 400							
1976	3 600	2 000						
1977	4 900		80	541			12	
1978	6 500	2 500	125			300	21	415
1979	9 100						30	
1980	14 250	3 400	371	1 255	580	454	58	795
1981	21 000	4 700	713	2 300	790	691	242	950
1982	31 857	6 250	1 152	3 500	1 385	1 143	361	1 400
1983	46 757	9 387	1 753	4 800	1 920	1 850	514	1 600
1984	67 300	14 550	2 623	6 600	2 750	2 585	860	1 900
1985	93 000	20 000	3 017	8 800				
Robot penetration (1984)	5.553	0.751	0.476	0.878	0.580	0.615	1.126	3.565

Source: Tani (1989: 193 (Table 1)).

Despite their advantages, the adoption of robots has been slow, except in Japan, which has led robot adoption from the mid-1970s, and Sweden (Table 6.4). Robotization has been greatest in the automotive and electrical/electronics industries, which together account for over 50 per cent of applications in most countries (Tani 1989). Most Japanese robots are assembly robots, and these are mainly found in the electrical and electronics sector, where the most intelligent robots are used, as well as in plastic moulding (Ishitani and Kaya 1989; Tani 1989; Yonemoto 1986). (Japan is the only country which classifies pick-and-place machines as robots, a fact which inflates its figures on robot penetration.) Japanese adoption of robots is also striking in their wider adoption within firms. It is not so much that there are more Japanese robot users, but that adopting firms have many more of them (Mansfield 1989). The same was true of the previous generation of production technology (Hicks 1986).

The adoption of robots and FMS by firms in these sectors is very different from the situation in most other countries, where they are used predominately in the motor car industry – characterized by giant firms – for welding tasks (Tani 1989). The diverse demand in Japan for new computerized equipment, in turn, stimulated development for specific small markets of machinery firms (Friedman 1988: 211–17). Small producers were able to utilize best-practice flexible machinery and to become internationally competitive. By contrast, the American lag in adopting robotics and other new flexible production technology is a result of adherence to traditional, short-term financial analysis which thwarted experimentation with new technology (Dean 1987; Hayes, Wheelwright, and Clark 1988; Mansfield 1989).

Table 6.5 Industrial robots and numerically controlled machine tools (NCMTs) in NICs

Country	NCMTs	Industrial robots
Argentina (1985)	500	
Brazil (1985)	1 711	
India (1985)	1 178	
Korea (1985)	2 680	55
Mexico (1984)	500	
Singapore (1985)	700	313
Yugoslavia (1983)	1 232	32

Source: Edquist and Jacobsson (1988: 130, 153 (Tables 9.1, 10.1)).

In the NICs, numerically controlled machine tools have diffused rapidly, but this is less true of robots (Table 6.5). Robots are most widely used in Singapore, a major centre of semiconductor production, almost entirely in semiconductor assembly operations of foreign-owned firms (Edquist and Jacobsson 1988). Flexible manufacturing systems are extremely rare in the NICs, except in Korea, giving evidence that only these two countries are comparable to the OECD countries. Second-tier NICs, such as Malaysia and Thailand, are far behind Korea and Singapore in the use of new technologies (Edquist and Jacobsson 1988:

171–89). Edquist and Jacobsson (1988) suggest that three factors influence the diffusion of new technologies in the NICs: information, especially from suppliers, and education and training. These, along with labour costs – which are also highest among the NICs in Korea and Singapore – seem to be behind the international patterns observed thus far.

Far less is known about the distribution of industrial robots within countries. The findings on the adoption of new production technologies in the UK and the USA suggests that traditional core-region locations with skilled labour and company R&D activities will be favoured (Gibbs and Edwards 1985; Hicks 1986; Oakey, Thwaites, and Nash 1982; Rees, Briggs, and Oakey 1986).

Flexible manufacturing systems – in effect, integrated clusters of robots – are far less prevalent. There are about 400 such systems installed in the world, in the same principal sectors as robots: automotive parts, aerospace, and non-electrical machinery (Warndorf and Merchant 1986). The adoption of robots or FMS is no guarantee of competitiveness and profitability for firms. Much depends on how these are used within a production system (Krafcik 1988). For example, in a comparison of Japanese and US factories where FMS was utilized, Japanese firms used them to produce nearly 10 times the number of different products. The US firms had opted for a larger-scale, lower-scale production (Jaikumar 1986).

The geographical impact of automation

In any event, both automation technologies and shorter production runs push assembly activities back to industrialized countries, breaking down the international division of labour (Kaplinsky 1984b; Sayer 1986; Sanderson *et al.* 1987). However, the situation is complex indeed, as Japanese firms retain at home the capability in R&D and technology, often opening only assembly plants to get around market protection in Europe, North America and elsewhere (Linge and Hamilton 1981: 35–6; Morgan and Sayer 1988: 72–4; Reich 1984). In virtually all technologically advanced industries, internationalization by large firms and their networks is dominated by advanced country locations, primarily because of the need for skilled labour in R&D, design, quality manufacturing, and marketing (Morgan and Sayer 1988). The 'triad' of Europe, Japan, and the USA is encroached upon only slightly by the NICs, Hong Kong, Korea, Singapore, and Taiwan. The division of labour *within* the Third World divides the more developed of them from others which remain dependent on cheap labour (J Henderson 1989;

Lipietz 1986; Salih, Young, and Rasiah 1988; Scott 1987). Consequently, despite large production volumes in semiconductors in South-east Asia, few local linkages have been generated (Henderson and Scott 1987).

Within developed economies, the decentralization of industry to rural areas and to foreign sites common in the 1970s is no longer the norm. Production is being pulled back to the central plants where R&D and engineering and interaction with suppliers and customers can be best accomplished (Camagni 1988; Schoenberger 1987). In both the UK and the USA, plants which adopted new production technologies did so at significantly higher rates in traditional core-region locations where R&D was also conducted (Gibbs and Edwards 1985; Oakey, Thwaites, and Nash 1982; Rees, Briggs, and Oakey 1986). A strong market orientation resulting from product differentiation also pulls flexible production back to industrial countries, to the detriment of Third World locations (Kaplinsky 1984b; Schoenberger 1988). The agglomeration tendency is not complete, however, since markets largely drive the new strategies. Within the Third World, therefore, countries which attract foreign investment are more often large markets, such as Korea, Taiwan, Mexico, and Brazil, than sources of cheap labour (Hoffman and Kaplinsky 1988; Sayer and Morgan 1987).

Networks of suppliers on the Japanese model are also part of the restructured geography of production systems. In the car industry, it is noteworthy that these have retained a bias towards traditional car manufacturing regions, because of the skill requirements needed to maintain quality in both components and in the final product. Some Japanese firms themselves have combined a requirement that suppliers be located relatively near by for JIT to succeed. At the same time, several car firms and their 'transplant' suppliers – Japanese car parts firms which have located factories in the USA – appear to have chosen their locations in small towns to avoid as much as possible labour union organization, a strategy employed by General Motors as well (Clark 1986b; Kenney and Florida 1988; Mair, Florida, and Kenney 1988). In other words, the dispersion of car production which has been observed (Glasmeier and McCluskey 1987; Rubenstein 1986, 1988) may be little more than a 'local-scale dispersal' that retains proximity to the core region of North American car production but on greenfield sites (Mair, Florida, and Kenney 1988: 370; Schoenberger 1987). It should be noted that a sizeable proportion of components are still imported from the Far East and Japan into the USA, the UK, and elsewhere in Western Europe (Morris 1988b; Reich 1984). A hybrid JIT system, with only daily deliveries from a small number of mainly distant plants, is weak evidence that flexible production and Japanese firms are identical in all places (Morris 1988b). The pattern used by US car makers is another 'rather odd hybrid called a JIT warehouse', which merely pushes the inventory costs away from the manufacturer and its suppliers and places them on the warehouse (Flynn and Cole 1988: 136).

In other industries as well, flexible systems have reduced the appeal of low-wage labour pools in favour of concentrations of different functions which contribute to production (Schoenberger 1987; Storper and Christopherson 1987). Geographically, the impact of this form of technological and organizational change is to make industry concentrate in areas where specialized firms and skilled labour are abundant, and where unforeseen changes could also be accommodated.

Flexible labour

The difficulty of transferring Japanese productivity and efficiency to other locations may rest largely on the importance of flexible labour. The presence of fewer routine tasks and few long production runs from flexible production leads to a demand for a 'highly skilled, flexible, coordinated, and committed work force' (Walton and Susman 1987). This form of flexibility, identified above all with Japanese carmakers, is closely related to and indeed evolved from Fordism (Gertler 1988b). 'The Japanese translation of the Fordist system . . . was simple. Toyota was the great innovator here, taking the minds + hands philosophy of the craftsmen era, merging it with the work standardization and assembly line of the Fordist system, and adding the glue of teamwork for good measure' (Krafcik 1988: 43). The Toyota system evolved during the 1950s and 1960s, when the much smaller Japanese market stimulated a myriad of experiments in flexible production technology (Cusumano 1988). For motor car firms which flourished in the larger American and European markets under Fordist production, more immediate adjustments are needed as the basis of competition shifts from price to quality, and from standardization to variety. But production costs still matter. As the JIT system keeps inventory levels at a minimum, a flexible work team minimizes the need for utility and repair workers, since other team members fulfil these tasks as well as substitute for absent co-workers (Krafcik 1988). The utilization of flexible labour seems to be much more significant to productivity (hours per vehicle) than the use of robotic technology. In 38 automobile plants located in Japan, North America, and Europe, Krafcik (1988) found little correlation with the level of flexible technology.

Whether this is 'neo-Fordism' depends on the degree to which new forms of work organization are employed. Neo-Fordism within the workplace involves the integration of different productive subunits into a larger 'flexible manufacturing system' for the production of a variety of products. This requires that workers themselves be 'flexible' rather than rigid in the definition of the tasks for which they can be called upon (Blackburn, Coombs, and Green 1985). Even if viewed solely within the framework of an individual firm, the implications are immense. When one takes the whole of inter-firm relations into account, the entire array of flexible accumulation is more significant yet, with ramifications for the networks of firms and entrepreneurship (Lorenzoni and Ornati 1988).

'Neo-Fordism' or 'post-Fordism' refers to production which replaces the task fragmentation, functional specialization, mechanization, and assembly-line principles of Fordism with a social organization of production based on work teams, job rotation, learning by doing, flexible production, and integrated production complexes (Kenney and Florida 1988). These are largely work practices that evolved in Japan and which involve large firms. In a variety of respects, the 'Japanese model' of production is in sharp contrast to the Fordist model, but it also involves a blend of large, powerful companies and use of tight relations with supplier firms and teamwork in production (Hiraoka 1989; Junkerman 1987). This teamwork, usually embodied in 'quality circles', is among the most frequent aspects of the 'Japanization' of industry (Littler 1988). Gertler (1988b: 426) points out that the greater skills demanded of labour in such flexible production systems results in greater reliance on, and *less* flexible relationships with, labour. These work environments are transplanted to countries which represent major markets for Japanese producers (Hiraoka 1989; Morgan and Sayer 1988). Resistance from labour unions is intense, and prevents widespread adoption of Japanese labour–management relations (Hoerr 1989; Hyman and Streeck 1988; Willman 1987).

Neo-Fordism and post-Fordism are also used to describe the increased work load standardization that has accompanied IT, especially in office work. The proliferation of computer technology and word-processing capability has had several layers of effects. First, it has eliminated the prospect of employment for those whose entry-level skills do not include familiarity with computers, or sufficient literacy to solve non-routine problems.

The computer skills needed for 'back-office' activities of banks, insurance companies, data-processing firms and others are not great, and many of these activities have gone to distant locations. For example,

Citicorp established a credit-card processing centre in Sioux Falls, South Dakota, over 1000 miles from the firm's New York headquarters. Most back-office activities, which formerly were literally in a back office of the headquarters of such firms, have simply suburbanized. In large part, suburban locations are where the demand for high-quality, 'cheap but educated' clerical labour is best met (Baran 1985; K Nelson 1986). Relatively high levels of education, language, and communication skills, and middle-class manner and social values are prized for the more demanding nature of clerical work (K Nelson 1986). For these reasons, members of minority groups and others whose education-related skills do not meet the criteria are increasingly out of the running for such jobs (Baran 1985; Cyert and Mowery 1987; Stanback 1987).

A second way has been to demand more flexibility on the part of workers. In addition to qualitative flexibility demanded by automated production, firms also make use of numerical labour flexibility through the use of *contingent workers*, a phrase used to designate workers hired at less the standard wage, for less than the standard work week or work year, or with less than standard fringe benefits (Belous 1989). This group is particularly common in the service sector, where part-time work is more common, among 'homeworkers' and, increasingly, in manufacturing among subcontractors (Christopherson 1989; Christopherson and Storper 1989; Harrison and Bluestone 1988; Hyman 1988; Morris 1988b; Pollock and Bernstein 1986). Temporary help services, especially common in health care, business services, and finance and insurance, reduce workers' 'attachment' to, and expectations of, an employer. Although many such jobs are clerical, temporaries also include lorry drivers, assemblers, and professionals (Mangum, Mayall, and Nelson 1985; Pollock and Bernstein 1986; Posner 1989). Indeed, the percentage of workers who work neither full time nor year-round is nearly 45 per cent in the USA, and 65 per cent in service industries, facts which are masked by statistics on 'employment' and 'unemployment' (Belous 1989; Christopherson 1989; Ettlinger 1988). The lower wages paid to women may have had the effect of lowering US productivity, by reducing the incentive to invest in new technology (Bernstein 1989).

Koshiro (1987) suggests that the use of part-time, seasonal, and temporary workers began to replace lifetime employment in Japanese firms, beginning after the 1973 oil crisis, despite conventional wisdom to the contrary. Friedman (1988) and Kenney and Florida (1988) show that such a dual labour force, especially regarding female workers, is common in Japan. To a large extent, the dualism is found between large and small firms, the latter characterized by lower wages (63% of large firms) and lower capital–labour ratio (33% of that in large firms) (Kaneda 1980: 31). The presence of informal production in many countries is still a common reality (Portes, Castells, and Benton 1989). In addition, part-time employment, a growing portion of employment in all developed countries, is overwhelmingly (over 60%) female (Belous 1989: 50–2).

Amin (1989a) notes that this type of flexible labour force has been a major part of the success of the 'Third Italy' and some other new 'flexible regions'. Flexibility may refer mainly to

> an ability simply to survive, and on the basis of an artisanal capacity to respond to new designs and new market signals, as well as other factors such as self-exploitation and the use of family labour, the evasion of tax and social security contributions, low overhead costs, and the use of cheap female and young workers, especially in the area of unskilled work (Amin 1989a: 30).

At the opposite extreme, labour skills and flexibility are demanded. Small-volume batch production systems require less labour than do massive mass-production facilities, but workers must be sufficiently educated and skilled to perform their jobs on an ever-changing mix of products. The transfer of Japanese labour practices to North America and Europe has not been smooth, in no small part because of cultural factors and varying histories of industrialization (Harber and Samson 1989; Hoffman and Kaplinsky 1988: 337–41; Holmes 1987; Wood 1988). The complete Japanization of the labour process has been the most difficult part of systemofacture to adopt.

It remains the case, however, that mass production and labour-saving technological change still abound in some settings. Philips UK provides an illustration of this as it attempts to maintain and improve upon economies of scale in order to be competitive. As Philips UK undertook a vast restructuring from 1977 to 1983, the proportion of scientists, engineers, and technicians in the UK rose from 13 to 21 per cent (Peck and Townsend 1987).

In the electronics industry, in many ways a prototype setting for the product cycle, chip design in R&D is now much more capital-intensive than previously, as expensive equipment is needed to create the intricate products which are often of a semi-custom nature. This capital intensity renders labour a less important consideration for the firm even at the production stage. The short life of many electronics products means that successive cost reductions through improvements in efficiency are less likely than in longer life cycles, and equipment is usable over several product life cycles, reducing average costs (E Schoenberger 1986). As a result, non-routine, adaptive behaviour is demanded of firms, in contrast to the linear sequence seen in the various production-oriented life cycle models (Beije 1987). This has parallels with the feedbacks in the innovation process discussed in Chapter 4.

Inter-firm relationships

Networks of suppliers on the Japanese model are being utilized, and not all of these are based on 'just-in-time' rationales. Although JIT, one of the beacons of flexible production, is generally developed around short-distance deliveries from local suppliers, many of the linkages of Japanese car makers for parts and components are scattered throughout Asia (Hill 1989). Thus, as Gertler (1988b) stresses, the reagglomeration tendency of flexible production is not absolute (Holmes 1986). Despite the increased importance of information flows, IT permits both dispersal and agglomeration and it is not clear what this means in spatial terms (Hepworth 1989: 152).

In fact, a wide variety of approaches is being adopted to remain afloat and profitable in the face of growing production capacity. In addition to the automation of production systems and innovations in labour relations, various types of alliances and collaborations, including those with Third World partners, have become widespread (Hoffman and Kaplinsky 1988: 285–318). The 'strategic supplier' is one component of the set of strategic alliances reviewed in Chapter 5 (Spekman 1988). Outside the motor car sector, recent alliances between semiconductor and biotechnology firms also provide excellent examples where advantages are neither traditional nor immediately obvious, but are certainly proliferating (Cooke 1988; Molina 1989: 132–44).

What has emerged is a *network* model which affects both large and small firms (Antonelli 1988b). Amin (1989b: 118) explains: 'We are not talking about independent small firms in the traditional sense, nor about subcontractors for large firms, but about the development of an industrial system (almost a corporation) composed of interlinked but independently owned production units'. The use of telematics or informatics brings geography back into the picture, through the need for infrastructure, to the issue of centrality versus peripherality, and the importance of agglomeration (Hepworth 1989).

In fact, Kaplinsky (1984a) sees the trend in automation to be very different from that of the past 25 years, when peripheral areas and Third World countries were able to benefit from industrialization. In the future, little production will disperse in the manner of the recent experience. As Jaikumar (1986) puts it, 'Flexible automation shifts the arena of competition from manufacturing to engineering.' Perez and Soete (1988) are not so pessimistic, provided lagging countries enter into the new technological systems early, while everyone is still learning and while old leaders are unlearning old ways. If the entry of newcomers is late, or only partial, it cannot succeed.

Flexibility appears to reinforce agglomeration economies and cumulative causation processes to a degree unanticipated. The necessity for close integration of the highly skilled activities of professionals and engineers with production tasks puts the set of acceptable locations in the hands of the professionals, the primary labour force. It is these workers for whom geographical location is among the 'social status or prestige rewards' associated with a job, and it is a factor which firms cannot ignore (von Glinow 1988: 76–7).

Agglomeration economies can be 'created', especially within industrial complexes of the type engendered by JIT technology (Sayer and Morgan 1987: 32; Schoenberger 1987). In other industries, where production is more disintegrated and fragmented among many small producers, location in urban agglomerations is more likely to continue. However, the infrastructure demands of flexible production are great, both in communications and conventional transport infrastructure. Just-in-time supply systems, for example, require reliable, high-speed road networks (Janssen and van Hoogstraten 1989). In most respects, then, 'world cities' and those with equivalent attributes and infrastructure are the only likely locations of flexible production complexes.

The agglomeration tendency of flexible specialization, in particular, is striking (Piore and Sabel 1984). It suggests that the networks of small firms – even without large-firm control – are sufficient to compete against large firms. Artisan networks in Emilia Romagna and other provinces of the Third Italy, in industries as diverse as knitwear, gloves, shoes, and ceramic tiles, began to make impressive showings in world markets. The competitiveness of these industrial districts lies in their flexibility – first, in terms of labour, since they rely largely on family members rather than employees, and second, in terms of innovativeness based on artisanal skills (Brusco 1989; Piore and Sabel 1984: 213–16, 226–9; Russo 1985). Each of these industries was poised to take advantage of the shift to fashion as opposed to mass production.

The larger question that remains unresolved at this time is the degree to which flexible accumulation will be the province of small and medium-size firms, building wider forms of common services inspired to a large degree by large-firm models, in contrast to the web of subcontractors amassed by large firms as they attempt to replicate the advantages of industrial districts (Sabel 1989: 18–19). The subcontracting that takes place in many cities goes down several levels, incorporating illegal as well as legitimate activities (Beneria 1989; Benton 1989; Ybarra 1989).

Scott and Kwok (1989) suggest that subcontracting and agglomeration go hand in hand, each reinforcing the other, in the southern California electronic printed circuits industry, in the manner of an industrial district. Takeuchi (1987) describes a similar dynamism in the machinery industry in the Tokyo region. To a large degree, it is this dynamic network of local suppliers and subcontractors which determines the success of agglomerations based on high technology industry. Glasmeier (1988a) found that branch plant operations, no matter how technical in orientation, resulted in few linkages and few spin-offs in the local area. Similarly, Harris (1988: 369) suggests that 'innovations in firms that are externally-owned (and large) may have adversely affected the innovative capability of the peripheral regions' of the UK. At an international scale, it may be the window of opportunity has closed on the value of cheap labour.

Conclusion

This chapter has shown, by way of numerous examples taken from around the world, that several things influence the innovativeness and competitiveness of places and countries. First and foremost, technical skills stand out as key to relating to the process of technological change and competition (Maggi and Haeni 1986). Actual 'production-systems', as Walker (1988) calls them, are very complex, and centre around organizational skills and information (Hepworth 1989: 146–9). Yet, to quote Harvey (1988: 109): 'What is most interesting about the current situation is the way in which capitalism is becoming ever more tightly organized through dispersal, geographic mobility, and flexible responses in labor markets, labor processes, and consumer markets, all accompanied by hefty doses of institutional, product, and technological innovation.'

Second, urban agglomeration is a central fact of life, as it has always been, despite having spent a decade or two in the background as the last flurry of rural industrialization took place. Urban areas contain the complex or synergy of factors which smaller, and especially remote, places cannot attain (Stöhr 1986a). Especially in information and knowledge, urban areas sustain a level of creativity not easily found or generated in other settings (Andersson 1985).

'Post-Weberian location theory' attempts to explain the 'transition from Fordist to flexible accumulation' (Harvey 1988; Harvey and Scott 1989). In seeking to develop a broadly applicable theory, it often fails fully to take into account the specificity of industries and of places (Sayer 1985; Walker 1985, 1988). More importantly, it imposes a structure of a systemic and overriding nature on what remains fundamentally a human process. 'New forms of production' either work or do not work because of people and their actions, their abilities and skills, and their interpersonal contacts (Fadem 1984; Malecki 1989). Cooke's (1985a) notion of 'labour geography', Clark's (1986a) 'geography of employment', and Scott's (1985) 'post-Weberian location theory' all encompass this sort of thinking, albeit each only in a partial way.

A second theme in this ongoing transition concerns the role of government or the state. The French 'Regulationist' school identified the close relationship between the needs of firms under Fordism – social security, stable prices, and monetary policy (Aglietta 1979). Government roles are complex, and are frequently passive rather than active. The role of the state under the regime of flexible accumulation is less collective, more individualized, and guaranteeing markets through procurement (Table 6.2). The 'entrepreneurial' state captures the new role of governments as 'partners' of industry, but there are many other interrelationships with economic activity as well, and including awareness of the local or indigenous potentials of specific regions (Albrechts and Swyngedouw 1989; Eisinger 1988; Moulaert and Swyngedouw 1989; Moulaert, Swyngedouw, and Wilson 1988).

There is considerable variety in the evolution towards more flexible, 'disorganized' work practices (Boyer 1988a, b; Lash and Bagguley 1988; Leborgne and Lipietz 1988). The relative roles of labour, of firms (or capital), and of the state interrelate to define the structure of the new partnerships still being worked out. An increasingly entrepreneurial role of the state has been evident in the USA since the 1970s (Cobb 1982; Goodman 1979). As this has become an international enterprise, prominent examples such as Japan are also emulated. Whether there is a 'transition' to flexible accumulation deserves a sceptical eye, since it may be far from general or universal (Conti 1988; Thrift 1989). Large, global enterprises are far from eliminated, and they are often able to profit from their own size and organization as well as the flexibility present in other firms, organizations, and structures (Hoffman and Kaplinsky 1988).

In sum, whether or not flexibility represents 'the end of Fordism' or something less is not the important issue (Gertler 1988b; Hudson 1989; Lovering 1990; Rees 1989; Sabel 1982; Sayer 1989; Schoenberger 1989). What is important is that some very novel combinations of organizing production – including labour, materials, information, and markets – are being undertaken. These are reversing the tendency towards dispersal of production, to rural and other low-wage sites, seen for the past quarter-century. At the same time, dualism persists even in the most developed settings, and may well be the foundation upon which flexibility rests (Berger and Piore 1980; Gertler 1989).

The essence of economic development, the process of technological innovation, is being centralized in urban agglomerations, and in the wealthy countries of the world, to a greater extent than we have seen for some time. The 'entrepreneurial state' and attempts to attract and generate the conditions for development are the focus of Chapter 7. At local, regional, and national levels, efforts to create 'creative regions' are common and flourishing, if not always successful.

Chapter 7

Creating technological places: nations, regions, and localities

Earlier chapters have shown that regional economic growth is a complex process which, despite similarities among places, has its unique local twists and turns. The penchant to try to imitate successes observed in other times and places has been especially pronounced in the case of policies to *create* technological capability, or at least the appearance of such a capability. At the national level, this is manifested in industrial policies and science and technology (S&T) policies that attempt to protect traditional industries and to 'target' certain industries and technologies which are thought to have especially high potential for future growth. Industrial policy measures used to protect established sectors include subsidies, government procurement, tax preferences, and national product standards. Increasingly, trade policies of both tariff and non-tariff varieties are being applied to mature sectors as well as to new, 'infant industries' (Boonekamp 1989). Macroeconomic policies are also critical in high technology contexts (Porter 1990; Roobeek 1990).

High-tech industry (however defined) is a key part of the S&T policies present at national and local scales alike, if only because places feel compelled to compete for a share of innovative industry and jobs. S&T policies have long focused primarily, perhaps exclusively, on government-funded R&D, whether conducted at government laboratories or in universities. The linear model (Ch. 4) and its assumed link between overall R&D and employment growth in high tech industry, while often misunderstood, informed a thrust for S&T policy in many countries which has focused on in-house R&D and did not focus on firms and their technological capabilities (Mowery 1983b). Combined with the economic rationale of 'market failure', which justified a state role, government grants for R&D remain a popular policy, in keeping with the linear model (Tisdell 1981). However, as Macdonald (1986) has shown in Australia, R&D grants have rarely funded R&D projects that would not have been funded by the firms themselves in any event.

Currently, most nations see high technology as a logical direction for their economic and industrial development efforts as well. The link between science and industry now stretches the usual meaning of S&T policy 'downstream' into industrial activity, and it pulls industrial policy 'upstream' into the source of competitiveness, technology (Arnold 1987; Brainard, Leedman, and Lumbers 1988; Freeman 1987; Rothwell and Zegveld 1985: 108–57). This shift in the thrust of government S&T policy reflects both a realization that a linear process of technological innovation no longer applies (if it ever did), and a concern with a larger set of 'barriers to innovation' which R&D effort alone cannot address. These include barriers related to government policy, such as regulation and standards, as well as those relating to communication and mobility of information, and flows of capital (OECD 1980b: 76–81; Piatier 1984).

At the local level, local authorities or governments are most often placed in a passive or reactive role with respect to technological and economic change. They often learn of new technologies late in their development or after they have already begun their diffusion, and they only learn of economic change the 'hard' way – after investment and disinvestment are concluded. Thus, regions with declining industrial bases have attempted to refashion their economies by means of the same high-technology sectors that are sought at the national scale. Since competition among places is fierce, local efforts are constrained by the existing situation, and favour cumulative strengths found only in agglomerated regional economies.

This chapter begins the discussion of technology policies at the national scale, where they are perhaps most evident. Competition for technological and economic superiority is of course also highly visible at the regional and local levels. An assessment of these policies concludes the chapter.

National policies towards high technology

Government activity in technology has been commonplace only since the Second World War. The major role played by technology in that conflict and the continuation of military priorities in the electronics and information technologies led to a convergence of three traditional foci of public involvement. Support of scientific and technical education and research, public procurement (largely for military purposes), and general modernization policies are three areas in which government policy has been customary (Nelson 1984: 13). As technological capability has grown to be the principal basis of national competitiveness, 'the new "holy grail" of international competition', science policy has converged with industrial policy (Nau 1986: 9; Rothwell and Zegveld 1985). Consequently, a nation's firms and industries have become more significant than military might in the perceptions of leaders and citizens (Brandin and Harrison 1987; Kotler, Fahey, and Jatusripitak 1985; Lewis and Allison 1982; Neff, Magnusson, and Holstein 1989). National priorities concerning R&D and industrial policies are important arenas in which government policy are thought to affect a nation's status relative to other nations (Porter 1990: 617–82; Roobeek 1990).

Science policy and R&D priorities

One of the major established means which countries have for policy to influence industrial competitiveness is in priorities for scientific and technical education and direct government funding of R&D. The level of government support varies widely, but in general the wealthiest countries expend the most on R&D, as pointed out in Chapter 4.

A number of different goals are possible for government R&D. The OECD utilizes several categories of R&D: defence, space, research or advancement of knowledge, health and welfare, and economic development.[1] In particular, one sees from Table 7.1 that among the OECD countries only the USA and the UK direct the bulk of their public R&D resources to defence purposes, followed closely by France (Nelson 1984). These mission-oriented, 'big science' efforts tend to have limited spillovers into commercial technology (Ch. 5).

Diffusion-oriented policies, on the other hand, 'seek to provide a broadly based capacity for adjusting to technological change throughout

---

1. These goals encompass a number of subcategories. Economic development covers both agriculture and industry as well as energy and infrastructure (transport and telecommunications, urban and rural planning, and exploration and exploitation of the earth and atmosphere). Health and welfare covers health, protection of the environment, and social development and services. Advancement of knowledge includes public general university funds and other civil R&D funds sometimes referred to as 'advancement of research' (OECD 1986a: 75).

Table 7.1 Government-financed defence R&D, OECD countries, 1985

	Million $	Percentage of total government R&D
USA	32 339	59
UK	3 024	49
France	2 666	35
W. Germany	906	12
Italy	360	10
Sweden	241	24
Japan	217	3
Canada	166	6
Australia	129	9
Spain	58	8
Switzerland	53	10
Norway	44	10
Netherlands	37	2
Belgium	10	2
Finland	7	2
Greece	4	3
New Zealand	3	1
Denmark	2	1
Austria	0	0
Portugal	0	0
Iceland	0	0
Ireland	0	0
Turkey	–	–
Yugoslavia	–	–

Source: OECD (1989a: 111 (Table 30)).

the industrial structure' (Ergas 1987: 205). Agriculture and industry are priorities of smaller countries and those which depend on primary exports, such as Australia and Canada (Table 7.2). Energy and infrastructure and health aims together receive 20–40 per cent of total government R&D in most countries. However, in this typology, Japan stands out because of its high level of support for energy and infrastructure and low level of government funds for research (Ergas 1987: 214). The policies of each country reflect both priorities for the use of public funds and the relative strengths of various sectors. Where priority is on defence and space, for example, the aerospace industry tends to be strong. Generally, government policies influence corporate

Table 7.2 Government civil R&D in OECD countries by objective, 1987

	Economic aims			Health	Other civil aims			Civil R&D as % of total government R&D 1985
	Agriculture	Industry	Energy and infrastructure		Environment	Research	Space	
USA	6.1	0.6	17.1	41.0	4.8	11.4	19.0	36
UK	12.4	25.5	14.8	16.9	8.0	13.5	8.0	33
France	6.6	19.8	18.3	11.8	3.4	27.2	11.0	51
W. Germany	3.5	27.4	18.9	9.8	9.3	22.0	8.8	58
Italy	5.8	31.7	19.7	9.5	3.9	10.9	15.5	68
Sweden	5.1	8.1	23.4	15.1	5.3	37.3	5.6	31
Japan*	22.5	11.7	37.4	6.6	4.9	3.4	13.5	70
Canada	19.9	17.3	10.1	15.0	12.6	25.2	0	79
Australia	21.5	15.9	7.4	15.4	15.4	24.4	0	62
Spain	9.4	30.1	4.7	13.3	13.5	11.9	12.3	67
Switzerland (1986)	11.4	2.2	17.8	8.0	4.7	55.9	0	43
Norway	16.9	24.6	13.7	21.0	7.1	13.0	3.7	54
Netherlands	7.2	32.6	14.2	8.4	6.8	17.6	5.8	52
Belgium	9.9	16.7	13.4	4.5	6.9	30.8	12.8	98
Finland	12.8	38.9	10.1	13.4	11.3	13.5	0	67
Greece	35.9	15.5	5.4	20.4	12.8	9.2	0.6	69
New Zealand (1983)	39.4	14.8	8.1	12.1	19.8	0.2	0.0	77
Denmark	12.6	24.0	9.8	12.6	4.2	33.1	3.8	57
Austria	9.6	23.6	8.2	10.3	2.7	45.5	0.0	57
Portugal	20.4	8.9	21.9	2.1	17.9	0.7	0	61
Iceland	46.6	18.3	17.5	4.9	2.4	10.0	0.3	73
Ireland	31.5	35.4	6.9	18.1	2.0	3.1	3.0	75

*Includes R&D performed by government only, 1985.
Source: OECD (1989a: 113 (Table 33); 114 (Table 36); 117 (Table 42)).

R&D most in sectors which rely heavily on government research funds and procurement (Crow 1988). In these sectors, government R&D complements, rather than substitutes for, private industry efforts.

In addition to direct expenditure on R&D, governments also play a significant – and growing – role in protecting and aiding national firms and industries in global competition (Porter 1990). This 'reindustrialization' role of government marks the joining of industrial and science policies. The joining of S&T or innovation policies with industrial policies did not begin until the late 1970s. As recently as 1976, Pavitt and Walker (1976: 51) were able to conclude that 'the promotion of innovation has not been considered by governments to be the most important component of policies for industrial development'. This lack of attention to innovation and technology changed dramatically during the 1980s. Technological innovation, particularly in certain 'high-tech' industries such as computers, electronics, aerospace, and biotechnology, is increasingly seen as a major driving force behind national economic growth and competitiveness. This realization has led to expansion of traditional research and education programmes and creation of a host of new, untested programmes intended to stimulate industrial innovation and technology transfer (Roessner 1989: 310; Roobeek 1990).

A great deal of total national R&D is funded by business enterprises, particularly in developed economies. This is especially true in Japan and West Germany, which do not have the marked defence orientation found in the USA, the UK, or France. In Japan, for example, 68.9 per cent of total R&D in 1985 came from industry, in West Germany, 61.8 per cent. This may be compared to 47.9 per cent in the USA, 41.4 per cent in France, and 46.1 per cent in the UK (OECD 1989a: 102). The commercial inclination of R&D, which also reflects a tendency for businesses to invest heavily in R&D, is related to the success of Japanese and West German businesses in world markets.

The example of Japan

Of all Western nations, Japan is thought to have perhaps the most articulated technology policy, determining the path of economic and technological activity. Japanese industrial policy dates back to the 'reconstruction period' immediately after the Second World War, and its roots extend back to the nineteenth century. The Ministry of International Trade and Industry (MITI) was established in the late 1940s to direct the reconstruction of Japan's war-ravaged economy

(Magaziner and Hout 1980; Prestowitz 1988: 122–44; Tucker 1985). MITI's role in Japan's success may be overstated. The success of several Japanese export industries was unrelated to MITI (or any other) policy, and included cameras, bicycles, motorcycles, pianos, zip-fasteners, and transistor radios. From the mid-1960s on, a number of other industries developed without any reliance on industrial protection and promotion policies, including colour televisions, tape recorders, clocks, calculators, electric wire, machine tools, textile machinery, agricultural machinery, and robots (Komiya 1988: 7–8). (The inclusion of machine tools in the preceding list is significant. Beginning with the machinery industry law in 1957, Japanese machine tool producers have been an object of attention from MITI, but these policies were unable to co-ordinate the thousands of firms in that sector (Friedman 1988: 84; Tsuruta 1988).) Table 7.3 illustrates the broad range of products and industries in which Japanese firms held more than a 50 per cent share of world trade in 1985.

To achieve this dominance, Japanese firms devote large amounts of resources – far more than do American firms – to designing, constructing, and equipping manufacturing facilities. By comparison, US firms commit relatively more to R&D and to marketing-related costs (Hull, Hage, and Azumi 1984; Mansfield 1988b). Japanese strength in new technology as well is evident in US patent data (Narin and Frame 1989). The US market is large and technologically sophisticated, and a company wanting to earn a substantial return from its technology will patent it in the USA. Between 1975 and 1985, patents to US firms and individuals declined from 64.9 per cent to 55.5 per cent, while the Japanese share doubled from 8.9 per cent to 17.9 per cent. France, West Germany, and the UK together held 16.5 per cent of US patents in 1985, up slightly from 16.0 per cent in 1975. By 1987, the three top recipients of US patents were Japanese firms: Hitachi, Canon, and Toshiba. However, Japanese technological strength is found in a limited range of technologies; 80 per cent of the US patents held by Japanese firms are associated with 26 per cent of the possible patent classes. These tend to be in important industrial areas, such as motor car engines, photography, drugs, computers, consumer electronics, and photocopying (Narin and Frame 1989: 601).

In addition to being limited in scope, nearly all of the products which Japan dominates are a result of foreign technology imported and then creatively adapted and improved (Mansfield 1988a, b; Nelson 1984). As Rosenberg (1982: 273) notes, 'The Japanese have elevated to a fine art what they call *improvement engineering*. Although this kind of technical skill was not responsible for any major, original inventions, it has enabled them to draw upon a large inventory of foreign technologies and reshape them to their own requirements with a high degree of sophistication.'

Table 7.3 Industries and products led by Japanese firms (1985)

Shipbuilding
Zip-fasteners
Colour cathode ray tubes
Magnetron tubes
Cameras
Plain paper copiers
Video cassette recorders
Ceramic packaging for integrated circuits
Micro direct current motors
64K RAM CMOS memories
Radio remote control services
Electronic music generators (synthesizers)
Electronic wrist watches
Numerically controlled machining centres
Hi-fi audio products
Compact laser disc players
Ceramic capacitors
Some resistors, coils, switches and filters
Electronic typewriters
Electronic calculators
Robotics (in the widest sense)
Tape recorders
Microwave ovens
Satellite ground stations
Facsimile telecopiers (fax)
Artificial leather
Motorcycles
Pianos
Magnetic tape
Liquid crystal displays

Source: Adapted from Freeman (1987: 28 (Table 11)).

An illustration of MITI involvement is seen in the computer industry, an early 'strategic sector'. Responses to American dominance – especially that by IBM – did not develop until the 1960s (Flamm 1987, 1988;

Shinjo 1988). Japanese firms were far behind their American competitors, despite the fact that the electronics industry was designated as the core of future industrial development in Japan as early as 1957 (Flamm 1987: 129; Shinjo 1988: 342). Ten years later, after the IBM System 360 computer began to overwhelm all competition, major projects were organized to raise Japanese competitiveness (Arnold 1987; Roobeek 1990). Among them was the super high performance electronic computer (SHPEC) programme, a co-operative project that pooled the resources of government labs and private corporations. Such co-operative research, funded partially by the government, is the norm in Japan, with MITI funding accounting for 40–50 per cent of research association budgets, which is then divided among member firms (Flamm 1987: 140; Peck and Goto 1981). Government procurement policy also served to protect the computer industry when it was in its 'infant' stage, when an informal 'buy Japanese' policy prevailed (Flamm 1987: 143–6).

The result two decades later has been a phenomenal growth in R&D in the strategic area of IT (hardware and software). Although it began in large part as a government priority, industry funding steadily took over, and reached 88 per cent in 1984 (Flamm 1987: 138). Consequently, by late 1989, Japanese firms had overtaken IBM in the fastest-growing portions of the computer market – laptop and portable computers. This dominance is based on an accumulated monopoly of technology for batteries, screen, and disc drives – in addition to chips – used in such machines (Lewis *et al.* 1989).

In Japan, as elsewhere, government research funds are disproportionately important in basic research, whereas product development is left to industry. In particular, MITI and Nippon Telephone and Telegraph (NTT) have been notable for their willingness to sponsor speculative, long-term research, such as the 'fifth generation project' (Feigenbaum and McCorduck 1983; Flamm 1987). Some projects have been unsuccessful, such as an electric car project. Overall, however, the major funding disbursed to R&D has had a significant effect on Japanese competitiveness, in part because of attention to markets and international competitiveness associated with Japanese R&D, in contrast to the military priorities found in some other advanced countries (Goto and Wakasugi 1988; Gross 1989; Odagiri 1985).

Japanese industrial policy became critical in the wake of the oil crisis of 1973, when it was made apparent that high technology, rather than resource-dependent, industries were preferable (Uekusa 1988). The oil crisis merely accelerated a process of strategic selection that had begun in 1970, when MITI's 'Vision for the 1970s' was published. The core of this 'Vision', repeated and intensified in its successors, 'Long-term vision of the industrial structure' (1975) and 'MITI policy vision for the 1980s', was the shift from capital-intensive heavy industries (steel, shipbuilding, motor cars) to knowledge-intensive machining and assembly industries, including general machinery, electronic and electrical equipment, transport equipment, and precision instruments (Uekusa 1988). The 'Vision for the 1980s' bluntly 'called for developing an industrial structure with high technology as its core, which it referred to as a "creative, knowledge-intensive industrial structure"' (Uekusa 1988: 97). The evolution of just such an industrial structure was evident by the mid-1980s (Kash 1989; Rothwell and Zegveld 1985: 148–9). The key mechanism for achieving the Japanese 'Visions' has been 'the promotion of industry-specific R&D, directly and indirectly' (Patrick 1986: xiii).

An interplay between government policy and the banking system in Japan aids its technological competitiveness. For example, the participation of institutions such as the Industrial Bank of Japan (IBJ) or the Japan Development Bank in consortia to finance large, risky projects inspires the confidence of other participants, because it is seen as the practical equivalent of a government guarantee (Jéquier and Hu 1989: 43). The IBJ has its own research staff and monitors trends in industries and of individual clients. Through this 'technological intelligence' role, 'the IBJ has helped its clients to diversify into more competitive, higher-technology product lines' (Jéquier and Hu 1989: 119–20). The wide-ranging activities of banks in Japan, on the model of the Crédit Mobilier in the nineteenth century, is found also in Europe, especially in West Germany, where 'banks were technically equipped to assess and therefore to take risk, and to offer their clients more than just money' (Jéquier and Hu 1989: 31). In this active model of banking, banks are 'technological institutions' that should not be ignored when discussing technology policies (Jéquier and Hu 1989: 193). The Japanese trading companies, or *sogo shosha*, serve a somewhat similar function by co-ordinating investments and projects (Kojima and Ozawa 1984).

Japanese forecasts and 'Visions' cover a broad technological scope. Industries related to basic materials, such as biotechnology, composites, ceramics, and new energy technologies, have received considerable attention from major projects (Table 7.4). These core or generic technologies correspond to the Japanese definition of high technology (Ch. 5), which is more oriented towards technological paradigms than is the thinking in most other countries. Service industries illustrative of convergent IT and 'new media', such as telecommunications, communications processing, and information processing, also merit attention in the 1980s and 1990s (Imai 1986: 155; Uekusa 1988: 106).

The MITI 'Vision for the 1990s' emphasizes quality of life priorities and 'unstrategic sectors', such as resorts, interior decorating, and fashion (Schlesinger 1990). Whether this 'Vision' will have the impact of prior goals is uncertain.

Table 7.4 Large-scale Japanese industrial technology projects

Project name	Period
High performance computers	1966–71
Jet engines for aircraft	1971–81
Electric cars	1971–77
Pattern information processing	1971–80
Sunshine plan – alternative sources of energy	1974–2000
Moonlight plan – large-scale energy-saving technology	1978–
Fifth generation computers	1979–91
Optical measurement and control system	1979–85
Basic-chemical manufacturing using carbon monoxide feedstocks	1980–87
Manganese nodule mining system	1981–89
High-speed scientific computers	1981–89
Basic industrial technology for the next generation (ceramics, polymers, alloys, composites; biotechnology)	1981–90
Perfect crystals	1981–86
Ultrafine particles	1981–86
Automated sewing system	1982–89
JUPITER (juvenescent pioneering technology for robots)	1983–90
Resource exploitation observation system (satellites)	1985–90
SIGMA (software industrialized generator and maintenance aids)	1985–89
Aqua renaissance 90 (waste water processing and purification)	1985–91
Interoperable databases	1985–92
Advanced material	1986–93

Sources: Imai (1988: 213 (Table III); Tanaka (1989: 363–64 (Table 1)); Tucker (1985: 28–9)).

Freeman (1987) emphasizes that the Japanese approach to technological forecasting from which the 'Visions' have emerged is well suited to identifying new technological paradigms early. Like many other processes in Japan, forecasts are developed by formal and informal consultations with thousands of experts, rather than relying on published statistics. This permitted Japanese policy-makers to recognize the crucial importance of IT at an earlier stage than in most other countries (Freeman 1987: 55–90). Canada has experimented with a similar broad consultation approach (Elzinga 1987).

The deliberate attempt to develop technological capability in a technology and in linked industries is exemplified in the case of aerospace technology. Japan, an undisputed economic and technological power by the late 1980s, has virtually no aerospace industry, but the roots of a strong industry predating the Second World War remain. As seen in the UK, France, the USA, and the USSR, aerospace technology is driven primarily by the military market (Nelson 1984). Japan could continue to import or co-produce military aircraft from the USA as it has done in the past. Co-production, while it transfers know-how, omits the 'know-why' behind a system design (Samuels and Whipple 1989: 49). Instead, Japan has chosen to co-develop with the USA a new fighter, the FSX, at considerably higher cost than alternatives.

Jet fighters are among the highest forms of aerospace technology, another designated priority of MITI for the next decade. MITI officials believe that the technology, manufacturing processes, and industrial organization needed to produce a first-class jet fighter present valuable new opportunities for Japanese aerospace (Samuels and Whipple 1989). Jet fighters are complex machines which comprise numerous sophisticated components and subsystems: structure, propulsion, avionics, and armaments. In addition, the design and manufacture of such aircraft depends on 'systems integration', which involves managing the operation of putting the subsystems together. 'Systems integration is the most challenging aspect of aerospace production, and because full-scale programmes are infrequent, opportunities to acquire these skills are rare' (Samuels and Whipple 1989: 47; Walker, Graham, and Harbor 1988).

Developing such skills also can help the 'roots' and 'fruits' of aerospace to grow, especially through its linkages and spin-offs to other sectors (Fig. 7.1). The 'roots' are technical processes and management skills; the 'fruits' are product innovations that will find uses in other sectors, such as motor cars, leisure industries, shipbuilding, energy industries, and housing (Samuels and Whipple 1989: 50). Space and satellite technology and commercial jet aircraft are also likely Japanese beneficiaries of this effort (Valéry 1989). It is perhaps ironic that, despite its unquestioned technological accomplishments, achieved without the military connection found in other countries, Japan is now drawn into military technology.

Aerospace is not the only sector targeted by Japanese policy. Biotechnology, photovoltaics, high-definition television (HDTV), as well as a continuing interest in electronics are all cited as next-generation technologies to which considerable corporate R&D and government co-ordination is currently devoted (Brown and Daneke 1987; Norman

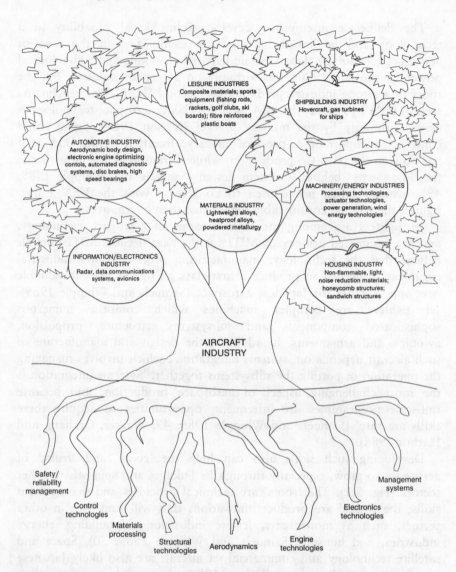

Figure 7.1 Japan's technology tree, with 'roots' and 'fruits' of aerospace technology. *Source*: Samuels and Whipple (1989: 50)

1989; Wysocki 1987; Yoshikawa 1988). Funding for R&D is going increasingly into science and basic research and less into technology (Arnold 1987: 259). Current Japanese achievements have come in part from acquisitions and investment in firms in the USA and elsewhere, bringing cries of plunder (Sun 1989). This continued reliance on foreign technology suggests that its openness to external ideas, rather than self-reliance, is Japan's chosen route to prosperity in the 1990s (Arnold 1986).

Other efforts, such as those behind HDTV, suggest corporate strategies oriented towards future possibilities, with 'roots' in basic technologies and branches and fruits in linked sectors (Valéry 1989: 9–12). More importantly, they highlight the fact that Japanese firms do not blindly follow MITI directives, but compete strongly, with high levels of R&D, both in the domestic market in Japan as well as internationally (Majumdar 1988; Oshima 1984). This competition, and the mix and balance of central organization and market dynamism, are the unique elements in the Japanese system (Okimoto 1986, 1989).

Europe: catching up

The science policies of other nations, partially as a reaction to the Japanese thrust, also began by the late 1970s to include investments in 'strategic' industries as a priority (Foray, Gibbons, and Ferné 1989). Unlike Japan, European governments reacted to IBM's dominance in computers by encouraging mergers and sheltering the resultant large national firms which might be seen as equal competitors to IBM. International Computers Limited (ICL) in the UK, Compagnie International pour l'Informatique (CII) in France, and Siemens in West Germany were 'placed on a high-powered diet of protection, procurement preferences, and subsidies' (Flamm 1987: 154). Government support for computer R&D rose steeply during the 1970s, but remained largely national in scope (Ballance 1987; Roobeek 1990).

The Japanese 'fifth generation project' in 1981 sparked a wave of co-operative R&D projects, both among European firms and as policy initiatives of governments and of the EC (Flamm 1987; Freeman 1987; Roobeek 1990). The decline of mainframe computers and the emergence of personal computers has altered the place of government in the industry. Small firms, even new start-ups, are frequently more influential than the industrial giants, and competition from NICs is very real, prompting a return to protectionism over R&D as the mode of policy reaction.

European policy-makers reacted to the rapid Japanese technological ascent in the 1970s with alarm. European alarm over a 'technology gap' *vis-à-vis* Japan and the USA revolved around such points as the following: '. . . the aggregate list of separate European national strengths is not impressive and the state of affairs is worrisome. From this point of view a technological gap exists and there may be truth in "Europessimism'" (Press 1987: 33). This gap is an overgeneralization, according to Patel and Pavitt (1987), and overlooks leading European positions in chemicals and nuclear energy, a lead over the USA in motor cars and metals, and over Japan in aerospace and technologies exploiting raw materials. Patel and Pavitt found a significant European gap only in electronics and software. Malerba (1985) identifies the problem in electronics as a lack of demand from the military or the computer industry. 'Insofar as there is a generalised process of economic "Eurosclerosis", its sources probably have little to do with any general lack of technological and entrepreneurial dynamism' (Patel and Pavitt 1987: 82; Peterson and Maremont 1990).

In the absence of any consensus but with the impression that something should be done, collaborative activity among firms and among governments, hitherto unthinkable, became the norm, both in individual countries and under the umbrella of the EC (Foray, Gibbons, and Ferné 1989). The first such programme was ESPRIT (European Strategic Programme for Research and Development in Information Technologies), which began in 1982. It has been followed by nearly a dozen others, many with intriguing names, in addition to R&D programmes in other broad areas (Table 7.5). A host of firm-to-firm collaborations have developed within EUREKA (Curien 1987: 63–4; Sharp and Shearman 1987: 72–3). It is yet unclear how the 'single market' of Europe will affect such collaborations in the future. To date, these collaborative research arrangements have taken on many different structures (de Woot 1990; OECD 1989b: 180–8). That they are still largely experimental rather than routine for both firms and governments is apparent.

Outside EUREKA, European governments have co-operated in high-technology ventures since the 1950s, when CERN (Conseil Européen pour la Recherche Nucléaire) was organized for high-energy physics research (Martin and Irvine 1984). Another project, the Airbus, continues as a means for Europe to maintain an aircraft industry despite being a financial burden.

Specific national programmes have also emerged, usually around a set of selected high-technology sectors (Rothwell and Zegveld 1988). Among the largest is the Alvey programme in the UK, which began in 1983 to improve the British position in microelectronics. It was the first British attempt to move beyond defence R&D and fundamental research conducted in universities and, more importantly, it involved the collaboration between firms and among industry, government, and universities and polytechnics (Freeman 1987: 120–38; Hobday 1988).

Table 7.5 European Community R&D programme, 1987–91

Purpose or activity	$US million
Information and communication	2610
Information technology (ESPRIT)	1840
Telecommunications (RACE)	630
New services (DRIVE, DELTA, AIM)	140
Energy	1350
Thermonuclear fusion (JET)	700
Fission	500
Non-nuclear energy	150
Industrial modernization	975
Manufacturing industries (BRITE)	460
Advanced materials (EURAM)	250
Standards and measurements (BCR)	215
Basic materials	50
Quality of life	430
Environment	300
Health	90
Radioprotection	40
Biological resources	320
Biotechnology (BAP, BRIDGE)	140
Food technology (ECLAIR)	120
Agriculture	60
Ocean resources	90
Ocean sciences	60
Fishing	30
General European co-operation	425
Total	6200

Source: Stevens (1990: 19 (Table 2)).

ESPRIT: European Strategic Programme for R&D in Information Technologies
RACE: R&D in Advanced Communications Technology
DRIVE: Dedicated Road Safety Systems and Intelligent Vehicles in Europe
DELTA: Development of European Learning through Technological Advance
AIM: Advances in Information in Medicine
JET: Joint European Torus
BRITE: Basic Research in Industrial Technology for Europe
EURAM: European Research in Advanced Materials
BCR: Community Bureau of References
BAP: Biotechnology Action Programme
BRIDGE: Biotechnology Research for Innovation Development and Growth in Europe
ECLAIR: European Collaborative Linkage of Agriculture and Industry through Research

Throughout Europe, individual national leader firms and strategic partnerships among those firms received massive funding for R&D ventures (Arnold 1987; Hobday 1989: 232). Critics of this approach contend that the 'top-down' approach and the focus on large 'national champion' firms common in Europe is an anachronism in a time of global firms and entrepreneurship (Marcum 1986; Woods 1987). In addition, Freeman (1987: 135–6) notes that the Alvey programme largely grew out of priorities in the defence area rather than in commercial markets, thus harming the development of suppliers of components and technologies for storage, display, printing, and communication needed for civilian markets. The dependence on Japanese suppliers for these items is the same as noted earlier concerning the American computer industry. Others contend that the nationalistic thinking behind British and other European policies is misplaced and that global strategies are still too rare (de Woot 1990). Moreover, comparisons made with Japan and the USA alone neglect the growing role of the NICs in world trade (Oshima 1987).

Sector-specific policies have emerged as well, especially in biotechnology, where the flurry of new firms in the USA and the naming of biotechnology as a next-generation technology by MITI in Japan, spurred the creation of European policies. The divergence between a market-led approach with no central direction in the USA and the co-ordination found in Japan has led to policies in Europe, such as the Dutch Biotechnology Programme, being 'between dirigism and laissez-faire' (Cantley 1986; Rip and Nederhof 1986; Sharp 1987). Linkages among firms and between academic and industrial research on the Japanese model are the preferred approach in European policy. The US model of entrepreneurship has been difficult to emulate, given the shortage of venture capital (Ch. 8) (Sharp 1987).

The American case

American policy is less explicitly interventionist, and has long held that government procurement for military and space programmes is a superior method of reaping the benefits of high technology (Abernathy and Chakravarthy 1979; Allen *et al.* 1978; Levin 1982; Molina 1989; Nelson 1982, 1984; Roessner 1987). This policy has recently been called into question, both because of the reduced potential for commercialization of military technology, and because of diminished concerns about military threats in comparison to economic ones (Neff,

Magnusson, and Holstein 1989). In addition, there is growing concern that 'the pattern of indirect commercial or economic benefits – spinoffs – is no longer providing the United States with enough technologies that are competitive in the international marketplace' (Kash 1989: 37). However, dependence on Defense Department funding has continued, for both software development and for advanced semiconductor manufacturing techniques (SEMATECH), and generally characterizes the US approach to S&T policy (Borrus 1988; Flamm 1987; Foray, Gibbons, and Ferné 1989). Beyond military demand pull, US S&T policy is only indirect and implicit, working through the regulatory and tax systems (Bean and Baker 1988; Rees 1987; Rothwell 1980).

In the wake of the Japanese 'fifth generation project' in the early 1980s, the US Congress passed legislation which allowed and encouraged firms to undertake co-operative R&D projects. Among these was the Microelectronics and Computer Technology Corporation (MCC), a consortium of some 20 companies, whose focus is on basic research, and the Semiconductor Research Corporation, for product development (Borrus 1988; Flamm 1987: 115–18; Peck 1986). Industry co-operation has not been sufficient to sustain other co-operative R&D ventures, however, especially in high-technology sectors, where the American eye for short-term pay-off prevails.

More common are one-to-one contractual joint ventures between firms of a more conventional form, where the costs and benefits are allocated in advance (Ch. 5). Most such collaboration has been for basic research, supplemented by participation in university research centres, some of which are supported by the US National Science Foundation (Arnold 1987; Fusfeld and Haklisch 1987). Many other collaborative arrangements have been formed between US and foreign firms, in an attempt to broaden the set of technological assets to which a firm has access, from research to marketing (Chesnais 1988; Mowery 1988a). Finally, decentralized applied research efforts at several universities, some of them also funded by the National Science Foundation, are intended to address the needs of industry (Flamm 1987, 1988).

Canada: avoiding technological dependence

The negative effects of external control and technological dependence particularly affect countries like Canada which are unable to maintain an 'offensive strategy' with respect to technology. Foreign ownership strongly influences R&D, both quantitatively and qualitatively. In the pharmaceutical industry, the degree of foreign ownership consistently decreased R&D intensity (R&D expenditure as a percentage of sales) in OECD countries (Palda and Pazderka 1982). The type of R&D undertaken by transnational corporations in host countries also tends to be different from that conducted at home. It is often oriented towards minimal fine-tuning and adaptation of a product – developed at a central laboratory for the home market – for a new market. Such *truncation* of a subsidiary's mandate, usually to production only, restricts it from developing an independent technological capability (Hayter 1982).

Canadian industry has been affected visibly by this impact of the international division of labour. R&D in Canada by subsidiaries of multinational corporations is virtually non-existent (Britton 1980; Britton and Gilmour 1978; Etemad and Séguin Dulude 1986a; Ondrack 1983). The development of 'world product mandates' at local subsidiaries depends critically, as in nearly all issues covered thus far in this book, on personality and human interaction. An active search for new products and identification of opportunity can create a world product mandate except perhaps in the most strictly controlled subsidiaries (McGuinness and Conway 1986; Science Council of Canada 1980). However, even in operations with world product mandates, patents resulting from R&D were most often applied for and held by the headquarters of key subsidiaries (Etemad and Séguin Dulude 1986b).

Because of the dependent nature of much of Canadian industry, Canada's 1979 S&T policy statement recognized the need for relationships between large 'core' Canadian companies and small firms. The encouragement of linkages among firms in Canada, aid to new innovative firms, and promotion of consortia and joint ventures was rather avant-garde for the time and foresaw these as critical issues for the 1980s (Science Council of Canada 1979). Steed (1982), working for the Science Council, proposed a more focused approach of supporting 'threshold firms' in the country, some of which if promoted could develop into core firms. Based on a concern over foreign ownership, the Canadian policy mood at present is to foster the growth of small indigenous enterprises in high-technology sectors. Steed's recommendation was to let the market determine which small firms survive and expand to the threshold size. In 1976, 165 threshold firms (employing 100 or more employees) were located across Canada (Steed 1982).

Further policy recommendations in 1980s continued to focus on enhancing entrepreneurship and industrial R&D (Science Council of Canada 1984). For S&T policy to have an impact on the Canadian economy is 'less of a long shot' than it had been before, in part because of the local initiatives which have become an important element of S&T policy in Canada, as in other countries (Steed 1989: 22). Canadian concerns over regional economic disparities have lessened, or perhaps policy-makers have become resigned to the realities (Ch. 3). Three metropolitan areas, Toronto, Montreal, and Ottawa, account for more than 60 per cent of Canada's total industrial R&D (Britton 1987; Steed 1989: 78). But Steed (1989: 80) concludes that 'any effort to reduce such concentration in central Canada would run against strong economic forces'. The importance of agglomeration economies or a critical mass in such places drew Lacroix and Martin (1988), analysing decentralization of R&D in Canada, to the same conclusion. For the 1990s the 'overriding issue' is competitiveness in a global setting of technology and innovation (Steed 1989: 15).

Industrial and technology policy in the Third World

Traditional 'push' and 'pull' forces for innovation are irrelevant for developing countries (Sagasti 1988). Markets are generally too small to stimulate innovation (other than imitation), and technological capabilities are too weak to create breakthroughs. As Chapter 4 pointed out, little R&D takes place in the Third World. 'The dependency of MNC subsidiaries and affiliates on R&D in their parent corporations applies in almost all sectors and countries, ranging from the bauxite industry in Jamaica to export-oriented electronics production in Malaysia' (Marton 1986: 64). When there is some locally oriented R&D by subsidiaries, 'they maintain very limited linkages with local technical and scientific institutions, mostly for confidentiality reasons' (Marton 1986: 76; Germidis 1977). This dependence on MNCs, and the small benefits which have resulted, set the context for policies in Third World countries regarding technological change.

Developing countries have generally responded in two ways to the rapid pace of technological change. First, most have shifted their industrial policies from import substitution to export promotion, with products aimed at markets in advanced economies in Europe and North America (Ballance and Sinclair 1983: 36–51; Dicken 1986). Second, there has been a move from labour-intensive products in which low labour costs are the prime basis of competition to more technology-intensive products (OECD 1988a).

This technological upgrading, even more than previous inroads in older industries, has threatened the developed countries whose comparative advantage has rested on technology-intensive exports. Protectionist curbs on imports, which earlier affected steel, textiles, and clothing, have spread to motor cars and electronics products, two sectors which appear to be priority sectors for NIC industrial policies. Partly to head off such protectionism, some major firms from NICs, such as Taiwan and South Korea, have begun to direct investment in North America (OECD 1988a: 71–2).

Some countries have followed the Japanese lead in formulating S&T-oriented industrial policies. Singapore, for example, has begun to focus on a set of industries which will lead it to become 'developed into a modern industrial economy based on science, technology, skills and knowledge'. Its industries earmarked for the 1980s were to be produced 'for the markets of the advanced capitalist countries, but increasingly in much higher value-added areas'. The industries included are among those which are on the lists of other countries as well, as Table 7.6 illustrates (Rodan 1989: 147–8). Considerable government investment is behind such an effort, including a Software Technology Centre in Singapore Science Park, the incorporation of Singapore Aircraft Industries in 1982, and the Singapore Technology Corporation, formed to develop and market Singapore-made high-technology products. Collaborative ventures with French, German, Japanese, and US partners are also aimed in the direction of high technology (Rodan 1989: 149–53). However, as Rodan (1989: 179) notes, policies alone cannot alter corporate investment patterns:

> Despite the best efforts of the Singapore state and notwithstanding some important technological upgrading, international capital had not committed itself to the sort of qualitative reassessment of Singapore's place in the NIDL [new international division of labour] that the [government] had hoped for. . . . The dominant form of industrial upgrading involved either the introduction of greater automation or higher value-added products rather than a shift away from the assembly process as such. The more conceptual stages of production still tended to elude Singapore, as evidenced by the limited progress in attracting engineering and design processes.

Japanese firms, in particular, seem unwilling to invest in Singapore's upgrading strategy (Rodan 1989: 180). Neighbouring Malaysia has noted the same lack of R&D on the part of both Japanese and American companies (Fong 1986: 75–6).

Table 7.6 Singapore's target industries for the 1980s

Automotive components
Machine tools and machinery
Medical and surgical apparatus and instruments
Specialty chemicals and pharmaceuticals
Computer, computer peripheral equipment, and software development
Electronic instrumentation
Optical instruments and equipment
Precision engineering products
Advanced electronic components including wafer fabrication
Hydraulic and pneumatic control systems

Source: Rodan (1989: 148).

The dominance of foreign technology throughout the Third World is thus largely a consequence of extremely low levels of local R&D. When R&D is absent, the only avenue for obtaining best-practice technology is from abroad. Molero (1983) shows that in Spain dependence on foreign technology has been unchanged since the 1950s in large part because of rising imports of foreign technology, especially machinery. By contrast, R&D is conducted by only a very small number of Spanish firms, and has increased little if at all since the late 1960s (Cuadrado Roura, Granados, and Aurioles 1983; Molero 1983). Some Spanish companies, however, have been able to export technology, particularly those in the aircraft, chemical, pharmaceutical, and electrical and electronic equipment industries, generally the most R&D-intensive sectors and firms (Sanchez 1988).

The lack of an R&D commitment on the part of MNCs has produced a long-standing interest in creating an indigenous technological capability in the Third World (Fuenzalida 1979). Government R&D institutes perform most of the R&D in developing countries, and little industrial R&D takes place (Crane 1977). The gap between R&D and production is enlarged further by the dualism between government and university scientists on the one hand and industrialists and other technology users on the other (Reddy 1979). A call for an indigenous S&T capability is 'a close cousin' of self-reliance, a more extreme concept that attempts to create autonomy independent of the industrialized nations (Urquidi 1986). For small countries, in particular, self-reliance is virtually impossible, and therefore both an R&D capability and links to knowledge in the world at large – such as learning by searching, discussed in Chapter 4 – are essential (Bagchi 1988; Freeman and Lundvall 1988; Walsh 1987).

A critical issue for less developed countries is what proportion of the total 'knowledge budget' should be devoted to indigenous S&T, rather than to imports of foreign technology (James, Street, and Jedlicka 1980). Lall (1980: 46) suggests 'confining dependence on foreign technology to activities where local technology is incapable of keeping up with science-based or very rapidly advancing technology abroad'. Given the pace and sophistication of technological change, assimilating and diffusing technology, and adapting and improving it are critical. These require education and training beyond that provided to élites, and demand technical education, not one focused on the humanities (Dahlman 1989). This focus may be shifting in some countries, but it persists in others (Bath and James 1979; Sen 1988). Education and training, as additional costs to technology acquisition beyond the cost of licensing discussed in Chapter 5, are often underestimated by developing countries (James 1979). Thus, implicit as well as explicit policies affect S&T, but this is not always recognized in the Third World. Broad economic policies – those affecting credit, trade, investment, and labour – and manpower policies, especially regarding education and training, often fail to support S&T activities (Dahlman 1989; Sagasti 1979; UN Office for Science and Technology 1980).

In most of the Third World, the gap between R&D and production looms large, and thwarts efforts by governments to be the 'chief technological innovators' (Thomas 1979). This gap may be responsible for the preference by firms for foreign-originated technology (James, Street, and Jedlicka 1980; Wionczek 1979). Another dimension of this issue is the 'self-confidence' of a country regarding S&T. Forje (1988) sees this as largely lacking in Africa, due partly to colonial policies, and cites India as an example of a self-confident nation. In this context, the 'recovery' of traditional technologies and crafts, and their ability to 'compete' with foreign technology, seem to be the major issues for African economies (Müller 1984; Wad 1984). Elsewhere, linkages with, and adaptations of, world best-practice technology seem less remote (Fransman and King 1984).

Informal technological activities are closely associated with formal R&D, in the sense that skilled and technically competent personnel are central to both sets of activities. A free flow of information and feedback must take place, and dualistic social structures inhibit this flow. R&D for monitoring or searching 'international technology intelligence' for alternative sources of technology is an important allocation for policy (Radnor and Kaufman 1988).

Evidence from India suggests that companies that do their own R&D are able to benefit more from technology imports. They were better informed about the technology market before becoming buyers, and they imported only those components they could not produce or purchase economically or fast enough. However, the more technically progressive firms are also the largest ones, and technology imports seem to restrict competition further (Desai 1980, 1985). Small firms are not generally involved either in importing technology or in conducting R&D (Desai 1985). Government research laboratories of the Council of Scientific and Industrial Research, which are expected to provide technology to small firms, are typical of the dualistic structure of research in developing countries (Ch. 4). They are generally removed from manufacturing and marketing, and thus do not do much to assist the competitiveness of small firms (Desai 1980).

In general, as suggested in Chapter 1, the technological lag between advanced countries and underdeveloped countries is growing larger. An inadequate base of skills and R&D infrastructure prevents necessary learning activities from taking place. The large-scale changes in techno-economic paradigms and families of related technologies make it difficult to envision a large number of developing countries becoming able to create and master new technologies (Perez and Soete 1988; Rada 1984). In most countries, S&T policy needs to focus on implementation and absorption rather than state-of-the-art technology as viewed in advanced countries. Different methods of technology acquisition and diversified sources may be a better means to reduce technological dependence (Ernst and O'Connor 1989). In addition, communication and information flow – from firm to firm, from government R&D facility to firms, and with foreigners – is critical to knowing about and learning technology. These have been strong points both in Japan and in the NICs during the past few decades.

Science and technology in the NICs

The problem of technological backwardness is not the case everywhere, however, especially in the NICs. Taiwan, a country with a significant presence of small firms, has seen considerable success in sectors favouring a high-tech industrial structure. The country's industrial strategy for the 1980s selected three 'strategic industries': information, electronics and machinery, and biotechnology. Specific products encouraged include precision instruments and machine tools, VCRs, telecommunication equipment, computers, motor cars and car parts (OECD 1988a: 39). In each of these, entrepreneurial and private-sector activity has been the norm, including the promotion of new firms serving market niches in high technology markets (Simon and Schive 1986).

The successful technology-exporting firms in the NICs have combined local technological capabilities with foreign technological elements. 'Successful technological development requires access to foreign technology' (Dahlman and Sercovich 1984: 42). After obtaining foreign technology, conscious effort is needed to assimilate, adapt, and make effective use of it. Technology exports facilitate technological development by permitting economies of scale and accumulation of experience, including access to additional foreign technology from foreign collaborators. Korea, perhaps more than any other country, has pursued an export-led development for these very reasons (Dahlman and Sercovich 1984: 42–4). In contrast to Taiwan, Korea's exports are dominated by four giant conglomerate firms, or *chaebol*, which control electronics and computers, as well as motor cars, shipbuilding, and steel. In Brazil, likewise, technological learning has generated a large number of spin-offs to other applications, either in the same firm or in new firms (Teubal 1986: 104–30).

A distinct openness to foreign technology has marked Korean policies and practices towards technology (Westphal, Kim, and Dahlman 1985). Both imports and exports of technology have allowed Korea to accelerate its technological development and industrialization. Buyers of outputs and suppliers of equipment and materials together comprise the most important sources of technology for Korean firms (Westphal, Rhee, and Pursell 1984: 285). The Korean Institute of Electronics Technology and all four conglomerates maintain R&D operations in the USA's Silicon Valley to keep up with developments in the industry. These facilities also allow it to hire US-trained Korean engineers and computer scientists (Chung 1986).

India, by contrast,

> has been more restrictive toward imports of technology and has not made very effective use of exports to broaden its technological capability. It has a broad base of technological capabilities in many sectors. But India's bias toward technological self-reliance has condemned large sectors of industry to technological obsolescence – there are limits to what developing countries can do on their own without periodic injections of new elements of foreign technology (Dahlman and Sercovich 1984: 45).

Indian self-reliance has resulted in little attention being given to world technology reconnaissance and learning, as well as a bias towards large, 'big science' projects (Dore 1984; Lall 1985). This focus has succeeded in military production and electric power generation, but has not generated widespread economic development (Morehouse and Gupta 1987). India's technological lag is particularly great in

> electronics and semiconductors, for which the country has relied largely on locally developed technology. As a result, there has been an adverse impact on production and technological developments in electronics equipment and products and on electronics applications in various other branches, especially capital goods manufacture, where, with its large pool of cheap skilled and semiskilled labour, India could have built up major export capability (Marton 1986: 238).

Indeed, the focus on the internal, rather than external, market has harmed India's potential in computers, since labour-intensive methods in offices and industries are the norm (Gupta 1986).

There is no guarantee that openness to foreign technology translates into local technological capability. In Malaysia, 'the opportunities for technology transfer through "learning by doing" have been substantially reduced because of several factors, in particular the heavy dependence on parent companies, lack of linkages with local producers, limited R&D activities, and private business interests superseding national interests' (Osman-Rani, Woon, and Ali 1986: 58). These structural features which are fairly common to multinational corporations, combined with a shortage of local skilled and technical personnel, keep technological capability weak in Malaysia and elsewhere in the Third World.

Stages in industrial development that generally correspond to the notion of technological learning are found in the South Korean model of industrial development (Enos and Park 1988; Linge and Hamilton 1981: 32–5). For example, industries considered critical for future prosperity are accorded special emphasis for promotion of exports (Table 7.7). In part, these correspond to the Japanese example of a regularly revised strategic selection of sectors. For example, South Korean firms are progressing towards a more significant role in aerospace (Nakarmi, Shao, and Griffiths 1989; Nolan 1986).

The Japanese model is also emulated in government procurement of output of 'infant industries' such as computers (Kim, Lee, and Lee 1987). This limitation is evident as well in the general tendency to emphasize imports of foreign technology while continually adapting it and improving it (Enos and Park 1988). Finally, in all the NICs, a

Table 7.7 Stages in South Korea's export-oriented industrial development

	1961–66	1966–71	1971–76	1976–81	1981–91
Infant industries	Textiles; clothing; footwear	Electronic assemblies; shipbuilding; fertilizers; steel	Motor vehicle assembly; consumer electronics; special steels; precision goods (watches, cameras); turnkey plant building; metal products	Automotive components; machine tools; machinery assembly; simple instruments; assembly of heavy elec. machinery; semiconductors	Aircraft
Industries becoming competitive		Textiles; clothing; footwear	Electronic assemblies; shipbuilding; fertilizers; steel	Motor vehicle assembly; consumer electronics; special steels; precision goods; turnkey plant building; metal products	Automotive components; machine tools; machinery assembly; simple instruments; assembly of heavy elec. machinery; semiconductors
Self-sustaining industries			Textiles; clothing; footwear	Electronic assemblies; shipbuilding; fertilizers; steel	Motor vehicle assembly; consumer electronics; special steels; precision goods; turnkey plant building; metal products

Source: Linge and Hamilton (1981: 33 (Table 1.9)).

determined effort is under way to increase R&D in order to be competitive in the early stage, not merely the mature production stage, of product cycles (James 1979).

Outside Asia, prospects are dimmer. In Latin America, for example, financial debt has forced some countries to retain an import substitution policy orientation, especially with regard to payments for foreign technology. In Brazil, R&D has been emphasized only in sectors, with export potential such as military goods, but little in other industries, such as computers (Frischtak 1986). In both Brazil and Mexico, import substitution and exports rely largely on the activities of foreign firms (Marton 1986; OECD 1988a). Mexico has made efforts to increase 'local content', or supply linkages with local firms, rather than to develop local technology. This has risks, not the least of which is that 'Mexico's ability to develop an export capability is dependent not only on a continuing stream of investment into the country, but also on conditions in potential export markets', notably elsewhere in Latin America (Miller 1986: 195). Export-oriented manufacture has also to meet international quality and price competition, standards which are difficult for local suppliers to meet (Marton 1986: 45).

Overall, it is clear that 'a strategy for technology is more than a policy for R&D' (OECD 1988c: 23; Roobeek 1990). Competition now revolves around complementary assets and collaborative strengths, with the best-practice capabilities residing within powerful global firms.

Technology transfer

If a nation has been unable to learn technology sufficiently to be able to compete, it can obtain technology through *technology transfer*. At one level, all of innovation diffusion and technical education are types of technology transfer, but it is more frequently a market process in which technology is bought and sold. A technology transfer 'package' consists of several pieces:

1. One or more indivisible technology modules, containing either 'core technology', that which is indispensable to a process or the use of a product or service, or 'peripheral technology', which can be transferred via technical documents, blueprints, demonstration, training, or technical assistance;

2. Permission to use various rights, knowledge, or assets under licence, franchise, or lease;

3. Hard goods, or embodied technology;
4. Soft goods, or disembodied technology, which may take the form of written documents, computer software, photographs, or oral transmission (Robinson 1988: 4–5).

Technology transfer also takes place within firms, through training sessions and experience; it also takes place among firms as information diffuses, through inter-firm personnel mobility, and as specific knowledge is passed on to suppliers and customers. Firms transfer technology internally as they attempt to incorporate new products, processes, and organizational forms (Fusfeld 1986: 221–44). Direct investment, for example, allows knowledge to flow relatively unhindered between the parent and the subsidiary, often at little or no cost. However, such flows are biased in favour of the organization over the perceived needs of host countries. As MacCharles (1987: 29) notes, 'Access to state-of-the-art knowledge and the ease with which intra-firm transfers of it take place helps to explain why subsidiaries adopt technical changes sooner and why they have higher productivity than their Canadian-owned counterparts.'

International technology transfer, taking place among global corporations and local enterprises in Third World countries, attempts to narrow the gap between global best practice and local technology. Technology is transferred internationally either 'bundled' or 'unbundled'. The latter refers to separation of various elements of a technology, such as R&D, construction, manufacturing, marketing, training, and co-production (Robinson 1988: 5–6). In essence, it is management skill which is needed 'to weld these various elements of knowledge into a viable productive effort' (N Clark 1985: 183). It is the ability to provide a complete 'package' that keeps a supplier in a strong bargaining position. The supplier has an incentive to keep technical knowledge relatively 'secret' in order to differentiate from that of competitors and to appropriate the benefits of that knowledge by selling it as a commodity (N Clark 1985: 183).

Although 'technology is not usually produced directly for sale', its transfer involves the transfer of production capacity to a different firm or country (OECD 1981a). Technology transfer also occurs when firms purchase products from each other, license each other's technology, trade patents, or examine competitors' products. Such commercial transactions entail the purchase of tangible goods and intangible know-how. The level of technological capability transferred is not equal to all recipients within a firm. For example, of 65 technologies transferred by US firms during 1960–78, the mean age of those transferred to subsidiaries in developed countries was 6 years, and to developing countries 10 years (Mansfield *et al.* 1982: 48).

Transfer agents include exporters (salesmen, consultants, foreign aid officials, engineers on contract) – diffusion agents in Brown's (1981) terminology – who specialize in the geographic relocation of technology. Their expertise depends on the absence of native experts. The cultural gap between foreign and local experts can delay rapid transfer (Scott-Stevens 1987).

There are two motivations for recipients of technology transfers. One is to increase the value-added or profitability of economic activity, such as from innovative products. Low-risk activities, such as production and marketing of established products, yield dependable but relatively small returns. Preliminary production and marketing of unproven products are riskier ventures, but potentially able to reap Markusen's 'super profits'. Finally, information gathering and R&D (learning by searching in Ch. 4) are the most risky, but possess the highest potential return (Robinson 1988: 48–9). Second, as one moves from production to design, a corresponding increase in the need for management and technical skills arises. This reflects the second motivation: to increase indigenous technological capability. As a result, technology transfers become partial as transferees require less complete packages of technology (Robinson 1988). Figure 7.2 illustrates the typical path of technology transfer, from complete to partial, as recipients acquire greater technological capability. In sum, 'creating an inventive capacity' is central to the development process (Amin 1986: 169).

Any technology transfer depends upon the absorptive capacity of the recipient (Frame 1983). It may demand a large amount of knowledge in order to use an innovation effectively. The initial amount of knowledge necessary may be quite small, as in the case of turnkey plants and their embodied technology, or the recipient might substantially define the technology. Rubenstein (1980, 1989: 362–87) has suggested the term 'embedded technology' to depict the knowledge and skills concerning materials, products, processes, procedures, and systems which influence their acquisition and improvement. Embedded technology 'is a residual when the formal R&D components of the overall R&D/Innovation process are removed'. One test for 'stand-alone' imbedded technology is, 'the appearance of innovation and technological capability in the absence of formal R&D or S&T [science and technology] institutions and roles' (Rubenstein 1980: 382). If this capability and 'accumulated experience of how to do things, and of what works and what doesn't work' is gathered, the specification of a technology is more likely than the passive receipt of existing technology, or what Robinson (1988)

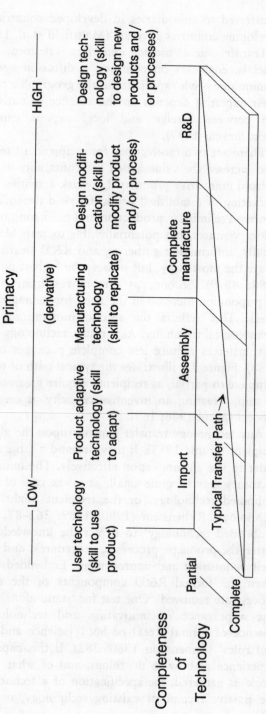

Figure 7.2 The sequence of technology transfer as technological learning occurs. *Source:* Robinson (1988: 18 (Fig. 2–2)).

considers partial technology transfer (Fig. 7.2). Direct import 'is the lowest level of technology acquisition. The product is imported as a "package" with the only substantial "local" content coming, perhaps, from the assembly process. The importing country will acquire some knowledge of the technology, particularly its operation and maintenance, but will only acquire a limited capability in design and manufacture' (S Thomas 1988: 327).

An increase in partial transfers, which have the potential to develop both local human capital and complementary technologies, is an important objective for companies and nations. The Brazilian firm Engesa simultaneously acquired and internally developed technologies for its military vehicle products, with which it has had marked success (Graham 1988; Lock 1986). As Brazilian arms development proceeded, the policy remained, in Catrina's (1988: 113) words: 'keep it simple, keep it cheap', for both externally purchased and internally developed technology. This defined the product niche to which Engesa was aspiring. A niche strategy also characterized Korean firms producing for civilian markets (Bae and Lee 1986; Kim, Lee, and Lee 1987).

The success of technology transfer depends more on the ability and willingness of the importing society to accept and absorb the technology than on the support of the exporter. Over the past century, technology has increased dramatically and technology transfer has become more difficult for recipients. Where craftsmen could formerly understand, copy, and improve technologies, this now requires knowledge of science and engineering, as well as literacy. As a consequence, underdeveloped countries must now turn to the advanced countries for their capital equipment, thereby losing the benefits – multiplier effects, backward linkages, externalities – of the capital goods industries (Headrick 1988: 11).

Without technology transfer, the difficulties which confront Third World countries in attempting to create competitive local industries are enormous. Firms must be able to recognize certain technologies as outdated and inadequate, and be able to identify technologies which need to be acquired from foreign sources rather than be developed (with greater delay) domestically (Katz 1982a, b). Moreover, products must meet international standards of quality, materials, workmanship, after-sales service, and delivery schedules (Rhee 1990).

Local or domestic technological capability is indispensable in order to alter, modify, and adapt transferred technology to local conditions. As Chapter 4 has shown, local learning by doing and informal skills are more important than R&D in mastering imported technology. Learning as a result of actual use of technology contributes to knowledge that can be used to change designs, production processes, and the way in which a technology is used. R&D can contribute only to transformations of designs. Experience in production suggests other changes in production methods and design alterations, but cannot address ways in which a product may be used (S Thomas 1988). Cohen and Zysman (1987) reiterate this point in their contrast between American R&D and strong Japanese production and organizational skills.

Technology transfer without development of technological capability has been especially common as developing countries attempted to increase their production capacities in a minimal amount of time. Headrick (1988: 13) points out that many factors favour complete, bundled technology transfer. Exporters of Western technology have an interest in selling machines, spare parts, and expertise. For importers, the acquisition of technology, while costlier, is faster and entails fewer risks than more complete transfer. Governments need foreign equipment and experts, but they tend to avoid the political dangers involved in cultural or social changes necessary to develop the knowledge, skills, and attitudes related to a particular device or process. Thus risk aversion leads to technological dependency. The 'cultural diffusion of technology', as Headrick calls it, is a much more difficult task. 'It takes a willingness to accept changes, a strong political cohesiveness, and a common vision of the future. Western societies, facing less culture shock, have readily imported industrial technologies. Among non-Western societies, only Japan had the requisite cultural and social base. Others have had to undergo political revolutions first' (Headrick 1988: 13).

An R&D capability and an educational system to support it is essential for making the 'technological leap' to sophisticated technology and skilled work. The presence of skilled technicians and engineers differentiates countries whose arms production is largely indigenous from those which continue to depend heavily on licensed technology (Brzoska and Ohlson 1986). Adoption and diffusion of new technology seem preferable to trying to create it, since it demands less traditional R&D, but they still require a high level of technical skills (Vickery and Blair 1989).

The question of whether transferred technology is 'inappropriate' for less developed countries arises frequently (Bhalla 1979; Clark 1985; Dicken 1986: 361–70; Stewart 1978, 1981, 1987). To a large extent, concern over appropriate technology stems from the interest in basic needs and 'bottom-up' planning which have influenced development planning since the 1970s, discussed in Chapter 3. Perhaps more importantly for Third World nations, technology transfer affirms their dependence on nations and firms from which they obtain technology. Control over the pace and form of technology remains where R&D and improvements in production process technology are ongoing (Buttel, Kenney, and Kloppenburg 1985; Kaplinsky 1984a). Thus, partially transferred technology is less inappropriate than complete turnkey systems. The 'image of modernity' and the conditions imposed by international sources of finance have also influenced the adoption of inappropriate technologies over indigenous ones. However, traditional or indigenous technologies also tend to be static and multi-purpose rather than quickly replaced by the next 'generation' of single-purpose, modern technology. The longer-term utility of local techniques may be a significant advantage, since mastery of the technology can become more widespread (Jéquier 1979).

The adoption of imported modern technology can destroy knowledge of traditional technologies and, because of their complexity, prevent learning by doing by semi-skilled and skilled artisans without sufficient formal engineering-related education (Bhalla 1979). In addition, modern technologies demand modern or imported managers and managerial techniques. And spare parts for repairs must come almost exclusively from the country – and firm – from which the equipment was purchased, thus extending technological dependence well beyond the initial decision (Stewart 1978).

A great deal of effort has taken place during the past two decades to 'blend' new technologies to the situation of small-scale and traditional activities in the Third World (Bhalla and James 1988). Many of these utilize microelectronic technology applied to education, health care, and social services; applications in production have thus far been limited to NICs such as Brazil, Hong Kong, and Singapore, where organizational flexibility and skill in management are present (Bhalla 1988). The shortage of managerial capacity, distinct from technical knowledge, is a critical constraint on industrial growth in underdeveloped countries (Clark 1985: 183). Managerial dependence is interdependent with technological dependence (Stewart 1978: 65), and can prevent both entrepreneurship and expansion of technological capability (Vakil and Brahmananda 1987). As Stewart and Nihei (1987: 151) put it in discussing Indonesia, 'the top three priorities for expanding absorptive capacity are management, management, and management'. As a result, the development effects of transferred technology are minimal: too few firms are formed, too little information is exchanged.

In developed countries as well, similar managerial gaps have been noted. Producing knowledge for in-house use requires highly trained and educated managers, and they are even less common in small firms

(MacCharles 1987: 30). Among large and small firms alike, there is a 'managerial gap' between Japanese managers and their Canadian counterparts, as measured by the slower adoption of quality circles, significantly less use of JIT delivery and other inventory control techniques, and a far lower proportionate use of robotics and CAD/CAM methods. Canadian managers tend to be older, less well educated and generally less experienced than counterparts elsewhere in the world (MacCharles 1987: 27).

Military technology in the Third World

One of the more familiar sectors in which developing nations choose to invest in technology development and acquisition is military or defence production. A military industry serves several purposes. It can boost national prestige and, at least in theory, be a means of both reducing imports and increasing exports (Catrina 1988). A further appeal of military technology is that it provides a vehicle for incorporating high technology into a nation's industrial base, as the Japanese jet fighter example illustrates.

The reasons for persistent national attempts to develop this particular type of technological capability rest largely on the experience of developed nations, whose industrial bases and electronics and computer industries have thrived in part because of connections to military production (Flamm 1988a, b). It is also an attempt to create an indigenous technological capability and to reduce the technological dependence associated with imported weapons (Catrina 1988; Brzoska and Ohlson 1986). However, arms production is not a particularly effective import substitution strategy. Generally, the largest arms producers also import massive amounts of weapons (Brzoska and Ohlson 1986: 27–8; Wulf 1986). The 'software' element of arms transfers, including services, technical training, continued technical support, and maintenance and repair, maintain dependence based on the know-how to maintain, reproduce, and use imported weapons (Catrina 1988: 9–15).

The level of arms transfers in the world is enormous, largely originating in the industrialized countries. Increasingly, however, several Third World nations, such as South Korea and Brazil, have developed arms industries as export sectors and bases of high technology (Table 7.8). The Third World production totals are low compared to those of the industrialized nations, but they comprised nearly 8 per cent of global arms transfers in 1982, falling to 1.5 per cent by 1986. Arms-producing countries have relatively high technological capability, but the military thrust dilutes research capacity in commercial and other scientific directions (Katz 1986).

Table 7.8 Leading arms exporters among developing countries, 1975–86 (in million constant 1984 US$)

Supplier	Cumulative total (1975–86)
South Korea	3 269
North Korea	3 310
Brazil	2 940
Israel	2 661
Egypt	961
Saudi Arabia	861
Libya	843
Pakistan	794
South Africa	375
India	320
Others	1 795
Total	18 128
% of total world exports	3.8

Source: Catrina (1988: 57 (Tables 3–5)).

Only a few countries have developed a diversified arms industry, encompassing all types of weapons, including aircraft, missiles, armoured vehicles, and ships. These include Argentina, Brazil, Egypt, India, Israel, South Africa, and Taiwan (Brzoska and Ohlson 1986: 16-17). Other countries, especially the ASEAN (Association of South East Asian Nations – Indonesia, Malaysia, the Philippines, Singapore, and Thailand), emphasize low-cost production of arms for export as part of broad industrialization policies (Ohlson 1986). The production technology in all Third World arms industries is licensed largely from American, British, French, and German firms; Soviet technology has been used in Algeria, India, and North Korea (Brzoska and Ohlson 1986: 26). In virtually all respects, arms technology is the same as any other, with learning and design capability essential to success, except in the uses to which it is put.

Technology acquisition under central planning

Central planning, which dominated Soviet, Chinese, and Eastern European governments since the Second World War, has been relatively ineffective in promoting innovation. Two principal shortcomings noted by Soviet observers themselves are ponderous bureaucratic structures to manage science and marked immobility of scientists (Sagdeev 1988). None the less, several areas of technology have been competitive with Western technology. Notable among Soviet technological accomplishments are space and defence equipment, welding equipment, cooling technologies for blast furnaces, electromagnetic casting of copper, and surgical instruments. Many of these are best-practice technologies and have been licensed to Western companies (Bornstein 1985: 40; Kiser 1989). Similarly, Polish and Czechoslovakian technology has been sold to both Western firms and other Eastern European enterprises (Monkiewicz and Maciejewicz 1986: 74–115).

At the opposite end of the spectrum of priorities, especially in services, the technological level has been poor. The routine use of the abacus in place of cash registers contrasts markedly with the industrialized world (Buck and Cole 1987: 134). Most significantly, the Soviet Union has essentially missed out on the revolutions in biotechnology and computers, the latter a result of a 1962 decision to eliminate the computer division of the Academy of Sciences (Galuszka *et al.* 1988). Lack of computer equipment still plagues Soviet researchers (Sagdeev 1988).

It is the absence of a market-pull effect in a centrally planned system which inhibits innovation (Shmelev and Popov 1989; Simon and Rehn 1988). Central priorities govern items and quantities to be produced, and determine sources of supply. Consumer demand, input prices, and competitors' inroads – the motivations for automated, flexible production – have played no role in the Soviet system. Thus, the impetus to automate in such economies is not as compelling as in competitive market economies. This is certainly the case at the factory level, where there is little incentive for managers to improve product designs, quality, or efficiency beyond that demanded by central planning authorities. A few consumer items in the Soviet Union, such as refrigerators, vacuum cleaners, clocks, watches, radios and televisions, and motor scooters, are of markedly higher quality than most, primarily because key components or the entire production originate in defence factories (Baranson 1987b; J Cooper 1986; Kassel 1989). 'The only way is to attract the attention of the army. The labs with no contacts with defence are beggars', says Yury Khronopulo, a Soviet physicist now living in

New York (Galuszka, Marbach, and Brady 1988: 86). This illustrates a dual economy – military versus civilian – not unlike that of Third World countries (Goldman 1987). A similar shift of production by military production facilities to consumer electronics items has begun to take place in China (Simon and Rehn 1988: 69–71).

The traditional stumbling block for Soviet innovation is not a lack of R&D, but weak links between research and production. Factories generally do not have engineering facilities, and R&D organizations typically lack production engineering capability or pilot production plants (Amann 1986; Kassel 1989; McHenry 1987). The interaction with suppliers and repeated incremental changes common in market economies, and embodied in the Kline–Rosenberg model (Ch. 4), are very difficult to effect under central planning (Bornstein 1985: 33). Indeed, interaction and information flow, especially with the international scientific community, have been limited and only by 1990 had begun to increase (Balzer 1985). The concentration of research in Moscow (and, secondarily, in Leningrad) permits some informal interaction there, but generally constrains mobility (Balzer 1985; Sagdeev 1988).

Chinese industry, like that in the Soviet Union, maintains a distinct separation between research and manufacturing, even in high technology industries such as semiconductors. The 'ends here' phenomenon refers to the tendency of research units to consider their work in isolation and to ignore manufacturing considerations (Simon and Rehn 1987: 269; 1988). The transfer of technology from defence to civilian sectors is, along with that from research to production and from coastal regions to the interior, among the recent reforms of the Chinese research system (Jin and Porter 1988).

Centrally planned economies, such as China, have also long had explicit policies to promote certain industries. China's National Science Plan of 1956 included the electronics industry, greatly expanded in importance, along with computers and telecommunications, in later national plans (Simon and Rehn 1988: 59–65). Electronics technology was seen in 1978 as the hallmark of modernization, and one of the 'four modernizations' – of industry, agriculture, national defence, and science and technology – proclaimed in that year (Simon and Rehn 1988: 14–15).

Through the 1980s, the Soviet Union relied on imports from the West 'for a small but critical percentage (about five percent)' of high-technology components (Baranson 1987b). It appears, however, that relatively little cumulative learning takes place in Soviet industry, in comparison to the pattern common in the West described in Chapter 4.

Imported technology is rarely improved upon or upgraded, even when it is put into use (Goldman 1987).

However, the market economy and new technology are increasingly important to the Soviets (Shmelev and Popov 1989). Automation and new technology are seen as a way to increase industrial output and to earn foreign exchange in world markets. In addition, the Soviet Union has wanted to maintain parity with the West in order to serve as a model of industrial development for the Third World (Baranson 1987a). In several respects, this has been the goal of Soviet design and engineering. Simplicity of design, incremental design changes, common parts and components, and manufacturing feasibility are more important in Soviet military goods than in the high-technology equipment which originates in the West, and this in part accounts for the popularity of Soviet arms (Baranson 1987b; Kiser 1989).

As *perestroika* and *glasnost* continue, joint ventures are being pursued in order to learn manufacturing and management skills that cannot be simply purchased (Galuszka *et al.* 1988). Deals to get access to Western technology are likely to proliferate in areas such as computers, machine tools, nuclear power, semiconductors, and telecommunications, as well as in major petrochemical projects in Siberia (Galuszka and King 1989). In addition, controls on exports of military goods from the West to the Soviet Union are slowly being eliminated as the Soviet threat appears to be disappearing. The US public now perceives the economic threat from Japan to be greater than the military menace of the Soviet Union (Neff, Magnusson, and Holstein 1989: 51).

Seeking out alternative suppliers and inputs, and any energetic interaction between users and innovators on the Western model, are unlikely but not impossible. The Ivanovo Machine Tool Amalgamation, which manufactures flexible manufacturing systems (FMS), has customers around the world and may be a model for additional Soviet enterprises to come. To service its customers in Switzerland and West Germany, it maintains a service department in Switzerland and pays attention to world standards of product quality. In return, the Ivanovo enterprise is permitted to keep 50 per cent of the earnings from its exports to developed countries (Kiser 1989: 151–61). This degree of awareness of the global market and of marketing is unusual in the Soviet Union and Eastern Europe (Shmelev and Popov 1989; Telesio 1990).

Reliance on entrepreneurship of some sort seems essential as a means of bypassing central planning, but the great majority of Eastern European firms have 'neither the R&D capability nor the motivation to seek to acquire cutting edge technology from the West' (Marer 1986:

208). Skowronski (1987) generalizes the problem as one of there being too few small enterprises, since it is these which have been most effective in technology transfer, as the Ivanovo example illustrates. Decentralized R&D, for example, remains low outside the Academy of Science and ministry structure, where the central planning system thwarts exploration by new groups or of novel problems, in favour of existing organizations and institutions and thrusts determined in the current Five-Year Plan (Fortescue 1985). Recent restructuring of R&D to establish 'interbranch science and technology complexes' still shows signs of providing insufficient interface between advanced scientific research and existing industry (Kassel 1989: 65).

Hungary has absorbed a large amount of Western technology, mainly to help its export industries to produce high-quality manufactured products. However, most of this has been in the form of embodied technology in machinery and capital equipment. Only small amounts (around 3% of total R&D) are spent on licences and imports of know-how, far less than is the case in market economies, where figures of 15–45 per cent are common. By comparison, Spain and Greece spend 200 per cent of their R&D budgets on such imports (Marer 1986: 118). Even recently, instead of licensing, various co-production arrangements are the most common form of East–West co-operation (Monkiewicz 1989: 116). Because cutting-edge technologies have been restricted from export to Soviet-bloc economies, a technology gap between East and West has persisted. Indeed, the technological level of these countries had fallen relative to NICs such as Brazil and South Korea (Monkiewicz 1989: 195–6).

Events in late 1989 which signal the end of post-war Eastern Europe suggest that a massive increase in joint ventures and other agreements to improve technology will move rapidly, but will be constrained by the inability to raise hard currency for imports. Otherwise, standards for determining the costs of inputs and the price of products remain difficult to deal with, and rely on Western – particularly West German – management and expertise. The technology gap throughout Eastern Europe *vis-à-vis* the West remains very large, especially in microelectronics-based and other high-tech products and processes (Amann and Cooper 1982; Marer 1986: 220–6; Templeman *et al.* 1989; Wienert and Slater 1986). Few such products are exported from, while many are imported into, the East (Monkiewicz 1989). Capitalist management and priorities for bringing technology to market will be needed to attain global competitiveness (Rossant, Galuszka, and Reed 1989).

Stifling bureaucracies in large enterprises and central planning continue to prevail, however, and suggest that innovation in Eastern Europe will not quickly be competitive (Brady 1989; Goldman 1987; Gumbel 1989; Schares 1989). The new inter-branch complexes in the USSR, like other parts of the Soviet economy, are inclined 'to upgrade traditional technologies, rather than develop and produce new ones' (Kassel 1989: 65). In addition, other socio-technical transformations will also need to be put in place. Computerization demands both fewer social controls and greater and better physical infrastructure. Present telephone systems are unable to transmit data, power supplies are unreliable, and repair and service facilities – rare for most products – will be essential (Balzer 1985: 43–5; Schares *et al.* 1989). These may demand the sort of institutional and managerial innovations demanded by radical innovations and techno-economic paradigms associated with previous long waves, as seen in Chapter 5.

Even the most industrialized countries are not completely successful at capturing all the influences on innovation in their technology policies. It is now clear that institutional and industrial structures also play a major part in conveying scientific and technical knowledge into the wider economy. A knowledge base of R&D needs to be complemented not only by strategic international collaboration involving a nation's firms, but also by cultural or societal cohesion about the nature and desirability of innovation (Roobeek 1990). Finally, the importance of 'sophisticated and demanding buyers', so absent under central planning, is also a key element in national and regional competitiveness (Porter 1990).

Policies for regional high-tech development

The allure of high technology has sparked a number of policies to create or generate innovativeness in peripheral areas or regions and to upgrade the technological capability of local firms through regional innovation centres (Brainard, Leedman, and Lumbers 1988: 25–30; Malecki and Nijkamp 1988; Schmandt and Wilson 1987; Smilor, Kozmetsky, and Gibson 1988a). There is, at the same time, a substantial local element at work, represented by independent policy initiatives on the part of local and regional governments. These are to some degree an outgrowth of the trend under 'flexibility' towards decentralization of government and a more entrepreneurial attitude on the part of local governments, authorities, universities, and other institutions (Ch. 6). Feller (1988) stresses the political nature of such policies in the USA, where high-tech development was a form of 'bureaucratic innovation' by state governors (Osborne 1988). Whatever the mechanism, any active policy to create an innovation in a place rather than to wait for 'spontaneous' growth to occur is very much like the growth poles of the 1960s (Stöhr 1988). This is taking place despite a belief by some that 'it is difficult to escape the conclusion that high technology policy is best left for nations rather than to regions or urban areas'. Much of the policy infrastructure, including capital markets, tax policy, and government procurement, operate only at the national level (Macdonald 1987: 368–9; Power 1988: 174–8). On the other hand, much of what supports the competitiveness of firms is local in character (Porter 1990: 154–9; Senker 1985).

Tremendous competition has ensued regarding the competition for high technology. Brainard, Leedman, and Lumbers (1988: 30) warn that 'the intense competition between regions places the less developed ones at a considerable disadvantage in that they are often unable to provide conditions and incentives comparable to those offered by more developed regions for drawing research-based organisations to their areas. This can widen rather than narrow regional disparities.'

The resulting situation – a gap that does not narrow – is typical in the USA where competition for economic development is itself a big business. All 50 states and thousands of cities, counties, suburbs, and smaller communities take part in what has been called a 'second civil war' (Pelissero and Fasenfest 1989; Ryans and Shanklin 1986). These governmental units are complemented by a group of private-sector interests, especially real estate agents and land developers, local chambers of commerce, utility firms, and banks (Levy 1981). Similar 'city-marketing' takes place in the Netherlands and elsewhere (Borchert 1987), but the USA has more individual entities competing for jobs and industry than any other country. The short-term benefits of such groups are generally small or difficult to identify (Humphrey, Erickson, and Ottensmeyer 1988). However, the larger, better funded industrial development groups, which tend to be those in large cities, are better able to create and maintain connections to other groups and individuals. These external connections and the knowledge they provide yield a better – even a state-of-the-art – mix of incentives, promotional advertising and ultimately, greater job creation (Humphrey, Erickson, and McCluskey 1989).

Competition for high technology has been a major fashion throughout the 1970s and 1980s (Eisinger 1988: 266–89; Office of Technology Assessment 1984). Research parks or science parks are created as potential cores of new Silicon Valleys, in an effort to emulate the early successful experience of Stanford Research Park in Palo Alto, California, or of Route 128 surrounding Boston. The attraction of R&D activities can be a high priority, or regions and communities can fail to differentiate within high-tech industry and embrace branch plants as well (Malecki 1984a; Miller and Coté 1987; Ryans and Shanklin 1986).

The designation with a catchy label of 'high-tech highways' or 'silicon strips' is a particularly popular way to add high-tech to conventional industrial recruitment (Table 7.9). These designated regions or corridors are intended to inspire perceptions similar to those elicited by Silicon Valley, Route 128, and the Research Triangle (Farrell 1983; Rogers and Larsen 1984). Many of them are wildly imprecise or optimistic; others are so large in area that any uniqueness of location is lost. (Some are also fleeting. 'Robot Alley' in central Florida was included in Farrell's (1983) compilation but had vanished two years later after an eminent robotics expert left the University of Florida and several firms gave up their branch plants in Florida to return to core facilities in the manufacturing heartland. It was still cited as a reality in 1988 (Premus 1988: 444).)

Table 7.9 Popular names for high-tech regions in the USA

Name	Location
Bionic Valley	Salt Lake City area
Electronics Belt	Orlando–Tampa, Florida
Golden Triangle	South-eastern New Hampshire (north of Boston)
Research Triangle	Raleigh–Durham–Chapel Hill, North Carolina
Satellite Alley	Montgomery County, Maryland (north-west of Washington, DC)
Silicon Bayou	Lafayette, Louisiana area
Silicon Beach	Fort Lauderdale, Florida, area
Silicon Desert	Phoenix area
Silicon Forest	Portland, Oregon, area
Silicon Mountain	Colorado Springs area
Silicon Prairie	Dallas–Fort Worth area
Silicon Valley East	Albany–Schenectady–Troy, New York, area
Tech Island	Long Island (east of New York City)

Sources: After Farrell (1983); Rogers and Larsen (1984: 242–51).

The experience of Silicon Glen in Scotland for other regions highlights the necessity of several conditions, most of which were noted in Chapter 6: large urban region, air connections, local university sources of workers, and patience. The long-term character of Scotland's high-tech development is striking; US firms set up their first branch plants there in the 1940s. R&D was not common until the 1970s, long after the first plants opened (Haug 1986). The Scottish semiconductor complex remains dependent on R&D and designs imported from the USA, and cannot yet be called a self-sustaining complex (J Henderson 1989: 153).

America's counterpart to Silicon Glen is North Carolina's Research Triangle, a planned technological region. The Research Triangle Park got its start even later than did central Scotland, in 1959. It was not until the early 1980s that it began to compete with other regions for major R&D facilities (Whittington 1985). However, the North Carolina region is not the equal of Silicon Valley or the Boston area in one important respect. It can still be identified as a region in which venture capital is scarce; consequently, very few spin-offs have occurred despite the area's attraction for the operations and R&D of large firms (Luger 1984; Malecki 1986b; Rogers 1986).

In each case, success has seemed to elude peripheral regions for reasons that, in light of the history of regional policy and of the nature of economic change, are somewhat easily identified. A prominent shortcoming of peripheral regions seems to be their low innovation potential, an outcome of the relative scarcity of R&D carried out there (Ewers and Wettmann 1980). Attempts to lure the R&D activities of large firms generally will have a larger beneficial effect than the attraction of routine production. However, 'technical branch plants', where both R&D and standardized production take place, are an increasingly common type of corporate facility in high-technology industries (Markusen, Hall, and Glasmeier 1986). Their contribution to local economic growth may be minimal, and hinges upon their degree of local linkages and encouragement of entrepreneurship (Glasmeier 1988a). R&D facilities alone, however, are not guaranteed to spark spin-offs, even in the long run. Some places with considerable R&D have not been able to generate new firms in significant numbers, as evidence from the UK and France as well as from North Carolina shows (Cooke 1985b; Stöhr 1988; Whittington 1985).

More importantly, not all policies are intended to spawn or support new firms, and instead only foster technological innovation in large corporations. However, the models created by California's Silicon Valley and Boston's Route 128, with their successions of spin-off firms, have inspired an assortment of policies to boost regional innovative capacity and entrepreneurial activity. At the local and regional level, as for nations, competitive clusters of small firms are vital for innovation and economic development.

The Japanese Technopolis plan

The most ambitious example of deliberate high-technology development is the Technopolis concept in Japan, a plan to transform all of Japan into a high tech archipelago by building a network of regional high-tech cities, shown in Fig. 7.3 (Tatsuno 1986; Toda 1987). Tsukuba Science City, the top tier of current Japanese policy, was planned during the 1960s and is now the home of 2 universities and 50 national research institutes. It serves as a prototype for 26 technopolises and research cores (Kawashima and Stöhr 1988; Onda 1988).

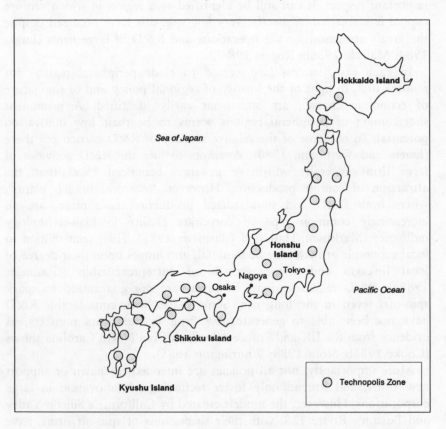

Figure 7.3 Japan's Technopolis sites. *Source*: Japanese Ministry for International Trade and Industry Industrial Location Policy Division (MITI) (1990: 13)

The Technopolis law in 1983 resulted in criteria used by MITI to select the locations. They include a typical set of urban attributes: proximity to a 'mother city' of at least 150 000 to provide urban services, proximity to an airport or bullet train station, an integrated complex of industrial, academic, and residential areas, and a pleasant living environment. These generally correspond to the location factors proposed as favouring R&D and technological activities.

The purpose of the Technopolis policy is to foster two types of R&D: transfer type, or manufacturing-oriented, R&D, and frontier or knowledge-oriented R&D (Imai 1986; Toda 1987). Manufacturing-based R&D and technology transfer among firms in specific industries in each local area have proceeded to some degree (Brainard, Leedman, and Lumbers 1988: 77–8). However, it is less certain that new 'frontier' R&D will take place. The centre of frontier R&D, Tsukuba, itself suffers from the typical problem of a government R&D centre: little non-government R&D is found there, largely because as it was planned manufacturing operations were restricted. Based on R&D like its model, Research Triangle in North Carolina, Tsukuba has been unable to spawn commercial spin-off firms (Glasmeier 1988b: 275).

The provincial technopolises are unlikely to reduce significantly the allure and advantages of Tokyo within Japan for important R&D and other key corporate activities. As occurred in the implementation of growth centre policies in virtually all countries, the intense competition among candidate regions has diluted the potential for diminishing the Tokyo region's dominance in Japanese technology (Glasmeier 1988b). From the original set of 14 technopolises in 1983, the list has grown to 26 (MITI 1990). The allure of the Tokyo agglomeration is unlikely to be diminished by the Technopolis plan, if only because of the concentration in Tokyo of residential amenities, top universities, and existing corporate headquarters and R&D (Glasmeier 1988b: 278–9). Its only real competition may be Osaka in western Japan (Morita and Hiraoka 1988). While acknowledging these problems, Nishioka and Takeuchi (1987) are more optimistic, if only in the fact that the Technopolis concept has aroused public opinion in Japan regarding regional development. A third tier of research cores furthers the decentralization of technological development to the local level, where many universities are located and others are to be built (Kawashima and Stöhr 1988; MITI 1990).

Other national high-tech policies for regional development

The critical importance of agglomeration is recognized in the Chinese decision to designate, in the country's Seventh Five Year Plan (1986–90), only a few key technologically advanced regions, mainly in the coastal areas (Shanghai, Jiangsu, Guangdong province near Hong Kong, and Beijing). This is in marked contrast to the previous strategy of establishing fully self-sufficient regional systems throughout the country (Simon and Rehn 1988: 64–5). Indeed, the creation of Silicon Valley-like sites in China acknowledged that an adequate technological infrastructure, an experienced and skilled labour pool, a well-developed industrial sector, and suitable environmental conditions are required (Simon and Rehn 1988: 134). As in Japan, the USA, and elsewhere, areas competed vigorously for designation as technology centres when the Silicon Valley idea was launched. Foreign investment has also concentrated in these centres (Aguinier 1988). However, even Shanghai, arguably the strongest agglomeration of research and production in electronics in China, still lacks sufficient supplies of power, a well-developed communications system, and a continuous supply of skilled labour (Simon and Rehn 1988: 139). Defence-oriented electronics remain in remote regions for security reasons, however, despite lower productivity and profitability (Simon and Rehn 1988: 135–6).

Political influence has been a dominant aspect in location decisions concerning government research facilities. In the USA, facilities may be placed in a location favoured by a member of Congress who has been, or continues to be, influential with the agency siting a facility. The National Aeronautics and Space Administration (NASA) complexes provided several opportunities for such 'pork-barrelling' during the 1950s and 1960s (Holman 1974). Since that time, the number of possible major facilities that could potentially land in a state or Congressional district has fallen, and slower growth in federal facilities has reduced the number of complexes to be sited, which has heightened competition. The most recent example of this competition was that for the US Department of Energy's superconducting supercollider. Over 30 states submitted proposals for the 30 square mile project, whose attributes included the employment of hundreds of scientists and engineers. It went in the end to a site near Dallas, Texas.

In the context of government research facilities, where profitability is not an explicit concern, subsidies could be expected to make the total cost of the scientific work undertaken in a location which requires subsidies somewhat more costly, and perhaps also somewhat less effective. The extra costs of the subsidy are likely to be long term, associated with attracting and retaining a labour force of mobile professionals, as well as costs connected with travel from a remote site to other research centres. When such considerations are taken into account, the decentralization of R&D is difficult to justify (Lacroix and Martin 1988).

The degree to which a major government research facility will generate wider regional economic benefits has received much attention over the years (Malecki 1982; National Academy of Sciences and National Academy of Engineering 1969). The general impact is small, being mainly associated with a concentration of relatively high income jobs. Any wider effects are difficult to identify, since they depend on the specific purchasing patterns of the facilities, frequently done via long-distance linkages. This is equally true of government facilities (Erickson 1977) and high-tech industry (Hagey and Malecki 1986).

A disturbing and complicating factor, when considering the effect of government policy, is the low degree of spin-off associated with government facilities (A Cooper 1986). A low incidence of entrepreneurship is especially likely from facilities devoted to basic research, such as the superconducting supercollider, from which entrepreneurial opportunities are less likely to present themselves than from other types of R&D which are spurred by market opportunities. Government research facilities may attract other research through their contribution to an area's technical labour pool. However, many government facilities are 'too large' and dominate their regions in the manner of company towns. The potential for entrepreneurship and diversified economic growth in this setting is slight.

Industrial recruitment and high technology

Many of the jobs which have located in rural areas and other 'new industrial spaces' over the past 30 years are a result of the various regional and local incentives offered. Other cost and accessibility factors have also contributed, such as lower land and labour costs in peripheral areas and air and highway transportation improvements which have improved access to many places previously considered remote. Although incentives have not been a dominant factor in firms' location decisions (Mulkey and Dillman 1976), the granting of subsidies and incentives as inducements to industrial location continues to be a major element of high-technology policies at the local level. Location incentives began in the USA during the 1930s, when 'the selling of the South' began, to use the title of James Cobb's (1982) book. Since that time, they have been widely imitated by other states, localities, and countries (Brainard, Leedman, and Lumbers 1988; McHone 1987). Like national governments, local governments want to build upon technological innovation, but do so without the background of science policy found at the national level, and without the full complement of policy tools with which to operate (such as tax policy and capital markets).

The use of incentives for industrial location has moved from the attraction of low-wage branch plants into the area of high technology (Eisinger 1988; Ryans and Shanklin 1986). The highly publicized location decision of the Microelectronics and Computer Technology Corporation (MCC) a few years ago showed how furious the competition among states and communities had become. Whether or not MCC's decision to locate in Austin, Texas, was strongly influenced by the state's package of land, tax breaks, and commitments of university upgrading is difficult to judge, but it showed other regions how location decisions can be influenced (Farley and Glickman 1986). The Austin experience illustrated the significance of co-operative institutional arrangements, especially among governments, firms, universities, and other groups (Smilor, Kozmetsky, and Gibson 1988b). Despite a lack of evidence that tax-based incentives affect location decisions, huge packages continue to be offered, especially as an inducement for large-employment projects, such as motor car plants (Elder and Lind 1987).

Regions and communities now engage in a new level of competition for jobs and economic activity. Although the programmes are often described in terms of 'science policy', in reality they really represent only a slightly higher level of industrial development policy. In this new competition, low-cost labour and land are secondary to pools of technical workers, world-class university research, air accessibility, and urban quality of life. Consequently, the tendency has been for regions already strong in high tech to grow stronger as new firms are formed and corporations locate new facilities.

Most state programmes in the USA for 'high-technology development' are basically extensions of existing economic development strategies (Luger 1985; Office of Technology Assessment 1984). 'Targeting' industrial recruitment towards high-tech industries is a common approach, aided by advertising in focused publications that narrow the readership to likely candidates. Several states have developed their own lists of 'next-generation' high-technology industries, at least partially in response to Japanese policies (C Thompson 1987). Of the 35 state programmes for advanced technology which were in place in the USA by 1985 (compared to only 4 in 1979), most were primarily concerned with focused industrial recruitment (Allen and Levine 1986; Office of Technology Assessment 1984). State recruitment advertising now emphasizes a combination of a state's 'high-tech traditions', skilled labour, and state universities, together with established local advantages such as low taxes, low wage levels, and other elements of a good business climate. A less standard aspect of current industrial recruitment is to attempt to interest firms to establish R&D facilities rather than merely branch manufacturing plants in high-tech sectors (Farley and Glickman 1986).

Infrastructure improvement

The regional infrastructure can help or inhibit technology-based economic development. Three different dimensions fall within the category of regional infrastructure. These include: universities and other research and technical training institutions, science or research parks, and other urban infrastructure. Several related programmes, to raise the availability of venture capital and to provide 'incubation' facilities for new firms, are dealt with in Chapter 8.

University funding

Three principal results are desired from the enhancement of universities. First, research findings lead to scientific and technological innovations, in accordance with the linear model, but also as a part of technological advance more generally. Second, development of a centre of excellence in a certain field can create or enhance a favourable public image and reputation. Third, training and education provide a pool of labour which can be important for regional recruiting (Angel 1989; Peters 1989; Shapero 1985). Boosting funding levels at state-supported universities is a relatively easy way to do something that visibly improves the high-tech status of the region; less frequently is any objective comparison made to universities or regions elsewhere.

Universities are found on most checklists of high-tech location factors, but their actual effect is far from clear-cut. Rogers (1986) stresses the interactions between university faculty and industrial firms in the vicinity, but his examples are not applicable everywhere. Segal (1986), drawing from British experience, makes the germane point that high-tech spin-offs take place in some locales – around the *best* institutions – without any direct university–industry connections being present. The Oxford example is a recent case in point (Lawton-Smith 1990). The

Stanford and MIT examples in the USA have probably distorted the perceived role of universities, but they do serve as (perhaps unrealistic) models for other regions (Howells 1986; Stankiewicz 1985). Faculty research and a pool of graduates provide different categories of attraction. Research – especially the best research – can be procured from a distance, as the many recent research support agreements in biotechnology by large drug and chemical firms have shown. To assume that universities function as 'knowledge centres' around which innovative firms will cluster is unrealistic (Howells and Charles 1988; Nijkamp and Mouwen 1987).

Despite the exaggerated expectations, the prominent success of Texas in rising to the top ranks of university research in the USA has shown that investment in universities can have discernible effects. Less frequently noted is the distinctive manner in which oil revenues have been almost the sole source of university funding. Once again, the duration or longevity of the state's investment is noteworthy, as is the fragility of the revenue base, as recent oil market developments have made apparent. Other states in the USA, perhaps recognizing this, have instituted more focused programmes, concentrating funding and promotion on selected high-tech fields. These include the microelectronics research units established at Arizona State University and at North Carolina, and biotechnology and medical research in Baltimore, Maryland, which were set up in order to appeal to a single sector of high-tech industry. A somewhat broader approach is that of New York, New Jersey, and Ohio, all of which have established advanced technology centres, each concentrating on a given field of technology, at universities within the state.

Targeting in a decentralized system runs the risk that all regions will pick the same small set of sectors. To some extent, this has happened in the creation of joint university–industry R&D centres in several states. By 1987, California, Massachusetts, Minnesota, New York, and Pennsylvania had all begun programmes in microelectronics and applications; Massachusetts and Pennsylvania had also focused on biotechnology (Schmandt and Wilson 1987: 269). In addition, a specialization in one or another research field may decrease the flexibility of universities, making it difficult to move into other fields or to promote interdisciplinary collaboration when the related specializations are located elsewhere (OECD 1981b: 56).

Because university research appears strongly to influence industrial R&D location decisions, several states in the USA have allocated substantial funding in order to improve their institutions relative to those in other states (Robinson 1985; Schmandt and Wilson 1987).

However, the states whose universities have improved most in recent years – California, Texas, and Virginia – are states which have strong bases in federal and industrial R&D alike (US General Accounting Office 1987: 14–17).

Regional policies to improve research draw heavily on the linear model's support for basic research. The state of Texas in the USA has noted two principal avenues of impact from such investment (Devereux *et al.* 1987). First, support for university research increases the competitiveness of the state's institutions for federal research grants, a conclusion supported by other recent work (Drew 1985; US General Accounting Office 1987). Second, such investment in higher education helps in the competition for industry currently taking place among a group of large or growing states, including Arizona, Massachusetts, California, Illinois, and Michigan.

The success of some areas and not others emphasizes once again the role of agglomeration. States in the USA which have several top-ranked universities, especially when they are located in major metropolitan areas, attract both government R&D funds and industrial R&D at disproportionate levels (Malecki 1985b). The significance of agglomeration contrasts with the experience and potential of states whose major universities are in small towns and rural locales. The three most-cited high-tech regions (Silicon Valley, Route 128 and Research Triangle) have both an urban setting and a cluster of at least three universities in close proximity.

One must not expect too much from policies to boost a region's universities. The importance of world-class research universities and a critical mass of technical people and technological activities are often noted. Shanklin and Ryans (1984) maintain that most areas of the USA have little or no chance to incubate or to attract high-tech firms, because they lack top-notch universities. In addition, university research must be prominent in both quantity and quality to attract a cluster of corporate R&D facilities. It is neither easy nor cheap to create and maintain the status of a top-notch university (Vaughan and Pollard 1986). Finally, a technological region needs more than university research and corporate R&D. It requires the dynamism of linked firms who produce best-practice goods and services.

Urban infrastructure
One of the most underappreciated location factors that attracts high-tech industry is a high level of accessibility to other locations. Firms and their professional workers depend on face-to-face communication with people in other facilities of the firm as well as with external sources

of information, and the presence of a major airport is a significant consideration (Ch. 6). In general, the issue of air accessibility and high-tech location reinforces the other advantages of large urban regions since they tend to be well served by a large number of airlines. For example, proximity to Heathrow airport near London is cited as critical to the location of firms (Morgan and Sayer 1988: 236–7).

'Quality of life' is another of the common attractions for high-tech industry. Professional and technical employees, in survey after survey, prefer locations with a mix of amenities that combine to mean quality of life. As a result, many cities are attempting to brighten up their images with new inner-city redevelopment efforts, public funding for parks and other civic improvements, and enhancement of local schools.

Regions and communities alike tout their cultural activities, park systems, and other publicly funded infrastructure and services. The paradox about amenities is that a desirable quality of life has its price – in the higher taxes needed to support public services, facilities, and education at levels competitive with those of other high-tech regions (Power 1988). And successful high-tech areas need continually to maintain and upgrade their services in order to keep their quality of life high. Not only R&D, but producer services as well, rely heavily on educational institutions in the local region. The education system, especially universities and other higher-education institutions, is the single most important factor in maintaining growth in producer services (Beyers, Johnsen, and Stranahan 1987).

Science parks

Given the clean office and research atmosphere of R&D facilities, many of them have settled into the suburban office and industrial parks that are now a commonplace location for economic activity in metropolitan areas. Whether called science parks, research parks, or technology parks, these more specialized developments cater to the preference for a campus-like setting with low-density, often dispersed, building sites. Onida and Malerba (1989: 163–4) suggest that for large firms in particular, science parks allow collaborative links to be established with a recognized 'centre of excellence' in a field, and to take advantage of an agglomeration of researchers and new graduates. The 'prestige factor' weighs heavily in the decisions of large firms, but does not necessarily indicate any linkage or interaction with the local universities.

Most experience shows that an urban environment is necessary for success. The UK's Cambridge Science Park – set up after the 'Cambridge phenomenon' was well under way – had nothing to do with the attraction of the Cambridge area for high-tech firms, although it now probably enhances it. Proximity to the London region is one of Cambridge's attributes. Instead, as in the prominent American examples, the science park, the most successful in the UK, now merely enhances the other attractions of the South-east region centred on London (Segal 1986).

Although a relatively recent phenomenon, science parks are now common throughout Europe, North America, and elsewhere (Gibb 1985). There are well over 100 science parks across the USA, but occupancy in existing American parks tends to be significant only in places where high tech has been successful for other reasons, such as Stanford Research Park, Princeton Forrestal Center, and New Haven Science Park near Yale University. Each of these is located within a major urban region and is affiliated with a world-class research university. State government funding has been central to the proliferation of American science parks, but they also represent an example of the growing trend towards 'public-private partnership' in economic development (Committee for Economic Development 1986). The state of Connecticut has backed the New Haven Science Park (centred on private Yale University) to the tune of $4.4 million. Joint ventures between universities and private developers are increasingly common, as at the universities of Florida and Maryland.

The partnership concept is also growing in France, especially in the context of technology centres (Perrin 1988a, 1989). French high-tech policy currently combines traditional top-down strategies with local initiatives (Dyckman and Swyngedouw 1988). Decentralization from Paris is still an objective, embodied in the development of Regional Centres for Innovation and Technology Transfer (CRITT). Local initiatives are also taking place, especially in the Grenoble and Lyon areas in south-eastern France. Sophia-Antipolis, outside Nice, is perhaps the most successful of the French *technopoles* (Dormard 1988; Perrin 1988a). Twelve such 'urban research parks' were in place by late 1989, out of a planned total of 37 – a number rivalling the Japanese Technopolis plan in magnitude and conceptualization. They are generally located near urban areas, including the Ile de France Cité Scientifique outside Paris (Lacave 1989; Scott 1988c: 60–77). The similarity between the *technopoles* and the *métropoles d'équilibre*, the French growth poles of the 1950s, is perhaps not a coincidence.

In the Federal Republic of Germany, technology parks did not appear until 1983. They have been, as in the USA, a product of local governments, and the parks preserve the regional imbalance which favours the prosperous South (Schamp 1987). This conclusion is that expected from prior observations (Brainard, Leedman, and Lumbers

1988). Technology parks have been only a part of a broad set of policies in West Germany, which include fiscal measures, infrastructure (including capital), and human capital (education and university research). Overall, however, as elsewhere, West German 'innovation-orientated regional policy . . . seems to be incapable of counteracting and balancing out present development in an área's economic structure' (Giese 1987: 250).

The Silicon Valley experience has spawned numerous other science parks as well, including that in Singapore mentioned earlier. A rather successful one is the Hsinchu Science-Based Research Park in Taiwan, which was originally conceived in 1969. By the end of 1984, 56 firms were operating in the park, including many leading international firms. Many of them promote a 'reverse brain drain' of skilled Taiwanese who were educated abroad (Simon and Schive 1986).

The conclusions from American experience are largely similar elsewhere. Joseph (1988) provides details of 16 technology parks in Australia, the most commercially successful of which by far is North Ryde outside Sydney with 90 firms. However, many of the tenants are multinational firms, similar to the situation at Research Triangle Park. Little, if any, interaction takes place with local universities other than purchases by the latter from the firms, and linkage to firms in the Sydney region is low (Joseph 1988: 186–7). Most of the firms had already been in the Sydney area previously, and chose North Ryde for its prestigious image.

Science parks or research parks, therefore, are an attractive but highly uncertain policy. They often present little more than a theme for real estate or property sales and occupancy. This may attract some firms, but parks themselves do not increase the propensity for new firms to form (Britton and Gertler 1986). Only metropolitan regions and their bundle of amenities and infrastructure are even potential locations for new firm formation (Draheim 1972; Goddard, Thwaites, and Gibbs 1986). The complex and dynamic advantages associated with urban size – face-to-face communication, pools of workers or the potential to attract and keep them – outweigh the largely aesthetic attributes of a science park. Policies have been unable to create the critical mass necessary to attract and keep professional workers, except in large urban areas. The synergy of amenities, accessibility, and agglomeration factors found in large urban regions cannot really be found in smaller regions (Andersson 1985; Dorfman 1983; Malecki 1987b; Martin 1986; Miller and Coté 1985).

It is not surprising, then, that so few science parks in the USA and elsewhere have been successful, nor that the most successful are in large urban areas (Abetti, LeMaistre, and Wacholder 1988; Cox 1985; Danilov 1972; Glazer 1986; Miller and Coté 1985). The advantages present in existing concentrations of high technology make it very difficult for new areas to supplant the established high-tech regions (Malecki and Nijkamp 1988). Given the 'right' conditions, however, science parks can add measurably to regional economic development (Goldstein and Luger 1990).

Directing technology

The strength of industrial and innovation policies is that they are based on a vision of what the future will comprise, both in technological trajectories and in industrial specialization. The greatest risk of such policies is not that they will fail, but that they will be so appealing that too many other regions and nations will be competing for the same goal in much the same way. It does not take high technology or frontier industries to sustain an economy. The true benefits of economic development – local linkages and continuing entrepreneurship – do not require high technology. They need a web of interactions and information flows within and into the region. Flows and linkages to the outside drain a region of ideas, talent, and control, unless they are balanced by a receptivity to new ideas coming from the outside. The successes of Japan and Korea reiterate this theme at the national scale (Enos and Park 1988; Otsuka, Ranis, and Saxonhouse 1988), while the Italian industrial districts confirm that localities can be receptive to outside information (Goodman, Bamford, and Saynor 1989).

It is clear from past experiences of high-tech regions that 'success breeds success' by forming a critical mass of workers, researchers, investors, and other supporting businesses and services. There can be considerable gain even if regional and local policies fail to attract high-tech industry. The comparative advantage of many places will improve from having upgraded local conditions, such as education, university research and amenities (Malecki 1984a). Policies at the regional level can provide a focus for regional advantages. Unfortunately, political considerations can dilute scarce funding by designating too many foci, or by selecting some that are inappropriate for high technology given the demands of firms and their workers. This may well be occurring in the peripheral technopolises in Japan (Kawashima and Stöhr 1988). It remains apparent that high technology will not appear in all places, despite policy efforts (Jowitt 1988; Malecki and Nijkamp 1988; Schlieper 1988).

Within regions, technology transfer depends critically on the level of information that is 'in the air' in a locale (Andersson 1985; Rogers 1982; Sweeney 1987). This quantity of information hinges on the firm size mix in the region: small firms much more willingly share information with each other, largely because this open flow has mutual benefits, as seen in industrial districts. Large firms must devote much of their information-oriented time to intra-firm communications. Intra-firm communication is likely to be particularly true of R&D, for which the demands for interaction both with headquarters and with production plants and sales offices are great.

Some regions are simply more open to new ideas just as they are to a diversity of imports. Regions where imports are diverse and state of the art are most likely to be innovative themselves. Imports, in turn, are likely to be innovative if learning by searching is taking place in the region in an active R&D environment, by a creative and knowledge-oriented labour force (Johansson and Westin 1987).

The easy answer to the failure to generate indigenous growth is the relative scarcity of R&D carried out in some regions and countries. R&D is important, but not all R&D has the same benefits for a regional economy. The significance of industrial R&D is simply that it indicates the level of technological progressiveness of a company and assists in giving it an open mind, reaching out for new information and being receptive to it (Sweeney 1987: 130). Technically progressive firms take part openly in information exchange, they continually search for information, and they maintain internal communication (Carter and Williams 1959; Sweeney 1985: 95–6). The major R&D laboratories within large, innovative corporations represent a sizeable component of the structure of high-tech activity, even if kept largely internal to the firm (Castells 1988: 67–8). Information flow is also a key element in successful exchange or transfer of research information from R&D to users. In such a contact system, users also become important sources of innovation, both of product designs and of production processes, and about the operational performance of innovations (Foxall and Johnston 1987; von Hippel 1988). In sum, regions and 'localities having high entrepreneurial vitality have characteristics similar to those of technically progressive firms, openness in giving as well as taking information, continuing effort in their own search for information and good internal communication flows. The vitality, once initiated, becomes self-reinforcing and sustaining' (Sweeney 1985: 95–6).

Conclusion

High technology has had some positive effects on national and local policy. It has prompted a more long-term perspective about economic development; and it has shown the significant advantages to be gained from investments in human capital, especially through education. Even if done only partially and half-heartedly, high-tech policies can have these effects. Interregional competition based on quality of universities, educational attainment, and quality of life is certainly an improvement over that centred on low wages, low taxes, and undervaluing of education (Cobb 1984). The creation of new high-technology complexes will not be an everyday, or everywhere, occurrence. Large urban regions, especially those where university research is abundant, are the prime locations, in no small part because of the entrepreneurial 'climate' and opportunities present. Other places, if they can attain some – even if not all – of the attributes of the large urban areas, might get at least a small development benefit. It remains to be seen whether far-sighted policies will win out over those aimed at short-term gains.

At the same time, technology is not fixed in either time or space. Scientific advances made in one location, whether a university or corporate lab, are typically shifted to other locations where further development takes place, and to yet other locations as commercial products enter into production. The aggregate economic impacts are thus dispersed to, and divided among, several locations. With large corporations, this internal process of technology transfer may have few benefits for any region or country. The same is true of most inter-firm technology transfer via licensing and other arrangements. In the end, corporate strategies and decisions regarding the location of activities determine where the benefits of science and technology end up.

There may be relatively little that policies as typically viewed can do to alter the situation (Goddard and Thwaites 1986). At national level, cultural or social difference remain extremely influential (Roobeek 1990). At the regional level, much of economic development continues to be characterized by cumulative causation (Ch. 3). Some possibility exists that a small region can become a technological complex, or that a declining city can revitalize itself around new technologies. More often, the choice must be among large, already rather prosperous, high-tech urban regions (Kutay 1989; Pottier 1985). The European resurgence of IT production, some of it under the umbrella of ESPRIT, continues to be concentrated in the advanced countries and, within them, in the major metropolitan regions (Gillespie *et al.* 1987). The fundamental problem for local and regional policy is that not all variables are manipulable by policy (Rubin and Zorn 1985), or are alterable only over the long run and after substantial investment. Over the long run, regions will be better off for having devoted money to education at all levels, to investments in public infrastructure (whether justified under the rubric of quality of life or not), and to thinking in terms of economic diversity and competitiveness.

In the context of most small, peripheral regions, however, innovative development may be 'too good to be true' (Martin 1986: 17; Brugger 1986: 42; Friedmann 1986b). The human capital of such regions is limited and mobile, so it can be expected that many talented people will simply leave. Historically, indeed, outmigration selects the most productive subset of the population of rural regions. In addition, the industrial and personnel structure of peripheral countries and regions is a result of their dependence on branch plants of large, multinational corporations. The evidence is clear that such plants tend to have lower levels of highly trained and skilled personnel, certainly lower levels of R&D, and little attention to non-routine activities or new products. Such a regional structure is unlikely to be innovative, especially in a setting of short product cycles, which require a continual flow of up-to-date information and knowledge.

It is the entrepreneurial spin-off process which is the principal local route of technology transfer. Indeed, new ideas enter an economy primarily through the identification of opportunities by entrepreneurs. The local nature of entrepreneurship poses great challenges, but it is just such a process which was the basis of the successes which national and regional S&T policies are trying to imitate. The process of entrepreneurship may be a more important one to regional and local economies than the process of technological change. The following chapter is devoted to entrepreneurs and new firms.

Chapter 8

Entrepreneurship and regional development

Entrepreneurship has acquired a new importance among the processes which affect regional economic change. The emergence and growth of new firms, and the presence of small versus large firms, vary considerably from place to place. This chapter examines the process of entrepreneurship, focusing on its role in regional economic development and spatial variation. Previous studies, for the most part, have emphasized the effect of spatial variables only through the occasional examination of local policies and, until recently, regional policies had generally ignored small firms. At the same time, regional research has found that small firms are more common and are more likely to emerge in some types of economic environments than in others, and national variations have attracted attention recently as well (Porter 1990; Roobeek 1990). The systematic study of entrepreneurship as a key process in economic change is just beginning, and is a critical need for understanding national, regional, and local development differences.

The emphasis in theory and policy on large firms over several decades lessened the importance of small firms in both conventional wisdom and governmental concern. The growth of giant corporations in most countries, including the UK, Germany, France, and the USA, took place for somewhat different reasons, but always with state support and favouritism for large firms. By contrast, 'micro capitalism' has flourished in several countries, especially in the industrial districts of the Third Italy and in Japan (Ch. 6) and in a more dispersed rural form in Denmark (Cooke and da Rosa Pires 1985; Kristensen 1989; Weiss 1988: 196–204). The entrepreneurial success of Japan or the Third Italy (Ch. 6) cannot be attributed to any set of cultural or social norms, such as a history of sharecropping and work in and as a family unit. Instead, policies in both Italy and Japan provide numerous privileges to small firms that would be lost if they were to grow (Friedman 1988; Weiss 1988: 204).

Much of the interest in entrepreneurship and the role of small firms originated in the study by David Birch (1979), which attributed 60 per cent of net job generation to small firms. This prompted a flood of research and policy initiatives concerning the small-firm segment of national and regional economies. Around the same time and using the same data source (the Dun and Bradstreet files), Jusenius and Ledebur (1977) found that differences in regional employment change were related almost entirely to differentials in the rate of firm births; regional closure rates were nearly identical. Later research by Birch, while not negating his earlier findings, lent a note of caution that was not embraced everywhere. For example, Birch and MacCracken (1984) found that in high-technology industries, large firms accounted for the bulk of new jobs. Kirchhoff and Phillips (1988), re-examining the evidence from 1976 to 1984, found that 74 per cent of all new jobs in the USA were accounted for by new firms.

The 'American job machine' via entrepreneurship and new firm formation is admired if not emulated elsewhere (OECD 1989c). Weiss (1988: 207) suggests that American policies are based on the underlying belief that small businesses are admired less for what they are than for what, if successful, they are bound to become. The industrial economics approach which lies behind such a belief generally sees firm growth as inevitable (O'Farrell and Hitchens 1988b). In the UK, where Mrs Thatcher's policies have mirrored those of the Reagan era in the USA, 'explosive' growth in new technology-based firms took place from 1970 to 1985 (Dodgson and Rothwell 1988). In France, the belief lingers that the conditions behind job creation in the USA are not readily transferred elsewhere and perhaps should not be (OECD 1989c).

At least some fraction of small-firm formation is a result of flexible policies by large firms to disintegrate their activities, such as subcontracting production and services to small firms (Ch. 6). This has the effect of increasing the apparent importance of small firms in an economy, and plays down the dominance of large firms. None the less, it has become standard thinking to place upon entrepreneurship responsibility for the bulk of employment generation within regions as well as nations, and to utilize it as a primary indicator of regional well-being (Coffey and Polèse 1985; Fischer and Nijkamp 1988).

From a research standpoint, there is a considerable distinction between *small firms* and *entrepreneurship*. Research on the latter is difficult at best, because national data bases are concerned primarily with the 'stock' of firms and of economic activity, and less with the 'flow' of new entrants into the economy. This partially explains the fact that employment gains and firm births may usually be identified, but that firm deaths and the resulting employment losses among small firms are rarely documented. The solution for researchers seems primarily to be survey research, which must be limited because of its expense (Aldrich *et al.* 1989). In fact, even firm formation or birth is not easily pinned down, because people often move into self-employment gradually (Johnson 1986: 12–15; Pickles and O'Farrell 1987). Bruno and Tyebjee (1982: 304–5), discussing the methodological problems associated with research on entrepreneurship, list data availability – on start-ups and failures as well as on geographical environments – at the top of the list.

It is difficult to identify actual failures, as distinct from more voluntary discontinuances or acquisitions (Bruno, Leidecker, and Harder 1987; Scott 1982). Of 250 firms studied over a 20-year period, Bruno, Leidecker, and Harder (1987) found that 39.2 per cent had merged or been acquired, 38.4 per cent had been discontinued, and 22.4 per cent were still surviving independently. Acquisitions were often by other small or medium-size firms, and took place within the first seven years after founding (Bruno and Cooper 1982).

Once identified, many factors are involved in the failure of young firms. Some prominent ones include product timing and product design, and an ineffective management team (Bruno and Leidecker 1987). As O'Farrell and Hitchens (1988a: 400) state: 'The relative ability to meet the key criteria required by the market is the principal constraint on small firm growth, and the lack of skills and inadequate training at managerial, supervisory, and shopfloor levels is the proximate cause of the production problems. Hence, getting production right is always a *necessary* condition of growth of *all* firms.'

This chapter reviews both the 'new firms' and 'small firms' components of regional economic change, although the considerable confusion between the two creates some problems for a systematic treatment (Storey 1982). Research on small firms varies greatly, if only because there is little agreement on the size of such firms, or how small is 'small'? A recent collection of European studies shows the difficulty, with small firms defined differently in nearly every country (Keeble and Wever 1986). In West Germany, no national data on small firms exist (Klaudt 1987). Incubators and venture capital as policies for new firms are briefly dealt with, primarily in the context of 'spin-offs' in a high-technology setting. Finally, some policy issues are touched on that address both small firm and new firm potential in regional and local development.

Entrepreneurship: the problem of definition

Entrepreneurship has been a topic of long-standing concern within economics and business, but there remains surprisingly little consensus on the behaviour and characteristics of entrepreneurs. The variety of interpretations has led most researchers to focus on 'less messy', more easily aggregated, factors of production, such as capital and labour. Since the time of Adam Smith, however, the distinction between capitalists and entrepreneurs has become blurred in both British and American economic thought (Kent 1984: 3). In recent reviews of the confusion with which 'entrepreneurship' has been viewed, there are four principal characteristics associated with entrepreneurs, including risk-taking, supplying capital, competent or perceptive management, and innovativeness. The supply of capital and the bearing of risk are traditional roles of the capitalist; competent (or superior) management is a more recent extension of the concept of entrepreneur (Hébert and Link 1982; Kent 1984). Indeed, in an effort to understand entrepreneurial behaviour, the personality characteristics of entrepreneurs have been by far the most common topic of research (Hornaday and Churchill 1987).

Despite perplexity over the definition, the Schumpeterian concept of entrepreneur remains dominant in most of the literature: the entrepreneur as innovator and source of disequilibrium (Leibenstein 1978; O'Farrell 1986a; Thomas 1987a). For Schumpeter (1934: 66) entrepreneurial innovation could occur in any of five ways:

1. The introduction of a new good or of a new quality of good;
2. The introduction of a new method of production;
3. The opening of a new market;
4. The conquest of a new source of supply of raw materials; or
5. The carrying out of new organization of any industry.

The disequilibrium of such entrepreneurial behaviour could result either from individuals or from organizations, as Nelson and Winter (1982) suggest. Casson (1982: 391) suggests that 'product innovation is the most important form of entrepreneurship, at least in a long-term perspective'. Even large corporations can be 'entrepreneurial' by attempting to instil the attributes of small firms into the organization (Johnson and Lucas 1986; O'Farrell 1986a). The term 'intrapreneurship' has been coined to describe this process (Vesper 1984).

Even if it is accepted that entrepreneurship implies disequilibrium, it may be important to draw a distinction between an emphasis on uncertainty and one on innovativeness. Uncertainty plays a major part in that the entrepreneur's perceptions are likely to be only partially correct concerning the decisions of others. *Knowledge* or information about the availability of resources (including the resource of technical knowledge itself) is a separate and crucial variable in the uneven and disequilibrating distribution of growth (Kay 1984; Kirzner 1984; Nelson and Winter 1982). This knowledge can come about either in an unplanned way or, more likely, at least partly as a result of deliberate investment in research (Kirzner 1984: 44–5; Bell 1984). 'Technological change is the major root source of information problems' (Kay 1984: 161).

Kirzner's three major types of entrepreneurial activity – arbitrage, speculation, and innovative activity – illustrate the distinction between entrepreneurs as innovators and as capitalists. Arbitrage, which consists of acting upon (not widely available) knowledge about the discrepancy between the prices at which goods can be bought and sold, calls for no innovation, risk bearing, or capital. Speculative activity is an arbitrage across time; speculators need not be innovative. Innovative activity, by contrast, consists in the creation (in the more or less distant future) of an output, method of production, or organization not previously in use. For such activity to be profitable, the innovation must display the same pattern of intertemporal price discrepancy that is identified in speculative activity. Both retain important parallels with the case of pure arbitrage. Innovation calls for the discovery of an intertemporal opportunity that cannot be said actually to exist before the innovation has been created. Recognition of an opportunity may be the 'core of entrepreneurship' (Timmons *et al.* 1987). All three kinds of entrepreneurial activity consist of taking advantage of price differentials; all are inspired by the pure-profit incentive constituted by the price differentials; all are made possible by less competent entrepreneurial activities (the errors of others).

The link between entrepreneurial competition and entrepreneurial discovery is *freedom of entry*, which spurs both incumbents and newcomers to seek new opportunities for profit (Kirzner 1984: 54). Although it places greater emphasis on non-innovative types of entrepreneurship, Kirzner's view of entrepreneurial behaviour is similar to that of Nelson and Winter's (1982) depiction of evolutionary economic change. In both views, however, entrepreneurs may represent, or work within, very large corporations, as long as they exhibit the qualities associated with entrepreneurship. Thus, firms, especially large firms, are thought to face a constant need to remain entrepreneurial and innovative (Maidique 1980; Maidique and Hayes 1984; Olleros 1986). The inclusion of large corporate entrepreneurs, however, conflicts with the popular notion of entrepreneurs as struggling founders of new firms or of small firms as a distinct subset within the economy (Mason and Harrison 1985).

In order to avoid such all-encompassing views, which would seem to include firms large and small, old and new, this chapter will focus on small firms, and will pay particular attention to those for which innovation has played a part in their founding. This narrow view of entrepreneurship based on innovation would not include actions and decisions (including firm formation) that merely respond to uncertainty and involve risk-bearing, but which are not truly innovative or novel. The Schumpeterian view follows that propounded by Morgan Thomas (1975, 1985, 1987a) concerning the crucial nature of innovation in regional change. Kent (1984: 4) concludes with the judgement: 'The entrepreneur is more than self-employed. Those who start businesses solely as an alternative to wage employment do not participate in the entrepreneurial event. Entrepreneurship requires the element of growth that leads to innovation, job creation, and economic expansion.'

Thus, not all small firms are entrepreneurships, but most entrepreneurship is found in small firms (O'Farrell 1986a). This interpretation is corroborated by the long-standing observation that innovation, especially more radical, 'leap-frog' innovation, is more likely to originate in small firms where older products and technologies are less entrenched than in large organizations (Birch 1987; Rothwell and Zegveld 1982; Sweeney 1985). One can also identify an intermediate category of firms, larger and older than those just emerging, but smaller than already dominant corporations. As Steed (1982: 40) puts it, 'Threshold firms develop out of the pool of small firms.' In a Third World context, elaborated later in this chapter, to focus on indigenous technological capability (Fransman and King 1984) underscores Schumpeterian entrepreneurship more than do conventional views (Leff 1979a).

Sectoral variations in firm formation

New firms are not equally likely to arise in all industries. Instead, they are likely to respond to the relative barriers to entry across sectors, and to the general level of opportunities presented in various technologies and markets (R Nelson 1986). The low entry barriers in retailing and services help to explain their abundance. The degree of innovativeness is particularly low in such sectors, where imitative and franchised establishments are standard (Wicker and King 1988). Outside retailing, sectors with low entry barriers are also common. In the Third Italy, textiles, leather goods, footwear and clothing, furniture, and metal goods are typical for small-firm specialization and new firm formation (Bellandi 1989).

The regional industrial mix is a factor which influences the degree to which new firms are likely to be founded in any particular place. Instability is characteristic of new industries, and it seems to diminish steadily as a sector develops (Bollinger, Hope, and Utterback 1983; Dorfman 1983, 1987; Gordon and Kimball 1987; Shearman and Burrell 1987). To a large degree, this change is related to an openness to learn and to seek out information, which were argued in Chapter 4 to be critical to technological competitiveness. From a regional perspective, the sectoral variation shows itself through the industrial mix and the propensity for new firms to arise in sectors already found in the area (Cooper 1972; Johnson and Cathcart 1979; O'Farrell and Pickles 1989; Sweeney 1985; Wever 1986). Regional clusters of threshold firms centred around certain industrial specializations are also identifiable (Steed 1982). At least one study has suggested that firms which survive for several years, rather than fail, are more likely to be closely related to the organization which the founder had left (Cooper, Dunkelberg, and Woo 1988).

Industries, and firms, which have limited markets for relatively small numbers of products, are less likely to provide opportunities for new firms. This is particularly characteristic of sectors which produce military products (Glasmeier 1988a). The local industrial mix therefore bears critically on the propensity for new firm births (Bollinger, Hope, and Utterback 1983; Dorfman 1983; Garvin 1983). In four high-technology sectors in the USA, for example, new firm formation was more frequent in locations where a sector was already established, rather than in new areas (Malecki 1985a). The profusion of spin-offs from high-technology firms has been a popular observation, based once again on the Silicon Valley experience (A Cooper 1973, 1986; Malecki 1981a; Rogers and Larsen 1984). In technically based sectors, high levels of education and occupation in technical and professional occupations are far more common than in other industries (Doutriaux 1984; Keeble and Kelly 1986; O'Farrell and Pickles 1989; Ray and Turpin 1987). However, industry mix alone does not account for the observed geographical variations in new firm formation; the relative size of establishments may play a greater role through the stifling effects of branch plants (Beesley and Hamilton 1986; Sweeney 1987).

Brusco's (1986) depiction of small firms in industrial districts emphasizes their innovative activities. These firms, whose markets are always national or international, are contrasted with the more dependent types of small firms – traditional local artisans and dependent subcontractors. To identify the *type* of firm, and thereby the potential for local development, one needs to know the extent of the market, the clientele (how many, how large?), the nature of the production runs (large or small), and the characteristics of the machinery. Brusco expects that entrepreneurship is highest where there is an accumulated managerial, technical, and commercial competence. Capecchi (1989) has observed that new firms which developed in the Third Italy were generally imitative but, at the same time, complementary to existing firms. Rather than competing directly, the new entrants tended to broaden the market further with slight variations on existing products. Other firms specialized on particular products or processes, subcontracting to existing firms. The Italian experience is similar to the aggregate pattern in the USA, where small-firm entry has been high in sectors where R&D is low but where small-firm innovation activity is high relative to that of large firms (Acs and Audretsch 1989).

Information channels are central to the transfer of knowledge from person to person, from firm to firm, and from region to region. Allen, Hyman, and Pickney (1983) found that information among firms in Ireland, Spain, and Mexico flowed mainly through informal channels – that is, through interpersonal contacts. Information channels among large firms and organizations, by contrast, are more formal. As Sweeney (1985, 1987) suggests, foreign subsidiaries obtain the greatest proportion of their technology from their parent firms. In fact, such plants have several channels of technology blocked to them, such as government-sponsored research, trade fairs, and industry associations, which are quite important to domestic firms. Firms in the same industry are also much more likely to be a source of technological knowledge to domestic firms. Allen, Hyman, and Pinckney (1983) conclude that domestic firms in many ways have easier access to foreign technology than do the subsidiaries of multinational firms, because of the more informal information flows among them.

There remain significant differences in communication behaviour from place to place, as Leonard-Barton (1984) found when comparing Swedish entrepreneurs with those from the Boston area. Swedish business people were less inclined to rely on informal sources, but tended to use a greater number of outside sources of all types, including publications and those formally connected to the firm.

Orientation towards national and international markets is what distinguishes among types of small firms. Of Brusco's (1986) three types of small firms (Ch. 6), those located in industrial districts are technologically sophisticated, because they work for a large number of clients, indicating that their level of quality and technical standards are high. Lorenzoni and Ornati (1988) suggest that firms in such 'constellations' are more willing to seek information from outside sources, such as consultants, universities, and other firms. The 'environmental texture' of such districts – suppliers and infrastructures particular to the production requirements of a group of firms – is especially supportive for new firms.

New firm formation, which also depends on flows of knowledge, varies geographically in part because of differences in innovativeness and in whether and how innovativeness is promoted. Large firms try to maximize information-intensive contacts by centralizing in large urban centres and around locales with innovative small firms (Pottier 1988). Around these cores are clustered service activities for whom large firms are the primary market.

The *local* nature of entrepreneurship bolsters the findings of the local specificity of economic development. 'Entrepreneurial vitality is very much a local phenomenon. . . . Prosperity and economic growth of regions and localities are strongly associated with the strength and vitality of the small firm sector in the region or locality' (Sweeney 1985: 94; Allen and Hayward 1990; Beesley and Hamilton 1986). Ganne (1989) describes the local network of small firms in the area around Grenoble, France, as far more significant a seedbed for new firms than the large technological organizations located there as part of the *technopole* policy described in Chapter 7.

Entrepreneurial capability relies on people, often in technological roles within organizations, who serve as resource persons to an external network. Not all people will be able to play the pivotal 'gatekeeper' role. For regions, the gatekeeper function is also critical, if informal, and largely relies on personal characteristics such as an inclination to interact with others (Falemo 1989). Within a locality or community, a key role is played by 'community entrepreneurs' for whom development of the community is a primary goal. In Sweden, for example, these individuals tend to have very large personal networks, but do not see themselves as entrepreneurs (Johannisson and Nilsson 1989).

Female entrepreneurs are generally similar to men starting new businesses. Women are somewhat more likely to form service or retail businesses, rather than manufacturing, but in most respects they face comparable challenges in seeking capital and expanding their markets (Brophy 1989). Women are somewhat less likely to obtain loans to start a business – in part, some studies have found, because they are less often perceived by bank officers to have successful entrepreneurial characteristics (Buttner and Rosen 1988). There are few actual differences between male and female entrepreneurs, and the major one is difficult to judge on paper or by quick evaluation. Women's social and professional networks tend to include few men, and therefore have also tended to include fewer professionals, such as accountants, who have access to broad business knowledge (Aldrich, Reese, and Dubini 1989; Hisrich 1989). Moreover, female entrepreneurs seem to be largely similar in different societies, including Italy, Sweden, the Pacific islands, and the USA (Aldrich, Reese, and Dubini 1989; Ritterbush and Pearson 1988).

This 'information rich environment' (Andersson 1985; Johansson 1987) and 'technical effervescence' (Miller and Coté 1987) are the principal characteristics of 'creative regions' or 'innovation know-how concentrations' (Perrin 1988b). Standing in sharp contrast, the predominant empirical finding about small firms in peripheral areas is their relatively low level of innovativeness (Malecki and Nijkamp 1988).

Innovation and entrepreneurship

The technological capability of a region or nation is concerned both with the stock of knowledge there and with factors which encourage or inhibit the formation, development, and reinforcement of technological capability. Sweeney (1987) has called this 'innovation potential'. The factors which make up the innovative potential of a region seem to be a complex mix of the following:

(a)   the sectoral and technological mix of the industry in a region;
(b)   the strength of the engineering sector;
(c)   the autonomy of decision-making in the industries and infra-structure of a region; or the dominance of branch plant employment;
(d)   the dominance of employment in one or two sectors, especially where these are of low use of best technical practice or are declining industries;
(e)   the strength or weakness of the quaternary information sector;
(f)   the technological orientation of the educational system or lack of it (Sweeney 1987: 131).

A full examination of this innovative capability calls for understanding of several elements of regional structure. First, the capability for R&D and technological activities, part of the design and creation of new technologies, must be grasped (Ch. 5). Second, the presence of a favourable or unfavourable industrial mix, domination by branch plants of large firms, a concentration of jobs lacking in skills, low levels of contact with outside networks and a general lack of urban agglomeration advantages, such as suppliers, markets, information, infrastructure, and capital. Low levels of education may be associated, in turn, with low levels of skills and business knowledge and a reduced range of contacts and information sources. However, many of these shortcomings can be overcome if education levels in the population are high; education is strongly related to a greater range of information sources and a wider set of customers and markets – in short, to an awareness of industry best practice (Ewers 1986; Miller *et al.* 1986; Stokman and Docter 1987).

In a detailed study of innovation in the Netherlands, Pellenbarg and Kok (1985) surveyed 461 firms of 100 employees or fewer. They defined four levels of innovation: *basic* (something completely new, and outside their study); *primary*, which is based on a basic innovation; *secondary*, which is new for the country but not the world; and *tertiary* – an adaptation or improvement. Four types of innovations used are: product, process, organization, and market. Of their 461 firms, 147 were engaged in one or more of these innovative activities. Only 52 ($\frac{1}{9}$) were engaged in primary or secondary innovation, and product innovations predominated. Small firms (less than 50 employees) produced fewer innovations than big firms, and young firms (less than 25 years old) fewer than older ones. To a large degree, the absence of R&D as a source of innovations in small firms is a major contributor to this situation. Instead, informal internal know-how and careful reading of market signals are more common elements (Docter, van der Horst, and Stokman 1989). Market information, the most important type of information, was best obtained from clients, journals, trade fairs, suppliers, and informal contacts (Kok and Pellenbarg 1987). Geographically, the peripheral Dutch regions were short of innovative firms, but not to the extent expected. In a similar study in West Germany, Meyer-Krahmer (1985) found that innovation potential – measured as the percentage of firms doing R&D – is unevenly distributed within the country. Higher levels of innovation potential (R&D) are found in more densely populated regions containing urban agglomerations. Davelaar and Nijkamp (1989a) similarly found that in urban agglomerations, firms are more likely to rely on external sources of information and innovation than are those located peripherally.

Even in urban core areas, small firms are not necessarily innovative.

For example, several of the firms surveyed by Markusen and Teitz (1985) had been affected by product-cycle developments which rendered their products less competitive than they had been previously. In response to these pressures, the Bay Area firms tended to be uninnovative, except to diversify their product lines. Process innovation was virtually non-existent, as was R&D. Much the same was found by Ajao and Ironside (1982) among firms in the Canadian Prairies. Innovativeness among small firms, then, is best explained by the dynamics which occur in some sectors – and some places – at the time when barriers to entry are low. As such, they do not respond well to policy initiatives at the local or regional levels (Shearman and Burrell 1987).

Entrepreneurship and local development

Coffey and Polèse (1984, 1985) have placed entrepreneurship at the centre of the process of local economic development, at least partially as a reaction to the well-documented consequences of 'external control'. Coffey and Polèse propose four stages of local development:

1. The emergence of local entrepreneurship;
2. The growth and expansion of local enterprises (to replace jobs lost by closed branch plants);
3. The maintenance of local enterprises under local control;
4. The attainment of an autonomous local control structure and of a local business service sector.

Their view of entrepreneurship appears at first to be an operationalization of the cumulative causation model (Ch. 3). It is, however, distinctly non-Schumpeterian, since innovation plays at most a minor role. Entrepreneurs start businesses and create jobs, but these jobs and new enterprises need not be centred around some innovative technology that allows for success or competitiveness outside the local region, much less globally. As Wever (1986) points out, most new entrepreneurs are in non-basic activities (shops, cafés, pubs, repair shops), where they most often simply replace fellow entrepreneurs. In addition, closure rates of new firms are often quite high and depend heavily on macroeconomic conditions such as interest rates and regulation (Rees 1987). In Wever's (1986) study, 50 per cent of new firms closed within 5 years; 60 per cent within 12 years. In a sense, the Coffey-Polèse model is very equilibrium-oriented, in that local entrepreneurs function mainly to counter the economic divergence brought about by external control. Other recent descriptions of endogenous regional development stress the necessity for political and institutional change that fosters entrepreneurship (Brugger 1986; Stöhr 1986b).

Entrepreneurial or 'creative' regions

In what regions and under what sorts of conditions, then, can we expect to find higher levels of entrepreneurship? Large urban regions have certain advantages over rural areas. In discussing 'creative regions', Andersson (1985) maintains that communication is a key to understanding the fundamental role of metropolitan regions in the creative process. His description of contact networks and communication is reminiscent of Silicon Valley as described by Rogers and Larsen (1984). Andersson (1985: 19) proposes that creativity as a social phenomenon primarily develops in regions characterized by: (1) high levels of competence; (2) many fields of academic and cultural activity; (3) excellent possibilities for internal *and* external communications (emphasis in the original); (4) widely shared perceptions of unsatisfied needs; and (5) a general situation of *structural instability* facilitating a synergistic development. Andersson's conditions for regional creativity can be translated into more conventional policy variables, but in general, they focus on three main elements: (1) the presence of professional and technical labour (competence); (2) urban agglomeration, or a threshold size of place, where cultural activity and communication will be heightened; and (3) conditions that promote synergy or instability. Andersson's emphasis on agglomeration economies has been echoed in recent research, both on entrepreneurship and on industrial location more generally (Oakey 1985; Scott 1988b, c). The search for information and willingness to interact with other firms in the local area are qualities also found in industrial districts, regions which have the features of development – linkages and entrepreneurship – vital to a self-sustaining economy (Ch. 6). The presence of small firms and of prior entrepreneurs is a key element of a dynamic agglomeration (Evans 1986).

Local information channels and knowledge are fundamental to entrepreneurs. Large firms try to maximize information-intensive contacts by centralizing in large urban centres. Around these cluster service activities for whom large firms are the dominant market.

However, localities which are dominated by large firms are not centres of entrepreneurial vitality, partly because there is a lack of openness in the flow of information from the large to the small. 'Large bureaucracies prefer to talk to large bureaucracies' (Sweeney 1985: 92–4).

That local conditions generate entrepreneurship is seen in the fact that new firms are rarely attracted from elsewhere. Cooper (1985) found that nearly all (84%) of firms are founded within commuting distance of where the founder previously worked. Cross (1981: 228–30) cites some Scottish evidence to the contrary. Diversity within the small-firm sector – not a dependence on a narrow range of technologies or markets – reinforces this vitality (Sweeney 1985: 95). As noted in Chapter 7, localities having high entrepreneurial vitality have characteristics similar to those of technically progressive firms: an openness regarding information and communication flows. 'The vitality, once initiated, becomes self-reinforcing and sustaining' (Sweeney 1985: 96).

A locality with high entrepreneurial vitality has a high birth rate and a high death rate of small firms (Gripaios and Herbert 1987; Jusenius and Ledebur 1977; Sweeney 1985: 97). Many new firms quickly die, and the rate of firm formation is critical in the difference between regional prosperity and disadvantage. However, other hallmarks of such a region may be harder to identify empirically: diversity and quality in the information flows, responsiveness in meeting the needs of entrepreneurs, and openness in provision of information (Sweeney 1987: 106). This 'information rich environment' is the principal characteristic of 'creative regions' (Andersson 1985).

Culture and entrepreneurship

The culture of some regions tends to result in significantly lower levels of entrepreneurship. Rural areas are often burdened by what can be thought of as life modes or cultural factors which suppress entrepreneurship, such as no small business class, a social structure dominated by church or a single industry (e.g. mining), and high rates of outmigration (Hjalager 1989; Illeris 1986). The new jobs that are created in rural areas, in tourism or branch manufacturing plants, are likely to provide only routine, part-time, and low-wage jobs (Corbeau and Sheridan 1988; Schaub 1984).

Demographic variables, including education, are not especially useful in identifying entrepreneurs, as Timmons, Broehl, and Frye (1980) found in south-eastern Kentucky. Attitudes about overcoming problems, travel outside the home county, and a greater concern for punctuality were the only variables consistently associated with success in simulations of new businesses. However, in a later study in Appalachia, education levels appear strongly related to entrepreneurial success, associated with a greater range of information sources and a wider set of customers and markets (Miller, Brown, and Centner 1987).

These are largely a product of prior work experience and accumulated expertise in technology, marketing, and management (Klaudt 1987; Utterback *et al.* 1988). In particular, experience accounts more for the success of new firms than does the 'optimism' and risk-taking propensity of the individual, factors often cited in research on entrepreneurs (Cooper, Woo, and Dunkelberg 1988). The importance of experience must also temper the attractiveness of education as a policy vehicle. As O'Farrell (1986b: 167) states: 'While the indirect influence of education in the development of skills, attitudes and values may be considerable, the decision to start a new firm usually arises largely as a consequence of the work and occupational experience during the post-education period of early adulthood.' In general, however, 'the success of the enterprise depends not on *where* it is located but on the skills of the management and employees' (O'Farrell and Hitchens 1988a: 413).

The cultural influences on entrepreneurship include the entrepreneurial culture of a locality – the examples of others, an already strong small-firm sector, the tradition of being self-employed (Delbecq and Weiss 1988; Illeris 1986; Martinez and Nueño 1988; Sweeney 1985: 91). The importance of prior entrepreneurs and other small businesses is that they demonstrate to others the achievements possible from taking risks and pursuing opportunities (Buss and Vaughan 1987). While it may seem to be a tautology to maintain that entrepreneurship depends on entrepreneurs or small businesses, it is just the case that the skills and experience needed to form a new business are developed best in a setting where such knowledge can be observed and learned.

The tendency for some regions to have a greater proportion of successful, more rapidly growing small firms may be more important than geographical differences in new firm formation. Again, social and occupational influences appear to be most significant in reinforcing existing spatial contrasts. Mason (1985) found the bulk of the most technologically and financially successful firms in the UK in the prosperous South-east. How do such regional variations come about? In a recent Dutch study, successful entrepreneurs (as opposed to those of failed or failing firms) were simply more prepared for starting a business: they had more starting capital, were more oriented to markets outside the local region, had a larger number of clients, and were somewhat more likely to have worked for a large firm previously (Wever 1986). Cooper and Bruno (1977), although not concerned with spatial variation, also found that successful firms are more likely to (1) have multiple founders (more possible in agglomerations), (2) have similar technology and markets to their parent firms (thus augmenting the industrial structure in certain sectors), and (3) have founders who have worked in large organizations. The advantage of prior work experience in large firms in these two studies shows the related importance of best-practice knowledge and information. A combination of large and small firms, operating in innovative pursuits, affects a local, as well as a national, culture in deep, probably permanent ways.

The ingredients of an entrepreneurial region

There is no shortage of 'ingredients' or regional conditions which foster high rates of entrepreneurship. Like those factors cited as supporting an innovative environment, the factors which comprise 'the environment for entrepreneurship' are generally those common to most, if not all, large urban regions (Bruno and Tyebjee 1982; Malecki and Nijkamp 1988):

- venture capital availability
- presence of experienced entrepreneurs
- technically skilled labour force
- accessibility of suppliers
- accessibility of customers or new markets
- favourable governmental policies
- proximity of universities
- availability of land or facilities
- accessibility to transportation
- receptive population
- availability of supporting services
- attractive living conditions

Bearse (1981) proposes eight less easily measured factors which contribute to the community climate for entrepreneurial activity:

1. The level of change/instability in the local economy resulting in gaps, imperfections, and market failures;

2. The level of uncertainty created by unexpected events and inter-firm competitive rivalry;
3. The degree of fluidity in the social structure and the degree of institutional resources to provide linkage of entrepreneurs with needed resources, especially information;
4. The level of diversification in the industrial, occupational, and social structure of the community;
5. The level of resources available such as (a) information, (b) capital, (c) specialized services, (d) space and (e) number of attractive non-entrepreneurial opportunities;
6. The presence of a critical mass of entrepreneurs and institutions involved in the gestation of new hard and soft technologies and an active, supportive venture capital community;
7. The cultural traditions of the relevant groups effecting the atmosphere for entrepreneurship; and
8. Government policy, including taxes, regulation, and economic management.

It is clear from these lists that several of the factors are simply conditions common to most, if not all, large urban regions. For example, availability of land and facilities, accessibility to transportation, and attractive living conditions tend to be attributes found in virtually any major metropolitan region. It is the other factors that appear to vary most among regions (Malecki and Nijkamp 1988).

Shapero (1984), in attempting to account for 'the entrepreneurial event', focused on urban characteristics in the environment for entrepreneurship. Previously (Shapero 1972), he suggested that a technical labour pool and a good financial community were the most important local characteristics. As he developed these ideas, Shapero (1984) suggested a dynamic process in which local investment propensities and an industrial base of small businesses together lead to a local economic environment that exhibits a readiness to lend or invest in new and different companies. Four qualities distinguish dynamic cities from all the rest: (1) resilience, (2) creativity, (3) initiative taking, and (4) diversity (Shapero 1984). It is evident that he was identifying the same, difficult to create urban characteristics specified by Andersson (1985).

The relationship between entrepreneurship and technological innovation is particularly important to high technology. The conclusions of the report of the US Office of Technology Assessment (1984: 7) suggest the links which are present: 'The most important conditions for 'home grown' (high technology development) are the technological infrastructure and entrepreneurial network that encourage the creation of indigenous high-technology firms and support their survival.' The elements of this regional technological infrastructure cited include: applied research and product development activities at nearby universities, federal laboratories, and existing firms; informal communication networks that provide access to information and technology transfer from those R&D activities; scientific and technical labour force, including skilled craftsmen, newly trained engineers, and experienced professionals (who also represent a pool of potential entrepreneurs); a network of experts and advisers (often augmented by university faculties) specializing in hardware, software, business development, and venture capital; a network of job shoppers and other suppliers of specialized components, subassemblies, and accessories; and proximity to complementary and competitive enterprises, as well as distributors and customers (Office of Technology Assessment 1984: 7).

Birch (1987: 140–65), describing job creation in the USA, lists five factors which enhance entrepreneurship – factors that are high quality in some places and not in others:

1. Educational resources, particularly higher education;
2. Quality of labour;
3. Quality of government;
4. Telecommunications;
5. Quality of life.

Birch's designation of educational resources is actually much more specific than it appears; he insists that research-based universities, at the leading edge of change, such as MIT and Oxford, are the necessary ingredient. Labour quality refers to skilled and adaptable workers, for whom training and adjustments are less costly, regardless of the wage level. Education at the primary and secondary levels (up to age 18) is often a key dimension of this adaptable workforce. Bartik (1989) found that market demand in a state and education level (proportion of high-school graduates) were among the most significant influences on small business starts. Quality of government refers to the balance between public services and their cost – 'tax-efficient' in Birch's terms. Bartik's (1989) study of small business start-ups in the USA 'indicate that some public services may encourage small business starts' (Bartik 1989: 1015).

Birch recognizes that quality of life is strongly associated with urban size, and enters into locational decisions for high-technology firms and their employees. 'Whatever it is, the quality of life has to do with "interesting pastimes" and an enjoyable atmosphere after work.

Whatever form it takes, "nice, exciting" places have decided advantages when it comes to appealing to executives of innovative firms for these are the areas where entrepreneurs and employees will want to settle and raise their families' (Birch 1987: 147–8).

Is quality of life important in entrepreneurship? Some empirical evidence exists, including that of Czamanski (1981), who found statistically significant effects of quality of life on a city's attraction of individual industries. Pennings (1982) specifically studied entrepreneurial activity, and found that only economic and health/educational dimensions were significant, even when urban population and the size of the local industrial base in the sectors studied (plastics, communication equipment, and electronics) were included; political, environmental, and social dimensions of quality of life were not. Placed in another light, 'a region's ability to attract and retain educated people is as important as its ability to attract firms' (Buss and Vaughan 1987: 445).

Dubini (1989a) recently studied six cities in Italy to determine their environments for entrepreneurship. *Munificent environments* contain:

1. A strong presence of family businesses and role models;
2. A diversified economy in terms of size of companies and industries represented;
3. A rich infrastructure and the availability of skilled resources;
4. A solid financial community; and
5. Presence of government incentives to start a new business.

By contrast, a *sparse environment* is characterized by:

1. Lack of an entrepreneurial culture and values, networks, special organizations or activities aimed at new companies;
2. Lack of a tradition of entrepreneurship and family businesses in the area;
3. Absence of innovative industries;
4. Weak infrastructures, capital markets, few effective government incentives to start a new business.

In short, access to other entrepreneurs, to consultants, and to sources of information is far more readily available in munificent settings than in sparse environments (Dubini 1989a).

The various conditions discussed thus far can also be classified into regional infrastructure, spatial concentration, and quality of life (M Thomas 1988). In Thomas's account, infrastructure includes capital, the economic–industrial base of technologically adept firms whose inputs and skills are transferable to new firms, a quality labour supply, technology networks, transportation and communication systems, and a sociopolitical structure of supporting institutions. Spatial concentration refers to the availability of input–output linkages and the synergy present in agglomerations. Quality of life is a less precise element that appears to operate through labour market choices of professional, especially R&D, workers. These comprise the elusive 'regional factor' which has proved difficult to define exactly (Thomas 1987b). However, local skills and entrepreneurship, unlike scientific knowledge, are more local phenomena, not readily transferable from place to place, and they depend on local economic structure as well as on the attitudes and culture of the region. Recent empirical work has shown that the specification of entrepreneurial environments is complex, not reducing to a few simple variables manipulable by policy (Moyes and Westhead 1990).

Venture capital and entrepreneurship

Venture capital appears in virtually every inventory of 'necessary' conditions for entrepreneurship, and is a high priority for policy initiatives in Western countries (Houttuin 1985; OECD 1985a). Most of our knowledge of the venture capital industry has developed since the mid-1970s, when Bean, Schiffel, and Mogee (1975) reviewed venture capital investments.

None the less, venture capital is also easily misunderstood, because it is not the same as conventional credit or loan funds. Venture capital involves *equity* investments, and profit through capital gains after stock is sold publicly. This equity may be lost entirely if the firm fails; usually, some assets remain. Venture capital also is high-risk, and may well be lost if the venture fails. Thus, there is a real distinction from relatively common public-sector capital programmes, which tend to lend money and to expect a full and regular payback. The contrast is such that purely state capital programmes in the USA are looked on with disdain by the venture capital industry (Wilson 1985).

The formal venture capital industry in the USA plays a role not found elsewhere. Venture capitalists in the USA 'sit at the center of multifaceted networks – which they actively help develop – comprised of financial institutions, large corporations, universities and entrepreneurs, and in doing so, forge important linkages between large and small institutions' (Florida and Kenney 1988b: 120). Thus, venture capital-financed innovation rests somewhere between the traditional dichotomy of entrepreneurial and corporate innovation (Florida and Kenney 1988b). Venture capitalists are financial intermediaries, managing capital provided by pension funds, corporations, insurance companies, philanthropic foundations, foreign sources, and families and individuals (Florida and Kenney 1988b; Y Henderson 1989). Banks are relatively unimportant either as sources of venture capital or as managers of venture funds in the USA. Venture capitalists play a unique role in the growth of new firms and industries and, while not part of Schumpeter's framework, provide a transition between the new, struggling firm and large, established companies in new industries (Kenney 1986).

Venture capitalists diversify their risk by holding stock in a portfolio of new ventures and by 'co-investing' with other firms (Bygrave 1988; Y Henderson 1989). This form of networking is especially prominent among firms in California (Bygrave 1988). Some of the syndicated investments are made to obtain information rather than simply to share risk, because venture capital firms have developed specializations in specific aspects of emerging businesses (Bygrave 1987; Robinson 1987). A further characteristic of US venture capital is substantial control over management decisions, usually via a seat on the board of directors, but including a broad range of involvement in firm management (Gorman and Sahlman 1989; MacMillan, Kulow, and Khoylian 1989; Rosenstein 1988). Investments, or 'deals', are made when profitability and market factors of a firm are attractive; deals are rejected because of poor market potential or inadequate management (Tyebjee and Bruno 1984a). The evaluation of management capabilities seems to be a central element in assessing the risk of new business ventures (Dubini 1989b; Rock 1987; Tyebjee and Bruno 1984b). However, deals are negotiated, and the share of a firm given up for financing can vary greatly (Bruno, Tyebjee, and Anderson 1985; Tyebjee and Bruno 1986).

Flows of risk capital vary greatly from region to region, and many regions of the USA are virtually without venture capital (Florida and Kenney 1988a; Haslett 1984; Leinbach and Amrhein 1987; Premus 1985; Silver 1980) (see Fig. 8.1). Although most venture capital is invested locally, the financial intermediary function of venture capitalists moves money from financial centres, such as New York and Chicago, to technology centres (California, Texas) where investments in new firms are made (Florida and Kenney 1988c). 'Availability' of venture capital in a region may mean simply that funds are gathered from financial partners, but invested through fund managers in other regions. Thus, California has attracted between one-third and one-half of all US venture capital for many years.

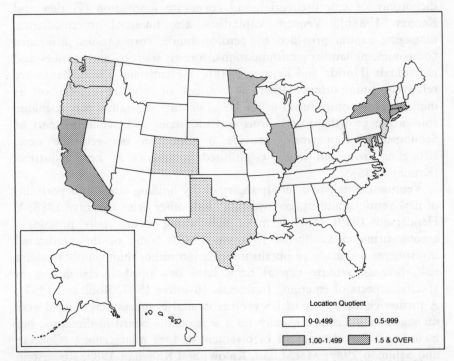

Figure 8.1 Concentration of venture capital investments in the USA, 1969–83.
Source: Calculated from Brody (1985)

In Canada, a similarly high level of concentration is found in Toronto, and secondarily in Calgary and Montreal (McNaughton and Green 1989). As in the USA, Canadian venture capital comes largely from pension funds (Brophy 1988; Macdonald and Perry 1985). About one-seventh of Canadian venture capital makes its way out of the country, mainly to the USA (Brophy 1988: 177).

A local network of financiers, entrepreneurs, and managers is also found, to a lesser degree, around Minneapolis and Boston (Florida and Kenney 1988c). Investment syndication and co-investment '"loosens" but does not eliminate the spatial constraints on venture capital investing' (Florida and Kenney 1988a: 42). In their stated preferences for investments, firms in Boston and San Francisco are the only ones to specialize in start-up capital, the most risky investments. The working of the local network of people can keep available capital local; without such an 'entrepreneurial climate', there is little to prevent it from leaving the area for more attractive areas.

Rural areas are at a severe disadvantage in terms of venture capital (Ewers and Wettmann 1980). In a study of new (less than four years old) small businesses in Wisconsin, Shaffer and Pulver (1985) focused on variations in the capital structure of rural and urban firms and those in the northern and the more industrialized southern part of the state. Firms in the northern region and those in rural locations reported an inability to find capital within 30 miles; they were also classified as more likely to be experiencing 'capital stress', defined as any of a variety of conditions related to insufficient capital.

Shapero and his students have long emphasized the significance of the local financial community on the potential for local entrepreneurship (Shapero 1971; Hoffman 1972). The appearance of the first, 'almost random' company formations in an area are the most difficult to account for, and may not always stimulate any follow-up of further entrepreneurship. There were substantial differences among localities in the degree to which local banks were willing to lend for new, untried ventures. If an area attains a level of sustained entrepreneurship, it is typically associated with the growth of a financial, legal, and service community to support it. Reid and Jacobsen (1988) confirm the central role banks play in business information networks in Scotland.

The source of initial capital and of the 'almost random' appearance of the first entrepreneurs is perhaps related simply to the availability of start-up financing. Financing for start-up firms can come from many sources, including personal funds, family and friends, local bankers, and outside lenders, but it tends to come from informal (non-institutional) sources that operate almost entirely via a network of personal, and *local* contacts (Bean, Schiffel, and Mogee 1975; Gaston and Bell 1987; Wetzel 1983, 1986). Prior to 1958, individuals were the primary source of small business financing (Florida and Kenney 1988c; Y Henderson 1989). The prospective entrepreneur will likely go to a lender with a reputation for backing new, unproven entrepreneurs and their start-up firms. Local bankers are unlikely to lend in such a risky situation – unless they have previously had successful experiences with firms of this type. In any event, banks are unlikely to make more than a small portion of their loans to new ventures, preferring to lend to firms which have already survived the start-up phase. However, once an area becomes known for its spin-off activity – in part a result of the willingness of local lenders – venture capital firms from other places may set up shop in order to profit from entrepreneurs in the area. This is certainly the experience of the Silicon Valley area, the premier technology-oriented venture capital complex (Florida and Kenney 1988c; Hambrecht 1984; Wilson 1985). Capital markets remain substantially local, despite the long-distance flows identified in research on venture capital (Rogers, Shaffer, and Pulver 1988).

Although there are few solid data on informal investors, there are a large number of them in the USA, providing small investments for new ventures (Haar, Starr, and MacMillan 1988). Collectively, they may provide more venture capital than that made available through the formal market, invest earlier in the life of new firms, and have longer 'exit horizons' than professional venture capital firms (Freear and Wetzel 1988; Tymes and Krasner 1983; Wetzel 1987). They may co-invest, but nearly always with other individuals (Aram 1989). In addition to their role as 'angels' in financing new firms, previous entrepreneurs act as role models and share a variety of routine business knowledge that is necessary to operate a business, such as regulations, taxes, accounting, suppliers, customers, and distribution. Some of the advice from previous entrepreneurs – hardly available from any other source – concerns mistakes made and lessons learned (Egge and Simer 1988).

Policies to increase venture capital

Policies to improve the availability of venture capital have been based on equilibrium concepts and assumptions that diffusion and market forces will cause a greater overall supply of capital to result in widespread benefits. Thus, they tend to ignore the human elements concerning networks of personal contacts, and historically rooted patterns of decision-making (Thompson 1989b). Public venture capital has been available in the USA for several years through the Small Business Investment Company (SBIC) and Minority Enterprise Small Business Investment Company (MESBIC) programmes. Although often classified as venture capital, capital available through such programmes tends to be debt or debt-like financing rather than the equity investments of conventional venture capital (Green and McNaughton 1989: 202). For the most part, the pattern of public venture capital is similar to that for private funds: Boston and San Francisco – two regions with an abundance of private venture capital – attract SBIC funds from other locations to invest in high-technology firms (McNaughton and Green 1989). These geographical and sectoral patterns of investment are intentional, and are stated as preferences of both public and private venture capital firms (Green 1989). Thus, federal government programmes have done little to alter the geographical disparities of capital availability for small business.

At least 20 states in the USA have created venture capital funds, in response to the overwhelming geographical concentration of investments by the private venture capital industry, and the perception that venture capital has indeed been a major reason for the growth of some regions (Ioannou 1985; Premus 1985). The objective of state venture funds is to favour firms within the state, because these firms might be unable to secure funding in the private sector. However, few of these public funds as yet have any track record (Fisher 1988). They may or may not exactly qualify as venture (i.e. high-risk) capital, but such programmes are based on an awareness of the capital needs of small business (Fisher 1988; Plastrik 1989).

Venture capitalists argue that government forays into venture investments are done by people with little knowledge or experience in venture capital (Wilson 1985). In response to this opinion, state venture capital funds in Michigan and Illinois are managed by private professional venture capitalists. Many state employee pension funds, such as in Ohio, Michigan, and Washington, now invest a small (1–5 per cent) portion of their funds in risky investments, often with, or as in Illinois, exclusively through private venture capital firms. In other cases, states match privately raised capital (Robertson and Allen 1986). Tax credits for a portion of individual investments in Indiana's venture capital pool are intended to create a larger supply of private funds. It is too early to know whether these attempts to alter the availability of venture capital will affect the patterns of new firms, but it is clear that they are addressing directly a major geographic inequality in the USA.

Venture capital in Europe

As recently as the late 1970s, venture capital was hard to come by throughout most of Europe. Private capital for high-risk investments, the mainstay of US venture capital, was particularly scarce (OECD 1986b). Tremendous changes have taken place to increase the pool of venture capital during the 1980s throughout Europe. The UK has been the leader, becoming entrepreneurial a bit earlier than other countries in Europe (OECD 1986b: 34–9). In 1984, the UK accounted for 75 per cent of all venture capital in Europe. One-quarter of the British total came from pension funds, an uncommon source of funds in Europe

(Tyebjee and Vickery 1988). However, growth took place rapidly in a large number of other countries, and the UK accounted for only about 40 per cent of the total venture capital pool in Europe in 1987 (Table 8.1).

Table 8.1 Estimated size of venture capital industries in selected countries

	Number of venture capital firms	Total venture capital pool ($m)
UK	110	4 500
France	45	750
Netherlands	40	650
West Germany	25	500
Denmark	14	120
Ireland	10	100
Sweden	31	325
Austria	11	50
Norway	35	185
Canada	44	1 000
Japan	70	850
USA	550	20 000

Source: Financial Times 29 April 1987; cited in Dodgson and Rothwell (1988: 236 (Table 3)).

Throughout Europe, a number of policy schemes, such as the Business Expansion Scheme in the UK, have greatly increased the availability of capital to small firms (Harrison and Mason 1988; OECD 1985a: 27–40; Rothwell 1985). Private capital has also expanded, but British investors, in contrast to those in the USA, have been less likely to invest in start-up firms or in those without a firm footing. Moreover, a large amount of British private venture capital (32 million in 1983) left the UK for American investments (Rothwell 1985). Banks are much more prominent in venture capital in Europe than in the USA, a pattern more similar to that in Japan (Jéquier and Hu 1989; Tyebjee and Vickery 1988). In Italy, West Germany, and Denmark, banks are the predominant source of capital (48%). Conversely, government agencies provide over 70 per cent of venture capital in Belgium and Luxemburg, Spain, and Portugal (Ooghe, Bekaert, and van den Bossche 1989).

An expected result of the growth of venture capital is its spatial concentration. In the UK, areas where venture capital firms locate and where 'good' investments are found coincide in the South-east region surrounding London (Martin 1989). This pattern merely reinforces the shortcomings of regions such as the North-east, which has failed to benefit from policies which focus on innovation and high technology (Amin and Pywell 1989).

A further policy initiative that has increased capital available to small businesses is the creation of an unlisted securities market, which permits small firms to sell equity under separate regulations from larger firms. In both the UK and Sweden, this is considered to have greatly expanded the pool of investment capital available (Dodgson and Rothwell 1988; Olofsson and Wahlbin 1984). Outside Europe, such as in Australia and Japan, little private capital is made available for venture capital (Wan 1989; Ooghe, Bekaert, and van den Bossche 1989; Smith and Ayukawa 1989).

The entrepreneurial climate

What is clear is that capital availability is not so much an institutional issue as a human issue. Networks of people who communicate with each other about business opportunities, procedures, and management are key elements of a positive regional entrepreneurial climate. These networks are important sources of key information. Knowledge about venture capital and its sources is less common in the USA, for example, among small banks and in rural areas, when compared to banks in metropolitan areas (Pulver and Hustedde 1988).

Capital is actually less critical to small business success than more simple, but less easily supplied, inputs. Capital is secondary to information, business knowledge, and management expertise – often of a rather simple nature. Technical education – available to more people than a university education – may do more to promote small firms than capital (Brusco 1989). With the provision of knowledge, human capital, invested as work instead of leisure, may have a more widespread impact (Brusco 1989; Cooke and Imrie 1989; Vartiainen 1988).

The *entrepreneurial climate* of a place is probably the greatest single variable influencing entrepreneurship as an ongoing part of local life. The climate 'includes the whole system of *values* in a society': attitudes towards science, towards economic and social change, towards private enterprise, and towards risk (Piatier 1984: 160). This climate as it affects entrepreneurs relies almost entirely on a well-connected network of investors, especially informal investors, previous entrepreneurs, and an aura of non-routine, innovative activity (Buss and Vaughan 1987; Dubini 1989a; Gruenstein 1984; Jarillo 1989; Miller and Coté 1985; Senker 1985; Shapero 1984; Sweeney 1985, 1987). The contrast between Boston and Philadelphia identified by Deutermann (1966) 25 years ago remains in force: Philadelphia is still an urban area in which venture capital is scarce. The effect of capital shortages on a region is reduced entrepreneurial activity. Some entrepreneurs may leave to set up new firms elsewhere; at best, a few may be able to get capital from investors in other cities (Freedman 1983).

The attitudes in the community as a whole are as important as solving the capital and other managerial problems of small businesses (Mokry 1988). Schell (1983) attributes much of the geographical variation in entrepreneurial activity to the developmental climate promoted by the power élite in a community. Schell suggests that some community and regional variables not related to city size, such as occupational and industrial differentiation, median income, education levels, and net migration, are related to entrepreneurship. He contends that one can group favourable government policy, receptive population, and attractive living conditions into a broader *sociopolitical* variable, which has a major influence on entrepreneurial activity in a community or urban area. This has obvious parallels with the 'cultural' influence on entrepreneurship discussed earlier in the chapter.

The influence of entrepreneurial 'role models' and attitudes in a region is especially important. Schell (1983) compared three urban areas in North Carolina and concluded that the élite in some urban areas failed to promote industrial development or to encourage local entrepreneurship. Freedman (1983) suggests similar problems in Philadelphia. The shortage of venture capital in North Carolina may be related to the fact that there are few local role models of entrepreneurship (Schell 1984). Prior entrepreneurs provide not only information and experience, but also capital which they are often willing to invest in the new firms of others. How to generate the first 'almost random' entrepreneur remains poorly understood.

Incubators and spin-offs

The 'incubator' hypothesis suggests that some locations, especially cities, serve to 'incubate' new firms, which may subsequently move to other locations in the urban region. Stretching the life cycle analogy to the 'birth' of firms, some of them may need extra attention during the first few fragile years of life. The attraction of current small business incubators is based on this assumption (Allen 1985). Davelaar and Nijkamp (1987, 1988) have reviewed the available literature and conclude that previous work is fairly ambiguous with respect to the hypothesis. It is also clear that the interpretation of most previous work refers only to a single urban area, rendering the findings irrelevant if one is interested in the wider regional scale most applicable for national policy. For example, Rees (1979) has suggested that the accumulation of a critical mass of production in an industry serves as an incubator for new firm generation. Peripheral regions, in turn, prove to have the weakest incubator potential in a study of 18 European regions (Nijkamp, van der Mark, and Alsters 1988).

Spin-offs are most common in large urban areas. It is in such places that a sufficient number of potential entrepreneurs are present, as well as the other 'environmental' factors that encourage entrepreneurship (Pennings 1982). Shapero (1971) studied 141 counties in the USA (of 300 000 population or less) which had technical companies formed between 1940 and 1968. The variables most highly correlated with firm formation tended to be those related to city size and agglomeration, such as manufacturing employment, educational expenditures, and income. More recent research by Armington, Harris, and Odle (1984) on small high-technology firms in the USA confirmed that they tend to form at higher rates in larger urban areas (Harris 1986).

A more common context for incubators of entrepreneurs is the spin-off of high-tech firms from R&D activities. To a degree unlike any other type of economic activity, R&D incorporates the potential for individuals to set up shop independently in a venture with high growth potential. The electronics and computer industries provide numerous examples of entrepreneurs literally becoming millionaires in their own firms. Examples such as these are what make high-tech industries so attractive to regional and local economic developers. In high-tech firms, the skilled professionals carry with them not only technical knowledge but also a variety of information and contacts that make entrepreneurship possible.

Technical knowledge is typically considered more important to such firms, thus prompting the conclusion that proximity to universities should be an especially important locational factor. However, universities have not been prominent sources of spin-offs, despite the examples of MIT and Stanford so frequently cited (Cooper 1972; Roberts and Peters 1983; Shapero 1972). Instead, universities provide a necessary resource to an area – technical personnel – as well as a pool of well-educated, potential entrepreneurs (Keeble and Kelly 1986; Segal 1986).

Rogers (1986) provides a good example of the reliance currently placed on research universities by policies hoping to promote the spin-off of high-tech companies. He reviews, largely by a series of anecdotes, the research university as the key institution around which high-tech growth occurs, relying on three principal examples: Silicon Valley, Route 128, and Research Triangle (the latter despite a lack of spin-offs). 'One seed for starting local entrepreneurial activities is to invest in improving a nearby research university, especially in such academic departments as electrical engineering, computer science, and molecular biology' (Rogers 1986: 177–8). Despite his optimistic tone, Rogers implicitly recognizes the importance of agglomeration: 'One trend is apparent; corporate money goes to the academic haves, not the have-nots' (Rogers 1986: 177).

Rogers (1986) acknowledges that the presence of an outstanding research university in a locale does not necessarily cause the development of a high-technology centre. Evidence on this point is provided by such excellent universities as Harvard, Columbia, Chicago, Berkeley, and Cal Tech, none of which has played an incubator role. The presence of a local research university is simply not enough to offset shortcomings in entrepreneurial climate or venture capital.

Generally, a university is but one of the factors bearing on entrepreneurship, especially in small urban areas. Only a small number of faculty found new firms, and they face the same problems as firms originating from other incubator organizations (Allen and Bird 1987; Olofsson and Wahlbin 1984). Myers and Hobbs (1985) studied nearly 300 high-technology firms located in university towns located more than 50 miles from the nearest major metropolitan area). For most, but especially the newer firms, the advantages of their locations suggest only indirect effects of a university: availability of technical personnel, presence of bankers knowledgeable about technology companies, and accessibility of airports and one-day package delivery services. Contacts between universities and *small* local firms are especially scarce, suggesting that policies to promote local technological growth through universities may well favour large firms (Bishop 1988; Corsten 1987; Rothwell and Zegveld 1982).

Cooper (1973, 1985, 1986) has persistently made the case that firms, not universities or government facilities, tend to be the 'incubator organizations' of entrepreneurs. This is largely true even in high-technology sectors, in both West Germany and the USA (Cooper 1985; Feeser and Willard 1988; Kulicke and Krupp 1987). Such firms are typically small (fewer than 100 employees), but a significant number of founders do come from large firms (Cooper, Dunkelberg, and Furuta 1985; Klaudt 1987). It is through the location of branch plants that quality of life may operate, according to Cooper, by attracting operations that employ large numbers of technical workers. These would have to be 'technical branch plants' of the sort found in Scotland or in parts of the US sunbelt (Cross 1981; Glasmeier 1988a). In *some* of these branch plant locations, then, entrepreneurship begins, and few, if any, entrepreneurs change location at the time of founding. Shapero (1972: 33) believed that 'more companies are generated out of small businesses than out of large ones', a view at odds with the work of Cooper. However, Shapero's point seems to be that branch plants are unlikely sources of entrepreneurs, a point made by several researchers on the drawbacks of branch plants as development vehicles (Watts 1981).

The location decisions of large firms, especially for R&D facilities, are closely linked to locational variations in entrepreneurship (Ch. 6). R&D workers belong to the social group likely to have the knowledge, contacts, and information conducive to spin-off from other firms (Massey 1984: 300–1). In addition, large firms tend to locate in urban environments which are also favourable for entrepreneurship. The locational needs of R&D favour large urban regions, for reasons related to agglomeration economies. For the professional workers, however, large cities can also be differentiated on the basis of their potential for entrepreneurship. Since the spin-off process is relatively common in several sectors, certain places acquire reputations among high-tech workers on the basis of their entrepreneurial climates.

Perhaps primary among the contacts that help to cultivate spin-offs are those with sources of financing for the new business. The success of previous entrepreneurs and the reputation of an area for venture capital add to the attractiveness of existing R&D concentrations for scientists and engineers, and especially for the more entrepreneurially minded. The cumulative flow of venture capital is heavily concentrated in a small number of places. Obviously not all new spin-offs will be successful, even in the hotbeds of high-tech activity. The presence of earlier local successes, however, greatly reduces the complete uncertainty and risk inherent in spin-off both for entrepreneurs and for investors. A place which has a reputation for entrepreneurial success will be relatively more attractive to the mobile professionals on whom large firms rely (Senker 1985). Table 8.2 illustrates the different sorts of factors that might be thought to dominate the location decision for branch plants, for R&D facilities, and for start-up firms. Although R&D facilities rely more on professional labour than do manufacturing plants, their dependence on transportation accessibility is rather traditional, and results in a tendency towards agglomeration (Ch. 6). New firms require a very different set of local characteristics, especially capital, which develop in some places from historical and cultural causes and not in others.

Table 8.2 Location factors influencing manufacturing plants, R&D facilities, and new start-up firms

Manufacturing plants
Labour availability and costs
Highway and rail transportation
Accessibility of raw materials
Business climate

R&D facilities
Availability of professional talent
Air transportation
Quality of life
Highway transportation

New start-up firms
Seed capital
Entrepreneurial climate
Venture capital
Incubator space
Bank financing
Space for expansion

Sources: Browning (1980); Gruenstein (1984).

Just as not all R&D leads to successful innovations, not all R&D generates new spin-off firms. The state of the local industry's technology must be sufficiently unstandardized, preferably with multiple market niches, and the barriers to entry by new firms must be low (Bollinger, Hope, and Utterback 1983; Garvin 1983). Even so, it would seem that the European experience with branch plants and with public sector R&D in peripheral regions has led to very low levels of entrepreneurship (Cooke 1985b; Sweeney 1987).

Among the factors inhibiting the indigenous potential of peripheral regions, most attention has focused on the dependence of such regions on branch plants of large, multinational corporations. The evidence is clear that such plants tend to have lower levels of highly trained and skilled personnel, certainly lower levels of R&D, and little attention to non-routine activities or new products (CURDS 1979; Goddard and Thwaites 1986; Miller and Coté 1987; Sweeney 1987; Watts 1981). Such a regional structure is unlikely to be innovative, especially in a setting of short product cycles, and it is unlikely that new firms will emerge. Unless a local economy meets some fairly large threshold size, its base of potential entrepreneurs – and the likelihood that they as a group will be able to come up with successive rounds of innovations as the product cycle progresses – will be inadequate to compete with other regions (Martin 1986).

Policies to promote entrepreneurship

From a policy viewpoint, it is easier to prevent a firm from failing than to persuade someone to start a new firm (Wever 1986). Policies may also be more helpful in facilitating the growth of small firms than in creating them (Utterback *et al.* 1988). Inequalities will be difficult to eliminate, judging from the findings of recent research.

> In munificent environments, government funding is likely to scarcely enhance the diffusion of entrepreneurship, as the environment itself is stimulating and offers a wide choice of opportunities. . . . In sparse environments, the presence of money directly available to entrepreneurs – which is the typical form blanket policies assume – is likely to result in a fruitless effort, since there is no infrastructure to support entrepreneurial effort (Dubini 1989a: 25).

After the first 'almost random' firm is started, the Italian example of co-operation and sustained entrepreneurship will be perhaps even more difficult to generate through policy design (Piore and Sabel 1983; Rosenfeld 1989–90).

In spite of challenging competition from other regions, several types of policies have been implemented by nations or by local governments in recent years in an attempt to promote new, usually high-tech, firms, often in conjunction with a local university. The most common vehicles include: venture capital (discussed previously) and business incubators, often combined with science parks. Other policies address the technological needs of small business by subsidizing R&D activities

(Rothwell and Zegveld 1985: 180–2). The latter fall largely within the scope of science and technology (S&T) policy (Ch. 7).

Incubator facilities for new firms

Incubator facilities, as implied by their name, address the specific problems faced by new firms. Because new firm spin-offs are an expected outcome of high technology, the question of how to raise their life expectancy has been used mainly in connection with high tech. Incubators range from inner-city buildings to communal facilities in new science parks (Smilor and Gill 1986). Most incubators are a product of local, rather than of regional or national, policy, and many are available to virtually any new firm, rather than only to high-tech firms. Those intended exclusively for high-tech firms are often called innovation centres (Allesch 1985).

Incubator facilities feature shared-tenant services, such as word processing, a copy centre, receptionist, conference rooms, and personal computer rental. The range of services and facilities can be quite extensive. A study of incubator facilities in Pennsylvania found that at least half of the 12 incubators studied provided in-house consulting services and management assistance on government regulation and procurement processes, preparation of business plans, and relocation planning (Allen and Levine 1986). Pennsylvania had 30 incubators, the most of any US state (Osborne 1988: 56).

The attraction of such shared facilities to new firms is that they can significantly reduce the initial capital needed. The drain on a young firm's finances can thereby be significantly reduced, whether by subsidized rent or by the landlord (often a major university) taking an equity stake (30% is common) in new companies in lieu of rent or fees (Andrews 1986). This is the case in several US cities, including Minneapolis, Milwaukee, Salt Lake City, and Atlanta. A similar approach is taken by Rensselaer Polytechnic Institute in Troy, New York, where every tenant must give up 2 per cent of its equity in addition to paying rent.

The goal of incubators is to 'nurture' firms and allow them to expand in size and employment so that they can outgrow the incubator and relocate elsewhere in the local area. Some business incubators require periodic reviews of each firm by the centre's board, comprised of local entrepreneurs and business representatives to identify problems and to determine when each will be able to leave. This management-consulting function of incubators goes a long way towards improving the entrepreneurial environment of places which might otherwise be deficient (Dubini 1989a; Krist 1985; O'Connor 1985).

Outside the USA, university science parks as innovation centres have proliferated (Gibb 1985). In the UK, two such parks were established in the early 1970s, at Cambridge University and at Heriot-Watt University in Edinburgh. They were not imitated for over a decade, but by 1987, 36 science parks had been established with the participation of universities or polytechnics (Monck *et al.* 1988: 70–97). The conclusion is that they have done well as incubators for new technology-based firms, and that informal relationships with academics have increased and become more important than firms first anticipated (Monck *et al.* 1988: 244–5).

Elsewhere, Chalmers University, in Gothenburg, Sweden, has been a source of 98 spin-off firms, supported by several courses tailored to young, innovative companies (McQueen and Wallmark 1985). The similarly successful experience of Twente, in the eastern part of the Netherlands, has been aided by the network of businesses and consultancies which provide advice on international markets (van Tilburg 1985). It is clear that science parks are now an accepted vehicle for cultivating and encouraging small firms in most advanced countries, as well as in NICs (Ch. 7). They are able to reflect local specializations of their academic institutions as well as in many cases local policy initiatives.

Comprehensive local development

Pennsylvania's Ben Franklin Partnership combines conventional strengthening of the state's public-sector universities with decentralized technology centres that link with private firms. The goals of the programme are explicitly high tech and entrepreneurial and, unlike the programmes of some states, the Ben Franklin Partnership has been able to build from an already strong base of university research and industry R&D. It is focused around four advanced technology centres, distributed throughout the state, each centred on one or more local universities, such as the University of Pittsburgh and Carnegie-Mellon University in south-western Pennsylvania and Lehigh University in the north-eastern part of the state. Ohio's Thomas Alva Edison Program and New York's centres for advanced technology are similar to Pennsylvania's, with their designated technology centres having specific areas of technology on which they concentrate. But the Ben Franklin Partnership 'is probably the most comprehensive economic development institution in the country' (Osborne 1988: 56).

Pennsylvania's state research funds must attract private funding and must involve at least equal financial participation by non-state sources. The first $28 million in state funds was matched by $84 million from private and philanthropic sources (Robertson and Allen 1986). The visibility and potential raised by these centres have helped to improve venture capital availability in Pennsylvania, especially in the Pittsburgh and Philadelphia areas, by providing leadership and seed funding (Singerman 1989).

The prospects for policy in advanced countries

Regional and local policies will not be able to 'create' entrepreneurs. Entrepreneurship is largely a product of the local mix of industries, firms, and plants and of the socio-economic environment created by the local population (Malecki and Nijkamp 1988). It can, over time, respond to the results of industrial recruitment and of efforts to upgrade the skill and management levels in a region. Incubators and venture capital pools can encourage entrepreneurs to stay in the area rather than to seek more favourable local conditions.

Policy initiatives alone will not create networks of entrepreneurs and investors. Such networks are a uniquely personal and often a local phenomenon, and this critical element of a place's entrepreneurial environment is perhaps the least amenable to any public policy (Committee for Economic Development 1986; Miller and Coté 1987).

Nor can policies do much to speed up the process of entrepreneurial dynamism, which relies on the formation of informal networks, service activities, and human capital (de Jong 1987). In large part the obstacle is the necessity for a critical mass in a region (Malecki and Nijkamp 1988). Small regions, even those with a high-tech base, will have a harder time attracting professional workers than will large urban areas. The synergy of amenities, accessibility, and agglomeration factors found in large urban regions cannot really be substituted for in smaller regions (Andersson 1985; Martin 1986; Miller and Coté 1985). Although Norris (1985: 98) suggests that, in the US context, multi-state regional policies would be able to create 'a critical mass that no single state can achieve alone', this is unlikely to work on such a large geographic scale. Urban agglomeration is the critical ingredient. The suggestion has not yet really been tried, but it would seem to encompass regions too large to affect local environments for entrepreneurship (OECD 1987; Sweeney 1987).

On the other hand, regional and local policy efforts can facilitate the entrepreneurial climate in a region or locality. This can be done by bringing entrepreneurs together in incubator facilities, and by integrating the local university and technical institutes as agents of technology and change into the fabric of civic and economic life. It must be remembered that entrepreneurship is very much a local phenomenon (Buss and Vaughan 1987; Osborne 1988; Sweeney 1985).

Existing concentrations of R&D attract additional R&D, reinforcing the pool of professional and technical workers – and potential entrepreneurs. Entrepreneurs can emerge outside the context of technological innovation, but some elements in common with innovation must be present. Information, contact networks, and technical progressiveness are standard in areas where entrepreneurship is common. Simple and short-term solutions are not likely to work in a situation where contacts and progressiveness either are, or are not, part of the local culture (Sweeney 1987). In turn, successful regions (and nations) will be those where workforce skills are associated with all stages of production, not simply R&D (Maillat and Vasserot 1988). For this purpose – and because policies cannot really 'create' entrepreneurs – it is especially important to address the information and innovation needs of small firms.

A policy framework to address the problem of competitiveness needs to deal with the information about best-practice technology and products. Although many small firms are innovative, they typically undertake no R&D and are heavily reliant on their external networks for information. That these networks are most dense in urban agglomerations is not surprising, but information can be enhanced in less munificent environments as well. In particular, the provision of intermediaries or liaisons who are technically proficient can greatly raise the level of knowledge in a region (Britton 1989a; Kelley and Brooks 1989; Sweeney 1987: 239–62). Positive experience with such a liaison approach is available from Portugal and the UK (Andrade 1989; Britton 1989a, b), and has been proposed elsewhere (Malecki 1988). In most countries, the information needs of small firms were long neglected in favour of R&D policies which in theory provide technological information from which all firms benefit (Britton 1989b).

What incubators do on a small scale, larger policies can also do to help small firms. More than anything else, firms need a continual flow of information on the state of markets, the relative merits of new machinery, on access to trade fairs, and on the changing standards for products destined for export markets. Government centres for this sort of information would fill a gap which producers' associations are often unable to do because of expense and problems in co-ordination (Brusco 1989: 267–8).

It is less clear that policies can facilitate the entrepreneurial climate in a significant way, especially at the local scale (Dubini 1989a; Mokry 1988). Recent research suggests that policies have been beneficial for nations and larger regions, such as the two most prominent high-tech regions, California and Massachusetts (Committee for Economic Development 1986). The interaction between public and private sector in a locale, indeed, appears to be of considerable importance in creating and maintaining contacts, and in creating a supportive milieu for new firms (Donckels 1989; Fosler 1988; Miller and Coté 1987). This interaction must be localized, but may be similar both in new and restructuring regions, such as Pittsburgh (Sauer, Currid, and Schwab 1988) and Austin, Texas (Smilor, Kozmetsky, and Gibson 1988b). The emphasis of current local policy is to structure *local* growth, 'from below' with minimal central direction (Albrechts and Swyngedouw 1989; Vazquez-Barquero 1987).

Finally, much of the variation found in entrepreneurship is a direct product of labour market processes. To a large extent, the professional labour market for high-technology workers, for example, determines that only some places will be R&D or high-tech locations. Other labour markets provide experience and skills helpful – and sometimes critical – to new firm formation, as O'Farrell's research has shown.

The indigenous potential of regions, identified during the 1970s (Ch. 3), is a concept similar to that of entrepreneurial climate (CURDS 1979; Ewers and Wettmann 1980; Sweeney 1987). Relatively little serious effort has gone into attempting to identify the indigenous potential of regions and their ability to generate new firms. Davelaar and Nijkamp (1989b), in a major study, found that indigenous potential in the Netherlands did not correspond closely to geographical setting. In particular, firms in remote areas were not less innovative than those in more 'favourable' environments. On the whole, the internal R&D of firms and the external network of R&D in the region are more important than the regional setting. Davelaar and Nijkamp (1989a) suggest that firms in urbanized areas rely more upon the external R&D environment, while remote firms must depend on their own internal efforts (Dubini 1989a; Jarillo 1989).

Overall, there is a strong tendency for entrepreneurship to be strongest precisely in those regions which need it least, suggesting that to rely on new and small firms will not eliminate regional economic differentials (Keeble and Kelly 1986; Storey and Johnson 1987). The shortcomings are greatest in rural regions, where urban characteristics are normally lacking by definition, and in peripheral regions, located outside a nation's technological heartland. The constraints on innovation and entrepreneurship are great but not equal in all such places (Malecki 1988; Reid 1988). Financial problems in regions without capital outnumber problems related to workforce or markets (Popovich and Buss 1989). In part, this is because of the networks of firms in place – even in rural areas – to resolve marketing problems (Young and Francis 1989).

The emergence of new firms depends most 'on community and business leaders who are informed about current practices and are willing to help newcomers in the commercialization and improvement of crafts, skills, and industries in the local economy. This support will not emerge in all communities' (Malecki 1988: 23). If tailored to local conditions, policies will have a greater effect (Dubini 1989a; Johannisson 1987; Kirby 1985).

Entrepreneurship in the Third World

Entrepreneurship is not confined to advanced economies. It flourishes to a greater or lesser degree virtually everywhere, and 'accounts for a very large number of jobs in the Third World' (Norcliffe 1985: 271). However, the dominance of modern state-owned businesses and multinational corporations has constrained small (informal) businesses to a second-class status, if not to small, local markets (de Soto 1989; Harper 1984). An intermediate indigenous form of industrial organization, the 'group', also operates in the modern sector (Leff 1979a, b). In fact, then, the dualism outlined in Chapter 1 may be less accurate than a dependent relationship which always favours the formal sector over small, informal enterprises. This overriding context challenges Third World entrepreneurs far more than those in advanced economies (Broehl 1982). The resilience of small-scale producers, who use labour-intensive, simple technologies and local inputs, is behind efforts to upgrade the indigenous technological potential of such enterprises (Fransman and King 1984; Norcliffe 1985).

As Chapter 4 emphasized, considerable innovation takes place within small, indigenous businesses in the Third World. However, the barriers to competitive enterprise on a world scale are great indeed (Harper 1984). Some firms have succeeded, even in high-tech industries, such as a firm producing computer network boards in remote Iraputo, Mexico (Moffett 1990). Outside high technology, education is less essential to success than information and market opportunity. Experience overseas and exposure to the modern practices of operating a business distinguished entrepreneurs in Tonga and Western Samoa who succeeded from those who did not (Croulet 1988: 86; Ritterbush 1988: 154). The importance of such experience is not unique to the Third World, of course.

Formal education contributes directly to basic business skills, such as literacy, numeracy, record-keeping, and familiarity with telephones (Fairbairn 1988b; Freeman and Norcliffe 1985: 89–91). However, almost 50 per cent of all rural households in Kenya engaged in rural industry. Entrepreneurship is easier and more likely to be successful where role models and others with appropriate training and experience are accessible (Harper 1984: 66–104). Deep-rooted cultural norms, by contrast, have plagued indigenous businesses in the Marshall Islands, where there have been few private-sector role models and where reinvesting profits in the business is rare compared to simply spending earnings on personal consumption (Carroll 1988).

Elkan (1988) observes that industrial entrepreneurs in the Third World typically come from four sources:

1. People who have moved up from the informal sector, often in the same industrial sectors, such as metalworking, tailoring, and furniture making;
2. Former employees of large expatriate or Asian-run businesses in the same industry;
3. People who started out as traders or merchants;
4. Well-educated politicians and senior civil servants who have become part-time entrepreneurs.

The urban informal sector, like that in rural areas, is often a family operation, much like that of industrial districts (Norcliffe 1985; Peattie 1980). Customized production is commonplace in Kenya, and provide ample opportunity for small-scale producers to serve the needs of households in five primary areas: food, clothing, shoes, kitchen utensils (pots and pans), and furniture (beds and tables). Of these, the largest industrial sectors of activity in rural Kenya are in brewing, charcoal making, baskets and mats, and weaving, spinning, knitting, and dyeing (Freeman and Norcliffe 1985: 42–51).

Like businesses everywhere, raising capital is the principal problem of those starting an enterprise in the Third World (76% of those in Kenya) (Fairbairn 1988a; Freeman and Norcliffe 1985: 97). Most capital derives from family and friends, as opposed to the formal financial sector, but the latter is used to a greater extent by 'progressive' entrepreneurs in Pakistan (Altaf 1988: 196–202). A wide variety of informal and ingenious sources of capital is also used, often linked to supply, machinery, or customers (Harper 1984: 47).

Information sources in the Third World are identical to those in advanced economies. The Pakistani sports equipment industry involves the experience of players as users and testers of improvements in equipment (Altaf 1988: 210). As in advanced economies, networks of business people are central to success. As Aldrich, Reese, and Dubini (1989) found in Italy and the USA, women entrepreneurs in the South Pacific also tend to form networks primarily with other women (Ritterbush and Pearson 1988). In general, communication and information build links which develop into markets (Harper 1984: 134). These links can be developed by multinational firms, often creating both local suppliers and customers (Harper 1989). Once established,

> there certainly is no substitute to learning from the operative market system . . . the irresistible conclusion being that entrepreneur competitive strength and efficiency from the market is reflected in price, quality, and innovative technology. To survive the entrepreneur needs to be up and about, the information systems utilised being more informal, more personalised. . . . To stay ahead they need to achieve improvements in products (Altaf 1988: 165).

The humble clothing (or garment) industry has provided opportunities for entrepreneurs in Bangladesh and Togo to be trained in the ways of world-class business. International markets are large but not entirely open, and shortages of skills and marketing networks persistently crop up (Crook 1989: 45–7; Maren 1988; Rhee 1990).

The policy environment for entrepreneurship varies from country to country, and varies even among those with a common colonial past (Dana 1987). Indeed, it might not be harmful simply to leave the informal sector alone (Fall 1989). Removal of restrictions on the private sector and privatization of public sector enterprises can create opportunities for entrepreneurs, especially as new firms emerge to serve the needs of, and form linkages with, thriving firms (Crook 1989: 45–7; Elkan 1988; Harper 1984). Perhaps the most important improvement for entrepreneurs is that access to credit could be improved, especially by expanding the sources of finance beyond banks (Fairbairn 1988a;

Harper 1984: 60–2). In other cases, it is not that credit is not available, but the interest rates charged to informal producers are often in excess of 100 per cent per year. At the same time, it would be too easy for governments to eliminate such sources of credit without replacing them with useful substitutes (Fall 1989). Alternatively, lifting interest rate ceilings in government financial institutions might create a flow of savings to be made into loans (Elkan 1988: 180–1). The counterpart to small business incubators, industrial estates for small-scale producers, works against informal businesses. The latter may operate production and trade in the same location, and rely on a very localized network of customers and linkages. To banish them to locations on the urban periphery is rarely a positive move (Elkan 1988; Harper 1984: 152–65).

Conclusion

Entrepreneurial development is most likely to be successful in large urban regions, especially those where innovativeness and creativity, an entrepreneurial 'climate', and commercial opportunities are relatively abundant. Other places, if they can attain some – even if not all – of the attributes of the large urban areas, may well be able to promote significant numbers of entrepreneurs. The recent focus on entrepreneurship has had some beneficial effects on regional and local policy, by prompting a more long-term perspective about economic development. It has also shown the significant advantages to be gained from investments in human capital, especially through education and training.

The prospects for rural and peripheral areas, as Martin (1986) has stressed, are less optimistic and likely to change only slowly. In rural areas, branch plants provide most of the new jobs, while small firms tend to start up and survive in urban areas (Miller 1985). Information flow in such a setting is reduced and less effective than where independent businesses are common. Knowledge and experience of international standards of quality, design, delivery, productivity, and training are critical in any context, urban or rural (Harper 1989; Malecki 1988; O'Farrell and Hitchens 1988a).

Just as only some places can realistically be R&D or high-tech locations, only some of these, in turn, are likely to see large-scale high-tech entrepreneurship. Other places can share in the larger growth of new firms outside high technology, but only if the human potential of the area is fostered through education and training, which are essential to traditional or low-technology industries as well. The importance of these elements of economic development is the major concern of Chapter 9.

Chapter 9

Development in a fast-paced world: challenges and questions

In this final chapter, it is useful to recall the hierarchy of measures that can be used to gauge economic development. At the most fundamental level are *basic human needs*. These needs are not only for adequate food, shelter, and medical care, but go beyond that to include employment and political power (Streeten 1979a). From the individual's point of view, then, a job which helps to provide for and improve the lot of one's family is a key priority not always met in many of the world's economies.

At the level of the national or regional economy, the second level in the hierarchy of development measures includes the usual *growth indicators* of employment and income. A goal of this type, while certainly preferable to declining employment and income, is a fine short-term accomplishment but does not make much progress towards a self-sustaining economy. Structural change, including industrial diversification, leads to the third step in the hierarchy – the *formation of linkages* with other firms, especially within the nation or region. These linkages support jobs not only in the initial enterprise or sector, but also in those firms and sectors linked to it. The fourth stage in development is the *generation of new firms*, ultimately the highest stage of economic activity, because it represents self-sustainability of an economy over time. It may, in fact, be only the beginning of a process of local development wherein an economy controls productive activities in other regions as well (Coffey and Polèse 1984, 1985).

Regions and nations are not isolated, perhaps especially in the context of science and technology (S&T). Nations themselves are unable to contain knowledge and the fruits of R&D within their borders. 'Technological knowledge and capability are international rather than national' (Nelson 1984: 75). Although Reich (1987) has proposed 'techno-nationalism', as an attempt to keep a nation's technology out of foreign hands, regional borders are more open than national ones, rendering 'techno-regionalism' even less likely to succeed than 'techno-nationalism'. With the activities of transnational firms, building and maintaining a national technological capability are difficult. 'The knowledge gleaned by perfecting each generation of technology spills over into other areas of the economy, creating a national pool of talent and technological experience that can improve productivity overall. But because each new generation of technology builds on that which came before, once off the technological escalator it's difficult to get back on' (Reich 1987: 64).

Technology is not a natural resource which some nations are lucky enough to possess while others are not. Technology as knowledge provides wealth via entrepreneurship as other resources cannot (Ayres 1988). Simple models and theories, such as the neoclassical paradigm and the linear model of the innovation process, do not hold up well in the complex and dynamic concepts – including a spatial division of labour, regional technological capability and learning – which characterize the world economy. Likewise, nations (and certainly the regions within them) 'are far more vulnerable to external influences than was envisioned in conventional explanations of growth and decline. Decisions emanating from places like Tokyo and Riyadh as well as the political give-and-take in [national capitals] have exerted far greater influence on the economic fortunes' of regions than have the 'mobile' factors of production (Weinstein and Gross 1988: 16).

Technology and competitiveness

Concern over national capability and competitiveness has been raised frequently in recent years, and technology plays a pivotal role in international competition (Brainard, Leedman, and Lumbers 1988; Cohen and Zysman 1987; Dertouzos, Lester, and Solow 1989; Ergas 1987; Guile and Brooks 1987; Pavitt 1980; Porter 1990; Roobeek 1990; Rothwell and Zegveld 1985; Spence and Hazard 1988; Teece 1987; Vernon 1989). Regional and national fortunes depend on the activities of the firms operating within them and on the strength of institutions, including educational and research institutions and informal networks. National and regional attitudes and culture regarding education, investment, and entrepreneurship comprise a societal dimension, or a 'regional factor' which has proven difficult to define precisely (Roobeek

1990; Thomas 1986). Skills and entrepreneurship, unlike scientific knowledge and technology, are very local phenomena, yet these are especially critical to economic change and Schumpeterian creative destruction.

International competitiveness hinges more than ever before on technology and education, since the costs of raw materials and of capital have become less variable (Cohen and Zysman 1987). Just as industrial policies have merged with S&T strategy, policies regarding international trade have increasingly had to take into account technological change and the competitiveness of nations (Ergas 1987; Krugman 1986). Investment must be made in education, in infrastructure for new technologies, and in diffusing knowledge, all with the aim of enhancing national capabilities. These influences, while difficult to measure, continue to determine economic growth (Scholing and Timmermann 1988).

Production skills and manufacturing capabilities have become decisive in international competition. For nations, as for individual firms, the decision to invest or not to invest in new products and processes is critical. If made, such investments in people and equipment accumulate into a body of know-how, resources, strategies, and habits that represent assets complementary to innovation. Moreover, 'advantage in a national economy is embodied not simply in the capacities of specific firms but in the web of interconnections that establishes possibilities for all firms' (Cohen and Zysman 1987: 102). This view of interconnectedness among producing firms, their suppliers and their customers, familiar in the French-language literature as the *filière*, depends on manufacturing systems being carriers of knowledge, skills, and organizational techniques (Dosi and Orsenigo 1988; Lundvall 1988).

The accomplishments of Japanese firms is due to achievements which span sectors but which remain soundly based in competitive manufacturing prowess. The slogan of the current industrial transition, manufacturing flexibility, encompasses an ability that goes beyond conventional short-term ability to adjust to shifting conditions within a fixed product and production structure. It reflects dynamic flexibility, or the ability to increase productivity steadily through improvements in production processes and innovation in products. For corporations, it implies a capacity to develop technological advances in products and processes continuously and a willingness to take risks (Cohen and Zysman 1987: 131–3; Hayes, Wheelwright, and Clark 1988; Sabel *et al.* 1987; Piore and Sabel 1984). 'As the division of labor becomes more and more complex – and that, after all, is what the colossal growth of producer services represents – the productivity of any worker, or any firm, depends on that of workers in other firms' (Cohen and Zysman 1987: 233). A strong national economy consists of tightly linked sectors comprised of dynamically flexible firms that mutually benefit from each other's competitiveness. The notion of linkages traditionally has rarely been touched on in economic policy, but it has begun to be addressed in recommendations concerning national policy towards investment in education and training, corporate priorities and behaviour, and government policies towards infrastructure (Horwitch 1986; Porter 1990; Spence and Hazard 1988; Teece 1987).

The national scale

International competitiveness and stature depend on economic and industrial performance as they once did on military strength. Industrial competitiveness can be helped or hindered by national policies, however, and the latter are known for their lack of timeliness and responsiveness to changing events. State-owned firms have been notable for their inability to adapt to shifting demands, capabilities, and priorities. Whether a nation's firms and industries are internationally competitive depends heavily on the technical capabilities present. These capabilities, represented first by skilled, technical workers and, second, by the general educational level of the labour force, embody the ways in which the capabilities of people contribute to organizational and economic success. In the UK, for example, despite greater demands for technical and engineering skills, significantly fewer workers with such skills are available than in West Germany or Japan, presenting a persistent source of concern (Albu 1980; Prais 1988; Walker 1980). This has perceptible consequences in competitiveness even in low-technology industries such as clothing, where German workers are far more qualified than their British counterparts (Steedman and Wagner 1989).

'Local content' regulation, countertrade requirements, and insistence on the use of local raw materials can enlarge the number and magnitude of linkages within a nation's borders. While it may be the case that production and employment are localized, the activities of global firms insist that there is no assurance that inputs would be obtained from local firms. Local content restrictions are a second-order import-substitution policy, attempting to assure that inputs rather than final products are produced locally. These and other non-tariff barriers to trade are a growing part of international competition (Cohen, Ferguson, and Oppenheimer 1985).

Katz (1982b) suggests that the state and its S&T bodies can act as a 'talent scout' in the field of world-wide technology and as a centralized buyer of technology in world markets. A state role in imports and exports is increasingly common via various countertrade arrangements whereby technology is purchased subject to the seller taking specified products in return (Alexandrides and Bowers 1987; Cohen and Zysman 1986; Hammond 1990; OECD 1985b). These are particularly common in arms transfers, called offset or compensation packages, which soften the financial impact of arms imports by providing economic or technological benefits to the buyer (Ch. 7) (Catrina 1988: 38).

National culture also affects attitudes about economic development. At the most general level, national cultures differ along dimensions related to family values, roles of individuals, biases concerning inclusion and exclusion (Berry 1989; Hofstede 1980). Such cultural values, especially regarding inclusion and exclusion and access to information and education, deeply affect legal systems and industrial organization (Aujac 1982; Casson 1987: 260). Legal systems, in turn, significantly influence the appropriability of technology, the principal incentive for innovation. Cultural or societal values deeply influence the transmission of skills, industrial relations, the strength of government in economic affairs, and the relative positions of large and small firms (Roobeek 1990). These values and institutions strongly affect the degree of inter-firm linkage and the rate of new firm formation.

The regional scale

Regions, states, provinces, *départements*, or other subnational governmental entities have the ability to tailor policies, incentives, and activities to their economy and culture, which may differ significantly from a national 'average'. While recognizing that technological change in particular is unable to be kept within a region, regions with the 'right' environment will perform better than other regions. This suitable environment will not always be predictable in the long run, but will depend on the regional industry mix and how well local industries and firms fare relative to other sectors. The presence of products with growth potential, in contrast to mature products, is a simple indication of regional industrial performance. Similarly, the presence of R&D activities in a region suggests the technological capability present there. Such a capability will enhance employment and incomes in any place, and will be evidenced by the mix of occupations in a region (Hekman and Strong 1981; Thompson and Thompson 1987). Regions with capability are able to adjust to changing conditions, such as the dramatic shock which took place in the watch-making region of Switzerland when electronic technology replaced mechanical technology (Maillat 1984). New local *filières* were developed by firms within the region and its 'development from below' has been notable.

Beyond this simple product-cycle perspective, and at a higher level of development, the technological capability of a region is enhanced by interaction: inter-firm linkages and inter-firm communication. The vital nature of inter-firm linkages is widely recognized, and they represent an indicator of development more advanced than employment alone. Priorities concerning 'local content' of material inputs have emerged regionally as well as nationally, especially where regional governments may provide funding for products intended as public goods or social overhead capital. Public transport, computer systems, and other large government purchases are among the areas in which regional governments have some 'clout' and may be able to affect the location of production or assembly, if the purchases are large enough. Local linkages support other firms, which supply inputs to purchasing firms. The industrial base of a region or country is weak or strong depending on the strength of its own firms and workers.

Regional attitudes are reflected in the value placed on education in other ways as well. Salaries of teachers may vary from place to place, especially if local or regional resources financially support public education. Even in a situation of fixed salary scales, however, teachers and schools differ markedly. In areas where education is prized, teachers are held in higher regard in the social hierarchy. The schools and class-rooms in such places will get an array of informal support from parents, organizations, and others in the community. This support may include both financial enhancement and voluntary cooperation and participation in school activities. The benefits of such informal activities outside of explicit policies are seen not only in the educational arena, but also in the generation of an entrepreneurial climate in a locale (Ch. 8).

The local scale

The nature of human activity is such that local scale interactions are the most frequent and the most meaningful. Despite the presence of long-distance interaction, most contacts, especially of an informal nature, are within a short radius of one's home base. These interactions, especially the 'half-hour cycle time' for information suggested by

Sweeney (1987: 25), are behind the local nature of linkages and technological progress in industrial districts, urban agglomerations, and other localities. The general processes of economic growth and development – employment, linkages, and entrepreneurship – take place first (and perhaps only) at a local scale, and are then aggregated to create regional and national totals. Economic change does not take place in the same way, and policies will not be equally appropriate, in all places. Locally specific attributes, shortcomings, and histories all play a part in defining the future potential of a regional economy (Kirby 1985; Sweeney 1987). Although interaction cannot be created by policies, initiatives for public–private co-operation are first and foremost local, in the informal, face-to-face interaction between people.

The importance of skills

Technological change, especially the robotics revolution in manufacturing, dramatically increases the advantages of batch production over mass production, and economies of scope (product variety) over economies of scale (production volume), but only if highly skilled workers and adequate networks for telecommunications are available. Deliberate policy choices to make investments in education and infrastructure have given Japan an edge over American and European firms. American firms, perhaps more than their European counterparts, have also been shortsighted by moving production off shore rather than to stay and sustain and develop the linkages and expertise domestically needed to preserve competitive manufacturing (Cohen and Zysman 1987; Ferguson 1983).

Among the most critical variations among places are information, contacts, and innovativeness. Information and contact networks encompass both publicly available (albeit perhaps costly) information and scientific knowledge and technology as a set of abilities and skills embodied in people. Geographical variations are inherent in those human elements. Although people and their knowledge are mobile, and become key elements in technology transfer and entrepreneurship, there is a great deal of stability over time in the firms and locales where innovative ability has accumulated.

The issues facing new businesses of all types are identical in most respects. Low technology is far more abundant than is high technology, and both groups can benefit from awareness of the critical importance of people (Naylor 1985). Education, and particularly the enhancement of skills that can be used across a spectrum of jobs, will benefit businesses and workers alike for a long time. If people learn how to learn and to make the incremental improvements which are part of the innovation process, wider benefits will accrue to them and to their employers and in some cases to society at large.

Education and training

Access to education varies tremendously from country to country (Ch. 4). Because of differential access within countries, education beyond the secondary level is likely to be available primarily, or only, in major cities, such as national capitals. Beyond the issue of geographic access, social distance may keep some, even most, people out of formal education (Jones 1982: 154–72; Reddy 1979). Formal education is one important element contributing to national labour skills.

'Schooling has come to be seen as the basic mode of technical training . . . the means for generating the technical skills needed for economic growth' (Peattie 1981: 143). Schools are also an important part of the system of stratification in any society. Access to formal education has much to do with access to jobs and job security, and to the possibility of social mobility between jobs and occupations. In Japan and Korea, the preoccupation of middle-class parents with their children's school performance reflects the cultural value placed on schooling (Arai 1990; Choo 1990). It also indicates the functioning of a system in which corporations hire, at each status level, from the graduates of particular institutions according to their prestige, and in which a person's initial entry into the occupational system after graduation largely determines a worker's future career path through a system of promotion dependent more on seniority than on individual performance (Peattie 1981: 144). 'In this aspect of their functioning, schools may be seen as the institution that legitimizes and moralizes that part of the technical order that we call the "division of labor"' (Peattie 1981: 144). As Easterlin puts it, 'the proximate roots of the epoch of modern economic growth lie in the growth of science and the diffusion of modern education' (Easterlin 1981: 19). However, education clearly is not sufficient to explain development, as seen in the experience of thousands of scientists and engineers in India who cannot find work. 'The existence of schools will in itself make very little difference to a society that does not provide appropriate channels in which school-learning can be put to use' (Peattie 1981: 152).

Regional and local cultures influence educational participation, as Jones (1982: 160) refers to it. Attitudes, often of long standing, affect the value placed on education, on reading (influenced by the number of bookshops and libraries), and on the desire to pursue higher education. Jones (1982: 164) suggests that there are three categories of young people, between 15 and 24 years, in most countries:

Group 1. Young people of superior ability, or those of above-average ability who receive the advantage of superior teaching and family encouragement, who wish to undertake tertiary education, actually do so, and generally succeed. In most cases, they enter professional employment. Most in this group come from upper- and middle-class backgrounds.

Group 2. Young people with the capacity to undertake tertiary education and to succeed, but who for a variety of reasons have come to accept self-limitation and prefer to seek early employment and early marriage and family rather than postponing these goals to compete for a degree and then seek work in an uncertain future. In developed countries, most drop out between years 10 and 12; others complete year 12 and then end their studies. Many in this group, especially those from working-class backgrounds, may have been socially conditioned by the family, the school, and the environment to accept an unrealistically low level of aspiration.

Group 3. Young people who have no intention of going on to tertiary education and who would be unlikely to succeed. They generally drop out of secondary school at the minimum learning age (or even earlier). This group may include some young people of superior ability who are grossly disadvantaged by school or by their environment. This group is common in the working class, in rural areas, and in migrant families.

These groups reflect, in addition, broad patterns of labour force reproduction, which tends over time to reinforce the workforce needs of local employers. Flexible production and the prominent place of small firms force labour training to the forefront of local policy (Lloyd 1989). Flexible firms need skilled and flexible workers. This need not mean exploitation of labour; rather, it implies the flexibility of non-routine work associated with innovation, niche markets, and economies of scope (Doeringer, Terkla, and Topakian 1987). While large-scale projections of increased skill requirements are easily made, regional and local variations are prominent, suggesting that training and education policies must be locally based (Spenner 1988).

Technological change and worker skills

Technological change raises the required level of theoretical knowledge necessary to perform a job. High-technology industries, in particular, demand understanding of complex capital equipment, technical documentation, and integration of complex technologies (Bartel and Lichtenberg 1987; Zymelman 1982: 438–9). General education also provides important skills, especially literacy and numeracy, and awareness about phenomena outside of daily existence. A further, long-run effect of education is that the more education one receives, one's peer group and interpersonal relationships will comprise more informed and skilled people. Finally, a higher level of educational attainment transfers into ability to deal with non-routine activities.

Literacy and numeracy may themselves be transmitted into skills in writing and communicating and skills in problem-solving (Duffey 1988). The need for worker skills varies over the life cycle of a product (Carnevale, Gainer, and Meltzer 1988; Flynn 1988; Maier 1982).

Carnevale, Gainer, and Meltzer (1988) suggest a hierarchy of *basic workplace skills*, which go beyond academic skills of reading, writing, and arithmetic (the three R's of American lore):

1. Knowing how to learn;
2. Competence in reading, writing, and computation;
3. Oral communication and listening;
4. Adaptability: creative thinking and problem-solving;
5. Personal management: self-esteem, motivation and goal-setting, and personal and career development;
6. Group effectiveness: interpersonal skills, negotiation, and team-work;
7. Influence: organizational effectiveness and leadership.

These 'workplace basics' are likely to be personal, but they also are broadly cultural, and will vary according to the demands of local employers. The fullest range of skills will be found in non-routine jobs, but a number of these basic skills are required by routine work as well. A machinist, for instance, must have technical reading skills in order to interpret blueprints and use precision instruments. He or she must be able to compute, e.g. from fractions to decimals. In addition, problem-solving is needed to select the appropriate tools and machines for a particular task, and to tinker with and make adaptations for custom jobs (Carnevale, Gainer, and Meltzer 1988: 22; Doeringer, Terkla, and Topakian 1987: 88). In a setting where custom products and short product life cycles are the norm, many work tasks are non-routine, and thus depend on the largely personal and interpersonal skills and adaptability which are essential.

In the financial service sector, four kinds of skills are needed:

1. *Social,* for effective interpersonal communication;
2. *Product-knowledge,* for effective marketing and selling of services;
3. *Keyboard and diagnostic,* for interface with systems and resolution of problems that arise; and
4. *Entrepreneurial,* for ensuring the viability of individual cost or profit centres in an organization (Rajan 1987: 225).

The decline in the demand for unskilled, usually manual, workers has been seen as a result of technological change in all countries (Maier 1982: 24; Johnston and Packer 1987; Brainard, Leedman, and Lumbers 1988). In essence, entry-level skills are rising most rapidly, a fact which hits hardest those who are disadvantaged or are in Group 3 of Jones' (1982) categorization. In countries which have distinctive minority populations, it is these groups that are less able to acquire the necessary entry-level skills (Cyert and Mowery 1987). An illustration of the changing skill mix found in the labour market is seen in the projections for the year 2000 in the USA by the Hudson Institute (Fig. 9.1). Low-skill jobs, such as labourers, are expected to decrease more than 50 per cent. The most skilled sets of jobs, on the other hand, are expected to double as a percentage of the total (Johnston and Packer 1987: 98–101). The skill ratings in Fig. 9.1 reflect a composite of language, maths, and reasoning skills. To reinforce the point above concerning entry-level skills, Johnston and Packer (1987: 100) state that 'jobs that are currently in the middle of the skill distribution will be the least-skilled occupations of the future, and there will be very few net new jobs for the unskilled'. For those who do not have basic skills, the growing demand for multiple, or polyvalent, skills is even more difficult.

Changes in the worker qualifications demanded by employers are seen in other countries through occupational shifts. In Denmark, for example, unskilled workers are expected to decline from 35 to 10 per cent of the workforce between 1980 and 2000, largely because of a tripling in the need for technicians and a doubling of managers and professionals (Brainard, Leedman, and Lumbers 1988: 88). In the UK from 1978 to 1984, professional engineers, scientists, and technologists were the only category of occupations to show an increase in share of employment (Bessant and Senker 1987: 159).

Education and training programmes are the primary systems by which the human capital of a nation is preserved and increased (Johnston and Packer 1987: 116; Porter 1990). Post-secondary, non-university training

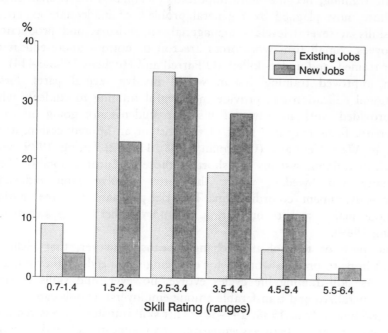

Figure 9.1 Increasing skill demanded of new jobs by the year 2000. *Source*: Johnston and Packer (1987: 100)

is especially weak in *laissez-faire* nations, such as the USA. 'Private firms underinvest in training because such investments can relatively easily be lost. In contrast to investments in capital goods, which can be bolted down, a worker can take his or her expensive training to another firm' (Spring 1989: v). In addition, small firms, which 'need people with a specific skill in twos and threes', are unable to justify full-pledged training programmes. The shift from large to small firms, increased subcontracting, and other dimensions of flexibility also mean that education and training in transferable skills, rather than employer-specific training, become more important (Seninger 1989). Small firms are those most plagued by a general problem of inadequate expertise and skills at several levels – managerial, supervisory, and production employees. 'Small firm workforces are top to bottom under-educated, under-trained, and under-skilled' (O'Farrell and Hitchens 1988a: 414).

An improved training system would involve several parts. First, vocational education can provide meaningful training to students who are provided work in private firms. For students not going on to a university, firms may pay most of the cost of an apprenticeship, as is done in West Germany (Osterman 1988: 110–32; Spring 1989: vii). American training systems fall short of those in Europe, notably West Germany and Sweden, where the latter involve joint industry–labour–government co-ordination. The US penchant for free market solutions makes such co-ordination voluntary rather than systematized (Spring 1989).

The pace of technological change means, however, that 'Schools cannot hope to prepare workers for emerging skill needs as they initially arise. As skill-training life cycles evolve, however, and skills become more generalized and transferable among employers, schools can provide such training' (Flynn 1988: 149–50). This skill-transfer process requires close collaboration between employers and educational and training institutions for identifying skill needs and standardizing training programmes in those skills. These skill needs are local because labour markets are locally unique, comprised of local workers, workplaces, and local histories (Flynn 1988). Central elements in local labour markets include the education level of the region and locality, which can vary considerably, and the diversity of employers, since greater diversity affords a variety of skills and experience (Browne 1981; W Thompson 1987).

Kuttner (1986: 22) asks a number of timely and relevant questions concerning policies that affect labour. For instance, how can industry (especially in advanced countries) be made more competitive other than by lowering workers' wages? If the idea is to 'accelerate transitions' from declining industries to emerging ones, how can this translate into decent jobs with good wages? What can be done to upgrade the quality and salary level of jobs in the service sector generally? These are important questions, because the effects of innovation on labour and employment are many and varied (Fischer 1990; Osterman 1988). Figure 9.2 shows that technological change affects not only the number of jobs but also work tasks, skill requirements, occupations, and work environments. The quantitative effects are the more typical objects of regional and local policy, and were the focus of Chapters 1 to 4. The qualitative effects on labour are the more complex, in part because they vary from workplace to workplace and from employer to employer.

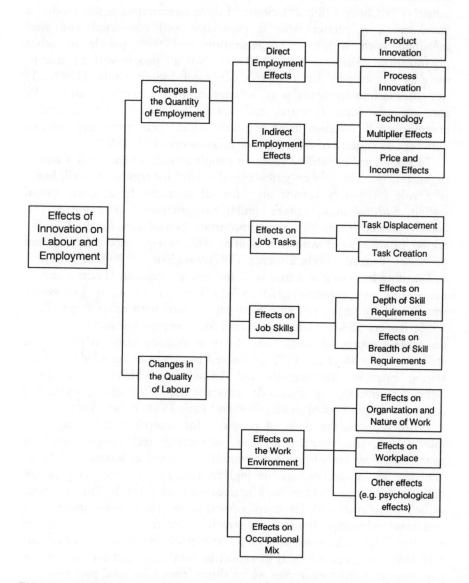

Figure 9.2 Effects of innovation on labour and employment. *Source:* Fischer (1990: 55 (Fig. 2))

The human element

The role of people throughout the process of technological change is critical. The 'most preferred and most effective mechanisms for technology transfer are "people intensive", rather than "paper intensive"' (Cutler 1989: 21). Some of these interactions centre around a 'gatekeeper', a person who is especially well connected and well informed both within an organization and with people in other organizations (Shapero 1985: 120–61). Not all people will be able to serve as 'gatekeepers' but their role is pivotal. Indeed, Cutler (1989: 21) describes technology transfer as 'a "contact sport": meeting with people, carrying new ideas forward, and joining individual effort toward a common goal'. Swedish firms regularly utilize their employees' contact networks as sources of information (Håkansson 1987, 1989).

Technological capability relies on people in technological roles within organizations. The role of 'gatekeepers' is vital for regions as well, but it is a role intensively reliant on informal contacts. In a more formal setting, *liaisons*, in a manner similar to agricultural extension agents, could connect firms with the appropriate consultants (located in the region or otherwise) who are familiar with best-practice products and processes (Britton 1989b; Sweeney 1987; Vaughan and Pollard 1986).

Learning by hiring is a means of technology transfer. 'Technology on the hoof', as Johnson (1975: 97–112) refers to it, is knowledge transferred by employees who formerly worked with other firms (Rhee 1990; Rogers 1982). Ideas transferred via a person joining a firm were the largest source of ideas (20.5%) from outside firms in the classic study by Langrish *et al.* (1972), followed by common knowledge (15%). Hiring provides information and knowledge which are otherwise difficult to obtain, and affects dramatically the ability of employers and nations to attain best practice (Enos and Park 1988; Ettlie 1980).

In all settings, the skills of people – for accepting and interpreting information, for improving and enhancing technology, and for generating new knowledge – ultimately determine economic outcomes. Whether in low-technology or high-technology circumstances, people define what organizations and nations can accomplish. An educated, skilled population has far more opportunities for change, innovation, and new endeavours than one where lives are trapped in a routine of poverty. While education and its advantages in the wider economy are typically considered national investments, local appreciation for learning continues to furnish examples which differ from a national pattern.

Infrastructure for development

Infrastructure, or social overhead capital, has always been a part of development (Hirschman 1958). In a technology-based economy, it is not enough for a region to have a lead in R&D; it must have the right socio-economic infrastructure (Hall and Preston 1988: 190–5; Roobeek 1990). In a sense, this infrastructure is the 'climate' of attitudes towards innovation and change found in creative regions. It consists in large part of the network of informal contacts stressed above. Information technology and telecommunications may be able to substitute for a local information environment in places where face-to-face contacts cannot readily be made (Lloyd 1989). It is not clear that the sorts of relatively routine business information needed by small firms can be found or provided outside the information infrastructure of core regions.

In a national context, physical infrastructures are more critical, because standards of international commerce demand movements of goods and information in certain forms and formats. Telephone service, regulated in most countries, now serves as a base for a wide range of information and telecommunications technologies. The infrastructure of the emerging industrial era is being based on the telecommunications networks of the past century, but must be modernized to replace electromechanical switching with digital technology (Antonelli 1989; Foray, Gibbons, and Ferné 1989: 69–77). To permit the transition to digital networks to affect small firms and backward regions will require massive, usually national, investments in telecommunications infrastructure (Arnold and Guy 1989; Cohen and Zysman 1987: 178–93; Hepworth 1989). The provision of this infrastructure will not be geographically uniform, as earlier innovations have not been. This means that regional inequities will persist (Gillespie *et al.* 1989; Hepworth 1989; Langdale 1983; Mansell 1988).

Other infrastructures allow nations to participate in global economic activity. Air and sea transportation demand modern airports and seaports. Investments must be made in air traffic control, security, and facilities in order to attract international investment. Innovations in port technology have insisted on containerized cargo, able to be shipped by sea, rail, or highway, and large vessels (Brookfield 1984; Hilling and Hoyle 1984; UN Conference on Trade and Development 1984). The decisions of ship operators about port selection for new services depend largely on state-of-the-art or best-practice facilities and operations being available (Willingale 1984). Once again, some places can be linked into the global network of 'world cities' while others are largely left out.

The challenge confronting the Third World

In an environment of unrelenting technological change, it is difficult to avoid a conclusion that cumulative advantages – 'the rich get richer' – condemn poor nations to at best a back seat in the world economy. Without sufficient capital for investment in new technology and infrastructure, many nations have taken on enormous debts to banks in wealthy countries (Fig. 9.3). The debts originated to invest in capital equipment for production, usually by state-owned enterprises, and to pay for oil imports (Peet 1987; Urrutia and Krueger 1989). Solutions, such as debt relief, are sorely needed to allow Third World countries to participate in the global economy (Sachs 1989).

There have been some remarkable cases of Third World economic success. By every conventional measure of economic success, the countries of East Asia are succeeding: growth, industrialization, export expansion, domestic saving, employment and wages, and income distribution (H Hughes 1988). 'There was a time when being well-endowed with natural resources was considered a distinct advantage. That was before the mineral-poor land-scarce East Asian NICs outperformed everyone, including the oil-exporting countries. Conventional wisdom was then turned on its head – lucky is the country that has no mining sector and few farmers' (Riedel 1988: 23). It also is argued that the East Asian countries succeeded because they were lucky to possess certain critical, non-reproducible assets or resources in greater proportion than other countries, such as cultural advantages.

> What many argue distinguishes East Asian countries, in particular the NICs, is the *quality* of their labour force. Diligence, loyalty, hard work and a strong appreciation for education are virtues which appear to be more abundant in East Asian NICs than elsewhere. Since these countries have common historical roots, the explanation has been found in culture, a factor relegated to the dustbin of development economics for more than three decades (Riedel 1988: 25–6).

One could also question whether the attributes of diligence, hard work, and educational attainment are exogenous or endogenous (meaning cultural). Attitudes towards work depend as much as anything on the availability of jobs and the reward given for effort, both of which are generally greater in East Asia than in most other developing countries. 'Educational attainment, the East Asian virtue most frequently referred to, is presumably as much supply as demand determined. Once

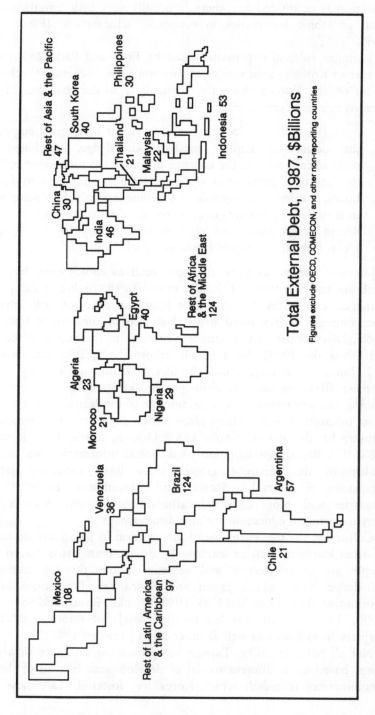

Figure 9.3 Total external debt of Third World nations, 1987. *Source: The Economist* 18 March, 1989: 92

education is accounted for, there is usually very little variation across racial or ethnic boundaries in economic achievement' (Riedel 1988: 26–7).

A similar cultural explanation used by Enos and Park (1988) in the context of Korea's rapid rate of absorption of petrochemical technology is that of 'the Korean character, particularly in the character of those Koreans drawn to engineering':

> . . . [V]aluable national traits are evident in Korea's engineers – the desire to acquire modern knowledge, combined with mechanical aptitude; the courage to undertake unfamiliar activities; the lack of resentment at working intensely and regularly for long hours; a spirit of cooperation; a willingness to subordinate private interests to collective goals; a rough, combative energy; and an urgent pragmatism – all tend to enable them to absorb foreign ideas, foreign techniques (Enos and Park 1988: 108).

Japanese success in some markets, such as motor cars, have been attributed to other forms of 'luck' – especially the timing of abrupt shifts in market conditions. On the other hand, an array of skills, many of them internal to individual firms, but also in the labour force as a whole, have contributed to Japanese mastery in the market-place (Near and Olshavsky 1985). For example, informal education has long gone on in Japan within large firms. The lifetime employment policy of many of these firms meant that firm-specific skills could be imparted by training and job rotation. A large amount of innovative activity – 'blue-collar innovation' – has taken place within Japanese firms through job mobility for decades (Okimoto 1989; Otsuka, Ranis, and Saxonhouse 1988: 97). Broad dissemination of technical information and the high quality of the Japanese labour force have enhanced Japanese competitiveness (Otsuka, Ranis, and Saxonhouse 1988: 111–12). Countries with more élitist educational systems and more restricted entry to technical education are at a disadvantage.

Culturally, the free exchange of information in Japan has led to more informal learning, such as learning by doing, than has occurred where people are overprotective and secretive about the transmission of knowledge. Once again, Japan and Korea are illustrations of such information flow (Enos and Park 1988; Otsuka, Ranis, and Saxonhouse 1988). This, in turn, has led to high levels of informal R&D and adaptations in Korea as well as other NICs (Amsden 1983).

Not all NICs are alike. Taiwan has grown slightly more slowly than Korea based on a different model of development. Instead of the large conglomerates (*chaebol*) which Korea has fostered, Taiwanese entrepreneurs are abundant and firms are small (Schiffman and Shao 1986). Taiwan's is more a case of local market-oriented development, whereas Korea exemplifies the benefits of greater central planning and direction (Ch. 7). Both have advantages, and Taiwan perhaps serves as a model of bottom-up development (Cook and Hulme 1988).

The regional context of technological change

We can usefully translate a recent summary of international competitiveness into the regional context:

> The . . . competitiveness of a [regional] economy is built on the competitiveness of the firms which operate within, and export from, its boundaries, and is, to a large extent, an expression of the will to compete and the dynamism of firms, their capacity to invest, to innovate both as a consequence of their own R&D and of successful appropriation of external technologies; but
>
> The competitiveness of a [regional] economy is also more than the simple outcome of the collective or 'average' competitiveness of its firms; there are many ways in which the features and performance of a [regional] economy viewed as an *entity* with characteristics of its own, will affect, in turn, the competitiveness of firms (Chesnais 1986: 91).

The structural deficiencies of peripheral regions, such as little local control, innovation, or inter-firm linkages, have led to proposals for local development based on entrepreneurship (Coffey and Polèse 1985). Policy proposals focus on the information gaps which are present in peripheral areas, because knowledge of the scientific and industrial state of the art is critical for the identification of potential innovations and entrepreneurial opportunities (Coffey and Polèse 1985; Sweeney 1987).

Despite numerous policy efforts to disperse new technology and innovative economic activities, they display a persistent tendency towards agglomeration and concentration (Clark 1972; Malecki 1981a, 1984a). Given our current understanding of non-routine economic activities in particular, there are sound reasons for this agglomeration tendency (Ch. 6). The places that offer conditions for 'regional creativity' have intensified levels of personal contacts (Andersson 1985; Törnqvist 1983). A major advantage of large urban agglomerations is that contact potential is greatest there, where the quality of information is high and the costs of obtaining it are relatively low. Although information levels can be enhanced by telecommunications, creativity requires personal contacts (Törnqvist 1983). For peripheral regions, their information links are mainly to large information centres, which maintains their dependent role in regional and international systems (Pred 1977; Sweeney 1987).

More than just agglomeration is necessary for regional development, although agglomeration encompasses some powerful forces. Other elements of regional structure, such as firm size and plant status and indicators of innovative capability and entrepreneurial vitality, may be present in some agglomerations but not in others. Regions dominated by large firms and by branch production plants are unlikely in any circumstance to have the level of information and of R&D, and knowledge of best practice and the state of the art needed to spawn new firms. Even outside high-tech sectors these aspects of technological capability are essential for regional growth to ensue.

The professional labour force, characterized by non-routine work, reinforces the agglomeration of creativity. The mobility of these workers and their access to information about employment in a large number of possible locations make it particularly difficult to foresee any reduction in their tendency to agglomerate in space. For these workers and for firms, large urban regions maximize their mutual benefits from the labour market: alternative employment opportunities for workers, a likelihood of obtaining a sufficient number of professional workers for firms.

The non-routine nature of their work in addition to their accumulated knowledge and contacts result in new firm formation at higher rates in areas where professional workers are concentrated. The formation of new firms in high-technology sectors is more dependent, relative to that of firms in other sectors, on the agglomeration of technical workers and on city size generally. In low technology, a very similar requirement of knowledge and innovation operates, separating successful from unsuccessful entrepreneurs (Malecki 1988). For example, beyond the need for starting capital, successful entrepreneurs are more oriented to markets outside their local region and have a larger number of clients (Miller, Brown, and Centner 1987; Wever 1986). Some regions have a history of entrepreneurship, accumulated over a long period of time, from which role models and experiences are made available to potential entrepreneurs through local information networks (Shapero 1971; Sweeney 1987). The tendency for entrepreneurship to be strongest in those regions which need it least suggests that a reliance on new and small firms alone will not eliminate regional economic differentials (Ch. 8).

Conclusions

The innovation process is lengthy and complex and all the necessary pieces may neither originate in a region nor always benefit that region's economy. There is no 'linear model' or simple one-way causality leading from S&T to regional economic development. Instead, there are strong, mutually supportive relationships which work best when S&T are part of the local culture, economy, and society.

The development of regional R&D and high-technology complexes anywhere takes long periods of time, and economic effects such as new local industry are highly unpredictable. Part of the unpredictability concerns the degree of local linkages by firms. Local purchasing varies greatly from firm to firm and from place to place, as do the characteristics of firms, whether labelled high tech or low tech. Consequently, the multiplier benefits of S&T – jobs, increased production, and new firms – will not remain entirely or even primarily in that region. The principal route by which the economic benefits stay within a region is new firm formation, as Chapter 8 has shown.

The (implicit or explicit) conclusion of most recent theory is that regions are helpless to do very much in the face of economic change, and that what they have tended to do in the past has been to assist rather than thwart capitalism in its quest for higher profits. Policies that arguably have had the greatest impacts on spatial structures have been aspatial policies – those whose goals have been macroeconomic, sectoral, or social in emphasis and whose spatial effects have been unintended, implicit, or indirect. Regional policies may simply be obsolete (Johnston 1986). On the other hand, if tailored to locally specific attributes, shortcomings, and histories, regional and local potential can be enhanced.

The spin-off process is ultimately the route of local technology transfer and economic growth. New ideas largely enter the economy through the identification of opportunities by entrepreneurs. The local nature of entrepreneurship poses immense challenges for policy at the subnational level, but it is just such a process that was the apparent basis of the California and Massachusetts successes which S&T policies elsewhere are trying to emulate. The process of entrepreneurship may be a far more important one to regional and local economies than the process of technological change.

For nations, it may be necessary to alter the national economic environment in major ways. This can be done over a period of time, but it cannot be planned entirely in the manner of selecting industries and jobs. Technologies change and new industries or market opportunities may arise which are best accommodated by combinations of large and small firms, in conjunction with local, regional, and national institutions. From a human standpoint, a well-educated and skilled population will be best able to capitalize on new opportunities and to take them into the future. Localities, regions, and nations are constrained by the imperatives of the market-place, but they learn, adopt, and adapt knowledge and technology only through the skills of their people.

References

Abe H, Alden J D (1988) Regional development planning in Japan. *Regional Studies* 22: 429–38

Abernathy W J (1980) Innovation and the regulatory paradox: toward a theory of thin markets. In Ginsburg D H, Abernathy W J (eds) *Government, technology, and the future of the automobile* New York, McGraw-Hill: 38–61

Abernathy W J, Chakravarthy B S (1979) Government innovation and innovation in industry: a policy framework. *Sloan Management Review* 20(3): 3–18

Abernathy W J, Clark K B (1985) Innovation: mapping the winds of creative destruction. *Research Policy* 14: 3–22

Abernathy W J, Utterback J M (1978) Patterns of industrial innovation. *Technology Review* 80(7): 40–7

Abetti P A, LeMaistre C W, Wacholder M H (1988) The role of Rensselaer Polytechnic Institute: technopolis development in a mature industrial area. In Smilor R W, Kozmetsky G, Gibson D V (eds) *Creating the technopolis: linking technology commercialization and economic development* Cambridge, MA, Ballinger: 125–44

Abler R (1975) Effects of space-adjusting technologies on the human geography of the future. In Abler R, Janelle D, Philbrick A, Sommer J (eds) *Human geography in a shrinking world* North Scituate, MA, Duxbury Press: 35–56

Abraham S C S, Hayward G (1984) Understanding discontinuance: towards a more realistic model of technological innovation and industrial adoption in Britain. *Technovation* 2: 209–31

Acs Z J, Audretsch D B (1989) Small-firm entry in US manufacturing. *Economica* 56: 255–65

Adám G (1975) Multinational corporations and worldwide sourcing. In Radice H (ed) *International firms and modern imperialism* Harmondsworth, Penguin: 89–103

Adam Y, Ong C H, Pearson A W (1988) Licensing as an alternative to foreign direct investment: an empirical investigation. *Journal of Product Innovation Management* 5: 32–49

Adas M (1989) *Machines as the measure of men: science, technology, and ideologies of Western dominance* Ithaca, NY, Cornell University Press

381

Adikibi O T (1988) The multinational corporation and monopoly of patents in Nigeria. *World Development* 16: 511–26

Ady R M (1986) Criteria used for facility location selection. In Walzer N, Chicoine D (eds) *Financing economic development in the 1980s* New York, Praeger: 72–84

Aglietta M (1979) *A theory of capitalist regulation* London, New Left Books

Agnew J A (1984) Devaluing place: 'people prosperity versus place prosperity' and regional planning. *Environment and Planning D: Society and Space* 1: 35–45

Aguinier P (1988) Regional disparities since 1978. In Feuchtwang S, Hussain A, Pairault T (eds) *Transforming China's economy in the eighties*, vol II: *Management, industry and the urban economy* Boulder, CO, Westview Press: 93–106

Ahlbrandt R S, Blair A F (1986) What it takes for large organizations to be innovative. *Research Management* 29(2): 34–7

Ajaegbu H I (1976) *Urban and rural development in Nigeria* London, Heinemann

Ajao A, Ironside R G (1982) Aspects of the life cycle of manufacturing firms in the Canadian Prairies. In Collins L (ed) *Industrial decline and regeneration* Edinburgh, University of Edinburgh, Department of Geography: 159–85

Albrechts L, Moulaert F, Roberts P, Swyngedouw E (1989) New perspectives for regional policy and development in the 1990s. In Albrechts L, Moulaert F, Roberts P, Swyngedouw E (eds) *Regional policy at the crossroads: European perspectives* London, Jessica Kingsley: 1–9

Albrechts L, Swyngedouw E (1989) The challenges for regional policy under a regime of flexible accumulation. In Albrechts L, Moulaert F, Roberts P, Swyngedouw E (eds) *Regional policy at the crossroads: European perspectives* London, Jessica Kingsley: 67–89

Albu A (1980) British attitudes to engineering education: a historical perspective. In Pavitt K (ed) *Technical innovation and British economic performance* London, Macmillan: 67–87

Aldrich H, Kalleberg A, Marsden P, Cassell J (1989) In pursuit of evidence: sampling procedures for locating new businesses. *Journal of Business Venturing* 4: 367–86

Aldrich H, Reese P R, Dubini P (1989) Women on the verge of a breakthrough: networking among entrepreneurs in the United States and Italy. *Entrepreneurship and Regional Development* 1: 339–56

Alexandrides C G, Bowers B L (1987) *Countertrade: practices, strategies, and tactics* New York, John Wiley

Allen D N (1985) An entrepreneurial marriage: incubators and startups. In Hornaday J A, Shils E B, Timmons J A, Vesper K H (eds) *Frontiers of entrepreneurship research 1985* Wellesley, MA, Babson College, Center for Entrepreneurial Studies: 38–60

Allen D N, Bird B J (1987) Faculty entrepreneurship in research university environments. In Churchill N C, Hornaday J A, Kirchhoff B A, Krasner O J, Vesper K H (eds) *Frontiers of entrepreneurship research 1987* Wellesley, MA, Babson College, Center for Entrepreneurial Studies: 617–30

Allen D N, Hayward D J (1990) The role of new venture formation/ entrepreneurship in regional economic development: a review. *Economic Development Quarterly* 4: 55–63

Allen D N, Levine V (1986) *Nurturing advanced technology enterprises* New York, Praeger

Allen J (1988) Service industries: uneven development and uneven knowledge. *Area* 20: 15–22

Allen T J (1977) *Managing the flow of technology* Cambridge, MA, MIT Press

Allen T J, Hyman D B, Pinckney D L (1983) Transferring technology to the small manufacturing firm: a study of technology transfer in three countries. *Research Policy* 12: 199–211

Allen T J, Utterback J M, Sirbu M A, Ashford N A, Hollomon J H (1978) Government influence on the innovation process. *Research Policy* 7: 124–49

Allesch J (1985) Innovation centres and science parks in the Federal Republic of Germany: current situation and ingredients for success. In Gibb J M (ed) *Science parks and innovation centres* Amsterdam, Elsevier: 58–68

Alonso W (1968) Urban and regional imbalances in economic development. *Economic Development and Cultural Change* 17: 1–14

Alonso W (1980a) Population as a system in regional development. *American Economic Review* 70: 405–9

Alonso W (1980b) Five bell shapes in development. *Papers of the Regional Science Association* 45: 5–16

Alonso W, Medrich E (1972) Spontaneous growth centers in twentieth-century American urbanization. In Hansen N M (ed) *Growth centers in regional economic development* New York, Free Press: 229–65

Altaf Z (1988) *Entrepreneurship in the Third World: risk and uncertainty in industry in Pakistan* London, Croom Helm

Amann R (1986) Technical progress and Soviet economic development: setting the scene. In Amann R, Cooper J M (eds) *Technical progress and Soviet economic development* Oxford, Basil Blackwell: 5–30

Amann R, Cooper J M (eds) (1982) *Industrial innovation in the Soviet Union* New Haven, CT, Yale University Press

Ameritrust, SRI Inc (1986) *Indicators of economic capacity* Cleveland, Ameritrust Corporation

Amin A (1985) Restructuring in Fiat and the decentralization of production into southern Italy. In Hudson R, Lewis J (eds) *Uneven development in southern Europe* London, Methuen: 155–91

Amin A (1989a) Flexible specialisation and small firms in Italy: myths and realities. *Antipode* 21(1): 13–34

Amin A (1989b) A model of the small firm in Italy. In Goodman E, Bamford J, Saynor P (eds) *Small firms and industrial districts in Italy* London, Routledge: 111–22

Amin A (1989c) Specialization without growth: small footwear firms in Naples. In Goodman E, Bamford J, Saynor P (eds) *Small firms and industrial districts in Italy* London, Routledge: 239–58

Amin A, Pywell C (1989) Is technology policy enough for local economic revitalization? The case of Tyne and Wear in the North East of England. *Regional Studies* 23: 463–77

Amin S (1974) *Accumulation on a world scale* New York, Monthly Review Press

Amin S (1986) Is an endogenous development strategy possible for Africa? In Ahooja-Patel K, Drabek A G, Nerfin M (eds) *World economy in transition* Oxford, Pergamon: 159–72

Amsden A H (1983) 'De-skilling,' skilled commodities, and the NICs' emerging comparative advantage. *American Economic Review* 73 (2): 333–7

Anderson A G (1969) The role of outlying laboratories. *Research Management* 12: 141–8

Anderson F J (1976) Demand conditions and supply constraints in regional economic growth. *Journal of Regional Science* 16: 213–24

Andersson A E (1985) Creativity and regional development. *Papers of the Regional Science Association* 56: 5–20

Andersson A E (1986) The four logistical revolutions. *Papers of the Regional Science Association* 59: 1–12

Andersson A E, Kuenne R E (1986) Regional economic dynamics. In Nijkamp P (ed) *Handbook of regional and urban economics*, vol 1: *Regional economics.* Amsterdam, North-Holland: 201–53

Andrade M H (1989) Promotion of industrial liaisons. In Commission of the European Communities (ed) *Partnership between small and large firms* London, Graham and Trotman: 348–54

Andrews D R, Tate U S (1988) An application of the stock adjustment model in estimating employment multipliers for the South Central Louisiana petroleum economy 1964–84. *Growth and Change* 19(3): 94–105

Andrews E L (1986) How much for a security blanket? *Venture* 8(2): 49–52

Andrews R B (1953a) Mechanics of the urban economic base: the problem of terminology. *Land Economics* 29: 263–8

Andrews R B (1953b) Mechanics of the urban economic base: a classification of base types. *Land Economics* 29: 343–50

Angel D P (1989) The labor market for engineers in the U.S. semiconductor industry. *Economic Geography* 65: 99–112

Ansoff H I (1987) Strategic management of technology. *Journal of Business Strategy* 7(3): 28–39

Antonelli C (1986a) The international diffusion of new information technologies. *Research Policy* 15: 139–47

Antonelli C (1986b) Technological districts and regional innovation capacity. *Revue d'Economie Régionale et Urbaine* 5: 695–705

Antonelli C (1987) The determinants of the distribution of innovative activity in a metropolitan area: the case of Turin. *Regional Studies* 21: 85–93

Antonelli C (ed) (1988a) *New information technology and industrial change: the Italian case* Dordrecht, Kluwer

Antonelli C (1988b) The emergence of the network firm. In Antonelli C (ed) *New information technology and industrial change: the Italian case* Dordrecht, Kluwer: 13–32

Antonelli C (1989) New information technology and industrial organisation: experiences and trends in Italy. In OECD *Information technology and new growth opportunities* Paris, Organisation for Economic Co-operation and Development: 81–99

Appelbaum E (1984) High tech and the structural employment problems of the 1980s. In Collins E L, Tanner L D (eds) *American jobs and the changing industrial base* Cambridge, MA, Ballinger: 23–48

Arai K (1990) Japanese education and economic development. In Lee C H, Yamazawa I (eds) *The economic development of Japan and Korea: a parallel with lessons* New York, Praeger: 153–70

Aram J D (1989) Attitudes and behaviors of informal investors toward early-stage investments, technology-based ventures, and coinvestors. *Journal of Business Venturing* 4: 333–47

Archer B H (1976) The anatomy of a multiplier. *Regional Studies* 10: 71–7

Armington C, Harris C, Odle M (1984) Formation and growth in high-technology firms: a regional assessment. In Office of Technology Assessment *Technology, innovation, and regional economic development* Washington, US Government Printing Office: 108–44

Armstrong W, McGee T G (1985) *Theatres of accumulation: studies in Asian and Latin American urbanization* London, Methuen

Arndt H W (1981) Economic development: a semantic history. *Economic Development and Cultural Change* 29: 457–66

Arndt H W (1987) *Economic development: the history of an idea* Chicago, University of Chicago Press

Arnold E (1985) *Competition and technological change in the television industry: an empirical evaluation of theories of the firm* London, Macmillan

Arnold E (1987) Some lessons from government information technology policies. *Technovation* 5: 247–68

Arnold E, Guy K (1989) Policy options for promoting growth through information technology. In OECD *Information technology and new growth opportunities* Paris, Organisation for Economic Co-operation and Development: 133–201

Arnold U, Bernard K N (1989) Just-in-time: some marketing issues raised by a popular concept in production and distribution. *Technovation* 9: 401–31

Arnold W (1986) Japan's technology transfer to advanced industrial countries. In McIntyre J R, Papp D S (eds) *The political economy of international technology transfer* New York, Quorum: 161–86

Arrighi G (ed) (1985) *Semiperipheral development: the politics of southern Europe in the twentieth century* Beverly Hills, CA, Sage

Arrow K J (1962) The economic implications of learning by doing. *Review of Economic Studies* 29: 155–73

Ashcroft B (1978) *The evaluation of regional economic policy: the case of the U.K.* Studies in Public Policy 12, Glasgow, University of Strathclyde, Centre for the Study of Public Policy

Ashcroft B (1982) The measurement of the impact of regional policies in Europe: a survey and critique. *Regional Studies* 16: 287–305

Ashcroft B, Swales J K (1982) The importance of the first round of the multiplier process: the impact of civil service dispersal. *Environment and Planning A* 14: 429–44

Aujac H (1982) Cultures and growth. In Saunders C (ed) *The political economy of new and old industrial countries* London, Butterworths: 47–70

Averitt R T (1979) Implications of the dual economy for community economic change. In Summers G F, Selvik A (eds) *Nonmetropolitan industrial growth and community change* Lexington, MA, Lexington Books: 71–91

Aydalot P (1981) The regional policy and spatial strategy of large organisations. In Kuklinski A (ed) *Polarized development and regional policy: tribute to Jacques Boudeville* The Hague, Mouton: 173–85

Aydalot P (1983) La division spatiale du travail. In Paelinck J, Sallez A (eds) *Espace et localisation* Paris, Economica: 175–200

Aydalot P (1988) Technological trajectories and regional innovation in Europe. In Aydalot P, Keeble D (eds) *High technology industry and innovative environments: the European experience* London, Routledge: 22–47

Ayres R U (1988) Technology: the wealth of nations. *Technological Forecasting and Social Change* 33: 189–201

Ayres R, Miller S (1982) Industrial robots on the line. *Technology Review* 85(4): 35–47

Ayres R U, Steger W A (1985) Rejuvenating the life cycle concept. *Journal of Business Strategy* 6 (1): 66–76

Bache J, Carr R, Parnaby J, Tobias A M (1987) Supplier development systems. *International Journal of Technology Management* 2: 219–28

Badawy M K (1988) What we've learned: managing human resources. *Research-Technology Management* 31(5): 19–35

Bade F-J (1983) Large corporations and regional development. *Regional Studies* 17: 315–25

Bae Z, Lee J (1986) Technology development patterns of small and medium sized companies in the Korean machinery industry. *Technovation* 4: 279–96

Bagchi A K (1988) Technological self-reliance, dependence and under-development. In Wad A (ed) *Science, technology, and development* Boulder, CO, Westview Press: 69–91

Bailly A, Boulianne L, Maillat D, Rey M, Thevoz L (1987) Services and production: for a reassessment of economic sectors. *Annals of Regional Science* 21(2): 45–59

Bailly A, Maillat D, Coffey W J (1987) Service activities and regional development: some European examples. *Environment and Planning A* 19: 653–68

Baily M N, Chakrabarti A K (1988) *Innovation and the productivity crisis* Washington, Brookings Institution

Bailyn L (1985) Autonomy in the industrial R&D lab. *Human Resource Management* 24: 129–46

Baker M J (1983) *Market development: a comprehensive survey* Harmondsworth, Penguin

Bakis H (1987) Telecommunications and the global firm. In Hamilton F E I (ed) *Industrial change in advanced economies* London, Croom Helm: 130–60

Baldry C (1988) *Computers, jobs, and skills: the industrial relations of technological change* New York, Plenum

Baldwin W L, Scott J T (1987) *Market structure and technological change* Chur, Harwood Academic

Ball N, Leitenberg M (eds) (1983) *The structure of the defense industry: an international survey* London, Croom Helm

Ballance R (1987) *International industry and business: structural change, industrial policy and industry strategies* London: Allen and Unwin

Ballance R, Sinclair S (1983) *Collapse and survival: industry strategies in a changing world* London, George Allen and Unwin

Balzer H D (1985) Is less more? Soviet science in the Gorbachev era. *Issues in Science and Technology* 1(4): 29–46

Bamber G (1988) Technological change and unions. In Hyman R, Streeck W (eds) *New technology and industrial relations* Oxford, Basil Blackwell: 204–19

Bandman M K (1980) *Territorial production complexes: optimisation models and general aspects* Moscow, Progress

Bar F, Borrus M, Cohen S, Zysman J (1989) The evolution and growth potential of electronics-based technologies. *Science Technology Industry Review* 5: 7–58

Baran B (1985) Office automation and women's work: the technological transformation of the insurance industry. In Castells M (ed) *High technology, space, and society* Beverly Hills, CA, Sage: 143–71

Baranson J (1987a) Preface. In Baranson J, Dolan J M, Heslin M, Leneman B, McHenry W *Soviet automation: perspectives and prospects* Mt Airy, MD, Lomond Books: ix–xi

Baranson J (1987b) Innovation within the Soviet economic system. In Baranson J, Dolan J M, Heslin M, Leneman B, McHenry W (1987) *Soviet automation: perspectives and prospects.* Mt Airy, MD, Lomond Books: 1–25

Bar-El R, Felsenstein D (1989) Technological profile and industrial structure: implications for the development of sophisticated industry in peripheral areas. *Regional Studies* 23: 253–66

Barff R A, Knight P L (1988) The role of federal military spending in the timing of the New England employment turnaround. *Papers of the Regional Science Association* 65: 151–66

Barker C E, Bhagavan M R, Mitschke-Collande P V, Wield D V (1986) *African industrialisation: technology and change in Tanzania* Aldershot, Gower

Barkley D L (1988) The decentralization of high-technology manufacturing to nonmetropolitan areas. *Growth and Change* 19(1): 13–30

Baron J N, Bielby W T (1980) Bringing the firms back in: stratification, segmentation, and the organization of work. *American Sociological Review* 45: 737–65

Bartel A P, Lichtenburg F R (1987) The comparative advantage of educated workers in implementing new technology. *Review of Economics and Statistics* 69: 1–11

Bartels C P A, Nicol W R, van Duijn J J (1982) Estimating the impact of regional policy: a review of applied research methods. *Regional Science and Urban Economics* 12: 3–41

Bartels C P A, van Duijn J J (1982) Regional economic policy in a changed labour market. *Papers of the Regional Science Association* 49: 97–111

Bartik T J (1989) Small business start-ups in the United States: estimates of the effects of characteristics of states. *Southern Economic Journal* 55: 1004–18

Bartke I, Lackó L (1986) Further considerations about regional policy in Eastern Europe. *Papers of the Regional Science Association* 60: 59–67

Bassett K (1986) Economic crisis and corporate restructuring: multinational corporations and the paper, printing and packaging sector in Bristol. In Taylor M, Thrift N (eds) *Multinationals and the restructuring of the world economy* London, Croom Helm: 311–43

Batey P J W (1985) Input–output models for demographic–economic analysis: some structural comparisons. *Environment and Planning A* 17: 73–99

Bath C R, James D D (1979) The extent of technological dependence in Latin America. In Street J H, James D D (eds) *Technological progress in Latin America: the prospects for overcoming dependency* Boulder, CO, Westview Press: 11–28

Batra R, Scully G W (1972) Technical progress, economic growth, and the north–south wage differential. *Journal of Regional Science* 12: 375–86

Batten D (1985) The new division of labor: the mobility of new technology and its impact on work. *Computers and the Social Sciences* 1: 133–9

Baumol W J (1967) Macroeconomics of unbalanced growth: the anatomy of urban crisis. *American Economic Review* 57: 415–26

Baumol W J (1981) Technological change and the new urban equilibrium. In Burchell R W, Listokin D (eds) *Cities under stress* New Brunswick, NJ, Center for Urban Policy Research

Baumol W J (1989) Is there a U.S. productivity crisis? *Science* 243: 611–15

Baumol W J, Blackman S A B, Wolff E N (1989) *Productivity and American leadership: the long view* Cambridge, MA, MIT Press

Bean A S, Baker N R (1988) Implementing national innovation policies through private decisionmaking. In Roessner J D (ed) *Government innovation policy: design, implementation, evaluation* New York, St Martin's Press: 75–89

Bean A S, Schiffel D D, Mogee M E (1975) The venture capital market and technological innovation. *Research Policy* 4: 380–408

Bearse P J (1981) *A study of entrepreneurship by region and SMSA size* Philadelphia, Public/Private Ventures

Beauregard R A (ed) (1989) *Economic restructuring and political response* Newbury Park, CA, Sage

Becattini G (1989) Sectors and/or districts: some remarks on the conceptual foundations of industrial economics. In Goodman E, Bamford J, Saynor P (eds) *Small firms and industrial districts in Italy* London, Routledge: 123–35

Becker G S (1964) *Human capital* New York, Columbia University Press

Beckmann M J, Thisse J-F (1986) The location of production activities. In Nijkamp P (ed) *Handbook of regional and urban economics,* vol 1: *Regional economics* Amsterdam, North-Holland: 21–95

Beesley M E, Hamilton R T (1986) Births and deaths of manufacturing firms in the Scottish regions. *Regional Studies* 20: 281–8

Beeson P (1987) Total factor productivity growth and agglomeration economies in manufacturing, 1959–73. *Journal of Regional Science* 27: 183–99

Beeson P E, Hustad S (1989) Patterns and determinants of productive efficiency in state manufacturing. *Journal of Regional Science* 29: 15–28

Begg I G, Cameron G C (1988) High technology location and the urban areas of Great Britain. *Urban Studies* 25: 361–79

Beije P R (1987) Innovation and production: two types of firm behaviour. In Rothwell R, Bessant J (eds) *Innovation: adaptation and growth* Amsterdam, Elsevier: 227–36

Bell D (1973) *The coming of the post-industrial society* New York, Basic Books

Bell M (1984) 'Learning' and the accumulation of industrial technological capacity in developing countries. In Fransman M, King K (eds) *Technological capability in the Third World* New York, St Martin's Press: 187–209

Bellandi M (1989) The role of small firms in the development of Italian manfacturing industry. In Goodman E, Bamford J, Saynor P (eds) *Industrial districts in Italy* London, Routledge: 31–68

Belous R S (1989) *The contingent economy: the growth of the temporary, part-time and subcontracted workforce* Washington, National Planning Association

Bender L D (1975) *Predicting employment in four regions of the western United States* Technical Bulletin 1529, Washington, US Department of Agriculture, Economic Research Service

Beneria L (1989) Subcontracting and employment dynamics in Mexico City. In Portes A, Castells M, Benton L A (eds) *The informal economy: studies in advanced and less developed countries* Baltimore, Johns Hopkins University Press: 173–88

Benton L A (1989) Industrial subcontracting and the informal sector: the politics of restructuring in the Madrid electronics industry. In Portes A, Castells M, Benton L A (eds) *The informal economy: studies in advanced and less developed countries* Baltimore, Johns Hopkins University Press: 228–44

Berentsen W H (1978) Austrian regional development policy: the impact of policy on the achievement of planning goals. *Economic Geography* 54: 115–34

Berentsen W H (1981) Conflicts between national and regional planning objectives: Austria and East Germany. *Papers of the Regional Science Association* 48: 135–48

Berentsen W H (1987) Relationships between trends in regional development and regional policy goals in central Europe, 1950–1980. *Environment and Planning C: Government and Policy* 5: 105–12

Berger S, Piore M J (1980) *Dualism and discontinuity in industrial societies* Cambridge, Cambridge University Press

Bernstein A (1989) What's dragging productivity down? Women's low wages. *Business Week* November 27: 171

Berry B J L (1972) Hierarchical diffusion: the basis of developmental filtering and spread in a system of growth centers. In Hansen N M (ed) *Growth centers in regional economic development* New York, Free Press: 108–38

Berry B J L (1973) *Growth centers in the American urban system* Cambridge, MA, Ballinger

Berry B J L (1989) Comparative geography of the global economy: cultures, corporations, and the nation-state. *Economic Geography* 65: 1–18

Berry B J L, Parr J B, Epstein B, Ghosh A, Smith R H T (1988) *Market centers and retail location: theory and applications* Englewood Cliffs, NJ, Prentice-Hall

Bertin G Y, Wyatt S (1988) *Multinationals and industrial property: the control of the world's technology* Hemel Hempstead, Harvester-Wheatsheaf

Bessant J, Cole S (1985) *Stacking the chips: information technology and the distribution of income* London, Frances Pinter

Bessant J, Haywood B (1988) Islands, archipelagoes and continents: progress on the road to computer-integrated manufacturing. *Research Policy* 17: 349–62

Bessant J, Senker P (1987) Societal implications of advanced manufacturing technology. In Wall T D, Clegg C W, Kemp N J (eds) *The human side of advanced manufacturing technology* New York, John Wiley: 153–71

Betz F (1987) *Managing technology* Englewood Cliffs, NJ, Prentice-Hall

Beyers W B (1976) Empirical identification of key sectors: some further evidence. *Environment and Planning A* 8: 231–6

Beyers W B (1979) Contemporary trends in the regional economic development of the United States. *Professional Geographer* 31: 34–44

Beyers W B (1983) The interregional structure of the U.S. economy. *International Regional Science Review* 8: 213–31

Beyers W B, Alvine M J (1985) Export services in postindustrial society. *Papers of the Regional Science Association* 57: 33–45

Beyers W, Johnsen E, Stranahan H (1987) Education and economic development: the producer services. *Economic Development Commentary* 11(4): 14–17

Bhalla A S (ed) (1979) *Towards global action for appropriate technology* Oxford, Pergamon

Bhalla A S (1988) Microelectronics use for small-scale production in developing countries. In Bhalla A S, James D (eds) *New technologies and development: experiences in 'technology blending'* Boulder, CO, Lynne Rienner Publishers: 53–64

Bhalla A S, Fluitman A G (1985) Science and technology indicators and socio-economic development. *World Development* 13: 117–90

Bhalla A S, James D (1988) *New technologies and development: experiences in 'technology blending'* Boulder, CO, Lynne Rienner Publishers

Bienaymé A (1986) The dynamics of innovation. *International Journal of Technology Management* 1: 133–59

Biersteker T J (1978) *Distortion or development? Contending perspectives on the multinational corporation* Cambridge, MA, MIT Press

Biggs S D (1983) Monitoring and control in agricultural research systems: maize in Northern India. *Research Policy* 12: 37–59

Birch D L (1979) *The job generation process* Cambridge, MA, MIT Program on Neighborhood and Regional Change

Birch D L (1987) *Job creation in America* New York, Free Press

Birch D L, MacCracken S J (1984) *The role played by high technology firms in job generation* Cambridge, MA, MIT Program on Neighborhood and Regional Change

Bishop P (1988) Academic–industry links and firm size in South West England. *Regional Studies* 22: 158–60

Bivand R (1981) Regional policy and asymmetry in geographical interaction relationships. In Kuklinski A (ed) *Polarized development and regional policies: tribute to Jacques Boudeville* The Hague, Mouton: 219–29

Black P A (1981) Injection leakages, trade repercussions and the regional income multiplier. *Scottish Journal of Political Economy* 28: 227–45

Blackburn P, Coombs R, Green K (1985) *Technology, economic growth and the labour process* London, Macmillan

Blackley P R (1986) Urban–rural differences in the structure of manufacturing production. *Urban Studies* 23: 471–83

Blair J P, Premus R (1987) Major factors in industrial location: a review. *Economic Development Quarterly* 1: 72–85

Bloomquist K M (1988) A comparison of alternative methods for generating economic base multipliers. *Regional Science Perspectives* 18(1): 58–99

Bluestone B, Harrison B (1982) *The deindustrialization of America* New York, Basic Books

Bluestone B, Jordan P, Sullivan M (1981) *Aircraft industry dynamics* Boston, Auburn House

Blumenfeld H (1955) The economic base of the metropolis: critical remarks on the 'basic–nonbasic' concept. *Journal of the American Institute of Planners* 21: 114–32

Boddy M, Lovering J (1986) High technology industry in the Bristol sub-region: the aerospace/defence nexus. *Regional Studies* 20: 217–31

Boehm W T, Pond M T (1976) Job location, retail purchasing patterns, and local economic development. *Growth and Change* 7(1): 7–12

Boisier S (1981) Growth poles: are they dead? In Prantilla E B (ed) *National development and regional policy* Singapore, Maruzen Asia: 71–83

Bollinger L, Hope K, Utterback J M (1983) A review of literature and hypotheses on new technology-based firms. *Research Policy* 12: 1–14

Bolton R (1966) *Defense purchases and regional growth* Washington, Brookings Institution

Bolton R (1982a) Industrial and regional policy in multiregional modeling. In Bell M E, Lande P S (eds) *Regional dimensions of industrial policy* Lexington, MA, Lexington Books: 169–94

Bolton R (1982b) Regional policy in the United States. *Canadian Journal of Regional Science* 5: 237–49

Bolton R (1985) Regional econometric models. *Journal of Regional Science* 25 (4): 495–520

Boonekamp C (1989) Industrial policies of industrial countries. *Finance and Development* 26(1): 14–17

Booth D E (1987) Regional long waves and urban policy. *Urban Studies* 24: 447–59

Borchert J G (1987) *City marketing en geografie* Nederlandse Geographische Studien 43. Amsterdam, Royal Dutch Geographical Society

Borchert J R (1978) Major control points in American economic geography. *Annals of the Association of American Geographers* 68: 214–32

Bornstein M (1985) *East–west technology transfer: the transfer of western technology to the USSR* Paris, Organisation for Economic Co-operation and Development

Borrus M G (1988) *Competing for control: America's stake in microelectronics* Cambridge, MA, Ballinger

Borts G H, Stein J L (1964) *Economic growth in a free market* New York, Columbia University Press

Botkin J, Dimancescu D, Stata R (1982) *Global stakes: the future of high technology in America* Cambridge, MA, Ballinger

Boudeville J-R (1966) *Problems of regional economic planning* Edinburgh, Edinbugh University Press

Boudeville J-R (1968) *L'espace et les pôles de croissance* Paris, Presses Universitaires de France

Bouman H, Verhoef B (1986) High-technology and employment: some information on the Netherlands. In Nijkamp P (ed) *Technological change, employment and spatial dynamics* Lecture Notes in Economics and Mathematical Systems 270, Berlin, Springer-Verlag: 289–98

Bourne L S (1988) Different solitudes and the restructuring of academic publishing: on barriers to communication in research. *Environment and Planning A* 20: 1423–5

Boyer R (ed) (1988a) Technical change and the theory of 'Régulation'. In Dosi G, Freeman C, Nelson R, Silverberg G, Soete L (eds) *Technical change and economic theory* London, Pinter: 67–94

Boyer R (ed) (1988b) *The search for labour market flexibility: the European economies in transition* Oxford, Oxford University Press

Boyer R (1988c) Defensive or offensive flexibility? In Boyer R (ed) *The search for labour market flexibility: the European economies in transition* Oxford, Oxford University Press: 222–51

Brady R (1989) *Perestroika* on ice. *Business Week* December 25: 44

Braham P (1986) Marks and Spencer: a technological approach to retailing. In Rhodes E, Wield D (eds) *Implementing new technologies* Oxford, Basil Blackwell: 123–41

Brainard R, Leedman C, Lumbers J (1988) *Science and technology policy outlook 1988* Paris, Organisation for Economic Co-operation and Development

Brandin D H, Harrison M A (1987) *The technology war: a case for competitiveness* New York, John Wiley

Braudel F (1982) *The wheels of commerce: civilization and capitalism 15th–18th century* vol. 2, New York, Harper and Row

Braverman H (1974) *Labor and monopoly capital* New York, Monthly Review Press

Breheny M J (1988b) Contracts and contacts: defence procurement and local economic development in Britain. In Breheny M J (ed) *Defence expenditure and regional development* London, Mansell: 188–211

Breheny M J, McQuaid R W (eds) (1987a) *The development of high technology industries: an international survey* London, Croom Helm

Breheny M J, McQuaid R W (1987b) HTUK: the development of the United

Kingdom's major centre of high technology industry. In Breheny M J, McQuaid R W (eds) *The development of high technology industries: an international survey* London, Croom Helm: 296–354

Brewis T N (1969) *Regional economic policies in Canada* Toronto, Macmillan

Britton J N H (1980) Industrial dependence and technological underdevelopment: Canadian consequences of foreign direct investment. *Regional Studies* 14: 181–99

Britton J N H (1985) Research and development in the Canadian economy: sectoral, ownership, locational, and policy issues. In Thwaites A T, Oakey R P (eds) *The regional impact of technological change* London, Frances Pinter: 67–114

Britton J N H (1987) High technology industry in Canada: locational and policy issues of the technology gap. In Breheny M J, McQuaid R W (eds) *The development of high technology industries: an international survey* London, Croom Helm: 143–91

Britton J N H (1988a) A policy prospectus on regional economic development: the implications of technological change. *Canadian Journal of Regional Science* 11: 147–65

Britton J N H (1988b) Economic change and the regional question. *Canadian Journal of Regional Science* 11: 167–79

Britton J N H (1989a) Innovation policies for small firms. *Regional Studies* 23: 167–73

Britton J N H (1989b) A policy perspective on incremental innovation in small and medium sized enterprises. *Entrepreneurship and Regional Development* 1: 179–90

Britton J, Gertler M (1986) Locational perspectives on policies for innovation. In Dermer J (ed) *Competitiveness through technology* Lexington, MA, Lexington Books: 159–75

Britton J N H, Gilmour J M (1978) *The weakest link: a technological perspective on Canadian industrial underdevelopment* Ottawa, Science Council of Canada

Brodsky H, Sarfaty D E (1977) Measuring the urban economic base in a developing country. *Land Economics* 53: 445–54

Brody H (1985) States vie for a slice of the pie. *High Technology* 5(1): 16–28

Broehl W G (1982) Entrepreneurship in the less developed world. In Kent C A, Sexton D L, Vesper K H (eds) *Encyclopedia of entrepreneurship* Englewood Cliffs, NJ, Prentice-Hall: 257–71

Brookfield H (1975) *Interdependent development* London, Methuen

Brookfield H C (1984) Boxes, ports, and places without ports. In Hoyle B S, Hilling D (eds) *Seaport systems and spatial change* Chichester, John Wiley: 61–79

Brophy D J (1988) The development of an integrated North American venture capital market. In McKee D L (ed) *Canadian–American economic relations* New York, Praeger: 153–79

Brophy D J (1989) Financing women-owned entrepreneurial firms. In Hagan O, Rivchun C, Sexton D (eds) *Women-owned businesses* New York, Praeger: 55–75

Brown A J, Burrows E M (1977) *Regional economic problems* London, George Allen and Unwin

Brown A L, Daneke G A (1987) The rising electric sun: Japan's photovoltaics industry. *Issues in Science and Technology* 3(3): 69–77

Brown L A (1981) *Innovation diffusion: a new perspective* New York, Methuen

Brown L A (1988) Reflections on Third World development: ground level reality, exogenous forces, and conventional paradigms. *Economic Geography* 64: 255–78

Browne L E (1978) Regional industry mix and the business cycle. *New England Economic Review* November/December: 35–53

Browne L E (1981) A quality labor supply. *New England Economic Review* July/August: 19–36

Browne L E (1983) High technology and business services. *New England Economic Review* July/August: 5–17

Browne L E (1988) Defense spending and high technology development: national and state issues. *New England Economic Review* September/October: 3–22

Browning H L, Singelmann J (1978) The transformation of the U.S. labor force: the interaction of industry and occupation. *Politics and Society* 8: 481–509

Browning J (1980) *How to select a business site* New York, McGraw-Hill

Brownrigg M (1980) Industrial contraction and the regional multiplier effect: an application in Scotland. *Town Planning Review* 51: 195–210

Brugger E A (1986) Endogenous development: a concept between utopia and reality. In Bassand M, Brugger E A, Bryden J M, Friedmann J, Stuckey B (eds) *Self-reliant development in Europe* Aldershot, Gower: 38–58

Brugger E A, Stuckey B (1987) Regional economic structure and innovative behaviour in Switzerland. *Regional Studies* 21: 241–51

Bruno A V, Cooper A C (1982) Patterns of development and acquisitions for Silicon Valley startups. *Technovation* 1: 275–90

Bruno A V, Leidecker J K (1987) A comparative study of new venture failure: 1960 vs. 1980. In Churchill N C, Hornaday J A, Kirchhoff B A, Krasner O J, Vesper K H (eds) *Frontiers of Entrepreneurship Research 1987* Wellesley, MA, Babson College, Center for Entrepreneurial Studies: 375–88

Bruno A V, Leidecker J K, Harder J W (1987) Why firms fail. *Business Horizons* 30(2): 50–8

Bruno A V, Tyebjee T T (1982) The environment for entrepreneurship. In Kent C A, Sexton D L, Vesper K H (eds) *Encyclopedia of entrepreneurship* Englewood Cliffs, NJ, Prentice-Hall: 288–307

Bruno A V, Tyebjee T T, Anderson J C (1985) Finding a way through the venture capital maze. *Business Horizons* 28(1): 12–19

Brusco S (1986) Small firms and industrial districts: the experience of Italy. In Keeble D, Wever E (eds) *New firms and regional development in Europe* London, Croom Helm: 184–202

Brusco S (1989) A policy for industrial districts. In Goodman E, Bamford J, Saynor P (eds) *Small firms and industrial districts in Italy* London, Routledge: 259–69

Brzoska M, Ohlson T (eds) (1986) *Arms production in the Third World* London, Taylor and Francis

Buck N (1988) Service industries and local labour markets: towards 'an anatomy of service job loss'. *Urban Studies* 25: 319–32

Buck T, Atkins M (1973) Regional policies in retrospect: an application of analysis of variance. *Regional Studies* 17: 181–9

Buck T, Cole J (1987) *Modern Soviet economic performance* Oxford, Basil Blackwell

Buhr W, Friedrich P (eds) (1981) *Lectures on regional stagnation* Baden Baden, Nomos

Burgan J U (1985) Cyclical behavior of high tech industries. *Monthly Labor Review* 108 (May): 9–15

Burgelman R A, Sayles L R (1986) *Inside corporate innovation* New York, Free Press

Burns L S, Healy R G (1978) The metropolitan hierarchy of occupations: an economic interpretation of central place theory. *Regional Science and Urban Economics* 8: 381–93

Burt D N (1989) Managing suppliers up to speed. *Harvard Business Review* 67(4): 127–35

Business Week (1976) The second war between the states. May 17: 92–114

Business Week (1980) The new defense posture: missiles, missiles, and missiles. August 11: 76–81

Business Week (1986) The hollow corporation. March 3: 57–85

Buss T F, Vaughan R J (1987) Revitalizing the Mahoning Valley. *Environment and Planning C: Government and Policy* 5: 433–47

Buswell R J (1983) Research and development and regional development. In Gillespie A (ed) *Technological change and regional development* London, Pion: 9–22

Buswell R J, Easterbrook R P, Morphet C S (1985) Geography, regions and research and development activity: the case of the United Kingdom. In Thwaites A T, Oakey R P (eds) *The regional impact of technological change* London, Frances Pinter: 36–66

Buttel F, Kenney M, Kloppenburg J (1985) From green revolution to biorevolution: some observations on the changing technological bases of economic transformation in the Third World. *Economic Development and Cultural Change* 34: 31–55

Buttner E H, Rosen B (1988) Bank loan officers' perceptions of the characteristics of men, women, and successful entrepreneurs. *Journal of Business Venturing* 3: 249–58

Bygrave W D (1987) Syndicated investments by venture capital firms: a networking perspective. *Journal of Business Venturing* 2: 139–54

Bygrave W D (1988) The structure of the investment networks of venture capital firms. *Journal of Business Venturing* 3: 137–57

Bylinsky G (1986) The high tech race: who's ahead? *Fortune* 114(8) (October 13): 26–44

Byun B-M, Ahn B-H (1989) Growth of the Korean semiconductor industry and its competitive strategy in the world market. *Technovation* 9: 635–56

Cainarca G C, Colombo M G, Mariotti S (1989) An evolutionary pattern of innovation diffusion. The case of flexible automation. *Research Policy* **18**: 59–86

Camagni R (1988) Functional integration and locational shifts in new technology industry. In Aydalot P, Keeble D (eds) *High technology industry and innovative environments: the European experience* London, Routledge: 48–64

Camagni R, Rabellotti R (1988) Innovation and territory: the Milan information technology field. In Giaoutzi M, Nijkamp P (eds) *Informatics and regional development* Aldershot, Avebury: 215–34

Caminiti S (1989) A quiet superstar rises in retailing. *Fortune* **120**(9): 167–74

Cantley M F (1986) Long-term prospects and implications of biotechnology for Europe: strategic challenge and response. *International Journal of Technology Management* **1**: 209–29

Cao-Pinna V (1974) Regional policy in Italy. In Hansen N M (ed) *Public policy and regional economic development: the experience of nine western countries* Cambridge, MA, Ballinger: 137–79

Capecchi V (1989) The informal economy and the development of flexible specialization in Emilia-Romagna. In Portes A, Castells M, Benton L A (eds) *The informal economy: studies in advanced and less developed countries* Baltimore, Johns Hopkins University Press: 189–215

Caporaso J A, Zare B (1981) An interpretation and evaluation of dependency theory. In Muñoz H (ed) *From dependency to development* Boulder CO, Westview Press: 43–56

Cappellin R (1988) Transaction costs and urban agglomeration. *Revue d'Economie Régionale et Urbaine*: 261–78

Carlberg M (1981) A neoclassical model of interregional economic growth. *Regional Science and Urban Economics* **11**: 191–203

Carlino G A (1982) Manufacturing agglomeration economies as returns to scale: a production function approach. *Papers of the Regional Science Association* **50**: 95–108

Carlino G A, Mills E S (1987) The determinants of county growth. *Journal of Regional Science* **27**: 39–54

Carlsson B (1981) The content of productivity growth in Swedish manufacturing. *Research Policy* **10**: 336–54

Carnevale A P, Gainer L J, Meltzer A S (1988) *Workplace basics: the skills employers want* Washington, US Department of Labor

Carney J (1980) Regions in crisis: accumulation, regional problems and crisis formation. In Carney J, Hudson R, Lewis J (eds) *Regions in crisis* New York, St Martin's Press: 28–59

Carroad P A, Carroad C A (1982) Strategic interfacing of R&D and marketing. *Research Management* **25**(1): 28–33

Carroll J J (1988) Obstacles to success: entrepreneurship in the Marshall Islands. In Fairbairn T I J (ed) *Island entrepreneurs: problems and performances in the Pacific* Honolulu, East–West Center: 111–36

Carter A P (1970) *Structural change in the American economy* Cambridge, MA, Harvard University Press

Carter C, Williams B (1959) The characteristics of progressive firms. *Journal of Industrial Economics* 7: 87–104

Casetti E (1981) A catastrophe model of regional dynamics. *Annals of the Association of American Geographers* 71: 572–79

Casetti E (1984) Manufacturing productivity and sunbelt–snowbelt shifts. *Economic Geography* 60: 313–24

Casetti E, Pandit K (1987) The non linear dynamics of sectoral shifts. *Economic Geography* 63: 241–58

Casson M (1982) *The entrepreneur: an economic theory* Totowa, NJ, Barnes and Noble

Casson M (1987) *The firm and the market* Cambridge, MA, MIT Press

Castells M (ed) (1985a) *High technology, space, and society* Beverly Hills, CA, Sage

Castells M (1985b) High technology, economic restructuring, and the urban–regional process in the United States. In Castells M (ed) *High technology, space, and society* Beverly Hills, CA, Sage: 11–40

Castells M (1988) The new industrial space: information-technology manufacturing and spatial structure in the United States. In Sternlieb G, Hughes J W (eds) *America's new market geography* New Brunswick, NJ, Center for Urban Policy Research: 43–99

Cater E (1988) The development of tourism in the least developed countries. In Goodall B, Ashworth G (eds) *Marketing in the tourism industry: the promotion of destination regions* London, Croom Helm: 39–66

Catrina C (1988) *Arms transfers and dependence* Philadelphia, Taylor and Francis

Chakrabarti A K, Souder W E (1987) Technology, innovation and performance in corporate mergers: a managerial evaluation. *Technovation* 6: 103–14

Chandler A D (1962) *Strategy and structure* Cambridge, MA: MIT Press

Charles D, Monk P, Sciberras E (1989) *Technology and competition in the international telecommunications industry* London, Pinter

Charney A H, Taylor C A (1983) Consistent region–subregion econometric models: a comparison of multiarea methods. *International Regional Science Review* 8: 59–74

Charney A H, Taylor C A (1986) Integrated state–substate econometric modeling: design and utilization for long-run economic analysis. In Perryman M R, Schmidt J R (eds) *Regional econometric modeling* Boston, Kluwer-Nijhoff: 43–92

Chaudhuri S (1986) Technological innovation in a research laboratory in India: a case study *Research Policy* 15: 89–103

Chesnais F (1986) Science, technology and competitiveness. *Science Technology Industry Review* 1: 85–129

Chesnais F (1988) Technical co-operation agreements between firms. *Science Technology Industry Review* 4: 51–119

Chiang J-T (1989) Technology and alliance strategies for follower countries. *Technological Forecasting and Social Change* 35: 339–49

Chilcote R H, Johnson D L (eds) (1983) *Theories of development: mode of production or dependency?* Beverly Hills, CA, Sage

Child J (1987) Organizational design for advanced manufacturing technology. In Wall T D, Clegg C W, Kemp N J (eds) *The human side of advanced manufacturing technology* New York, John Wiley: 101–33

Chisholm M (1987) Regional development: the Reagan–Thatcher legacy. *Environment and Planning C: Government and Policy* 5: 197–218

Choo H (1990) The educational basis of Korean economic development. In Lee C H, Yamazawa I (eds) *The economic development of Japan and Korea: a parallel with lessons* New York, Praeger: 171–83

Christopherson S (1989) Flexibility in the US service economy and the emerging spatial division of labour. *Transactions of the Institute of British Geographers* NS 14: 131–43

Christopherson S, Storper M (1989) The effects of flexible specialization on industrial politics and the labor market: the motion picture industry. *Industrial and Labor Relations Review* 42: 331–47

Chudnovsky D (1986) The entry into the design and production of complex capital goods: the experiences of Brazil, India and South Korea. In Fransman M (ed) *Machinery and economic development* New York, St Martin's Press: 54–92

Chung J S (1986) Korea. In Rushing F W, Brown C G (eds) *National policies for developing high technology industries: international comparisons* Boulder, CO, Westview Press: 143–72

Clapp J M, Richardson H W (1984) Technical change in information-processing industries and regional income differentials in developing countries. *International Regional Science Review* 9: 241–56

Clark C (1957) *The conditions of economic progress* London, Macmillan

Clark G L (1981) The employment relation and the spatial division of labor: a hypothesis. *Annals of the Association of American Geographers* 71: 412–24

Clark G L (1986a) Regional development and policy: the geography of employment. *Progress in Human Geography* 10: 416–26

Clark G L (1986b) Restructuring the US economy: the NLRB, the Saturn project, and economic justice. *Economic Geography* 62: 289–306

Clark G L, Ballard K P (1981) The demand and supply of labor and interstate relative wages: an empirical analysis. *Economic Geography* 57: 95–112

Clark G L, Gertler M S, Whiteman J (1986) *Regional dynamics: studies in adjustment theory* Boston, Allen and Unwin

Clark K B (1983) Competition, technical diversity, and radical innovation in the US auto industry. In Rosenbloom R S (ed) *Research on technological innovation, management and policy* vol 1, Greenwich, CT, JAI Press: 103–49

Clark L H (1988) Prosperity is tied more to where you live. *Wall Street Journal* October 6: A2

Clark N G (1972) Science, technology and regional economic development. *Research Policy* 1: 296–319

Clark N (1975) The multi-national corporation: the transfer of technology and dependence. *Development and Change* 6: 5–21

Clark N (1985) *The political economy of science and technology* Oxford, Basil Blackwell

Clark W A V, Freeman H E, Hanssens D M (1984) Opportunities for revitalizing stagnant markets: an analysis of household appliances. *Journal of Product Innovation Management* 1: 242–54

Clarke I M (1985) *The spatial organisation of multinational corporations* New York, St Martin's Press

Clarke I (1986) Labour dynamics and plant centrality in multinational corporations. In Taylor M, Thrift N (eds) *Multinationals and the restructuring of the world economy* London, Croom Helm: 21–47

Cobb J C (1982) *The selling of the south: the southern crusade for industrial development 1936–1980* Baton Rouge, Louisiana State University Press

Cobb J C (1984) *Industrialization and southern society 1877–1984* Lexington, University of Kentucky Press

Coffey W J, Polèse M (1984) The concept of local development: a stages model of endogenous regional growth. *Papers of the Regional Science Association* 54: 1–12

Coffey W J, Polèse M (1985) Local development: conceptual bases and policy implications. *Regional Studies* 19: 85–93

Coffey W J, Polèse M (1987a) Trade and location of producer services: a Canadian perspective. *Environment and Planning A* 19: 597–611

Coffey W J, Polèse M (1987b) Intrafirm trade in business services: implications for the location of office-based activities. *Papers of the Regional Science Association* 62: 71–80

Coffey W J, Polèse M (eds) (1987c) *Still living together: recent trends and future directions in Canadian regional development* Montreal, Institute for Research on Public Policy

Coffey W J, Polèse M (1989) Service activities and regional development: a policy-oriented perspective. *Papers of the Regional Science Association* 67: 13–27

Cohen R B, Ferguson R W, Oppenheimer M F (1985) *Nontariff barriers to high-technology trade* Boulder, CO, Westview Press

Cohen S S, Zysman J (1986) Countertrade, offsets, barter, and buybacks. *California Management Review* 28(2): 41–56

Cohen S S, Zysman J (1987) *Manufacturing matters: the myth of the post-industrial economy* New York, Basic Books

Cohen W M, Levinthal D A (1989) Innovation and learning: the two faces of R&D. *Economic Journal* 99: 569–96

Collier D W, Mong J, Conlin J (1984) Relationship between investment intensity and R&D intensity. *Research Management* 27(5): 11–16

Colombo U (1988) The technology revolution and the restructuring of the global economy. In Muroyama J H, Stever H G (eds) *Globalization of technology: international perspectives* Washington, National Academy Press: 23–31

Commission of the European Communities (ed) (1989) *Partnership between small and large firms* London, Graham and Trotman

Committee for Economic Development (1986) *Leadership for dynamic state economies* New York, Committee for Economic Development

Conroy M E (1975) *The challenge of urban economic development* Lexington, MA, Lexington Books

Conti S (1988) The Italian model and the problems of the industrial periphery.

In Linge G J R (ed) *Peripheralisation and industrial change* London, Croom Helm

Cook P, Hulme D (1988) The compatibility of market liberalization and local economic development strategies. *Regional Studies* 22: 221–31

Cook P, Kirkpatrick C (eds) (1988) *Privatisation in less developed countries* New York, St Martin's Press

Cooke P (1982) *Theories of planning and spatial development* London, Hutchinson

Cooke P (1983) Labour market discontinuity and spatial development. *Progress in Human Geography* 7: 543–65

Cooke P (1985a) Class practices as regional markers: a contribution to labour geography. In Gregory D, Urry J (eds) *Social relations and spatial structures* London, Macmillan: 213–41

Cooke P (1985b) Regional innovation policy: problems and strategies in Britain and France. *Environment and Planning C: Government and Policy* 3: 253–67

Cooke P (1986) The changing urban and regional system in the United Kingdom. *Regional Studies* 20: 243–51

Cooke P (1987) Research policy and review 19. Britain's new spatial paradigm: technology, locality and society in transition. *Environment and Planning A* 19: 1289–1301

Cooke P (1988) Flexible integration, scope economies, and strategic alliances: social and spatial mediations. *Environment and Planning D: Society and Space* 6: 281–300

Cooke P, Imrie R (1989) Little victories: local economic development in European regions. *Entrepreneurship and Regional Development* 1: 313–27

Cooke P, da Rosa Pires A (1985) Productive decentralisation in three European regions. *Environment and Planning A* 17: 527–54

Coombs R, Saviotti P, Walsh V (1987) *Economics and technological change* London, Macmillan

Cooper A C (1972) Incubator organizations and technical entrepreneurship. In Cooper A C, Komives J L (eds) *Technical entrepreneurship: a symposium* Milwaukee, Center for Venture Management: 108–25

Cooper A C (1973) Technical entrepreneurship: what do we know? *R&D Management* 3(1): 59–64

Cooper A C (1985) The role of incubator organizations in the founding of growth-oriented firms. *Journal of Business Venturing* 1: 75–86

Cooper A C (1986) Entrepreneurship and high technology. In Sexton D L, Smilor R W (eds) *The art and science of entrepreneurship* Cambridge, MA, Ballinger: 153–68

Cooper A C, Bruno A V (1977) Success among high-technology firms. *Business Horizons* 20(2): 16–22

Cooper A C, Dunkelberg W C, Furuta R S (1985) Incubator organizations: background and founding characteristics. In Hornaday J A, Shils E B, Timmons J A, Vesper K H (eds) *Frontiers of entrepreneurship research 1985* Wellesley, MA, Babson College, Center for Entrepreneurial Studies: 61–79

Cooper A C, Dunkelberg W C, Woo C Y (1988) Survival and failure: a longitudinal study. In Kirchhoff B A, Long W A, McMullan W E, Vesper K

H, Wetzel W E (eds) *Frontiers of entrepreneurship research 1988* Wellesley, MA, Babson College, Center for Entrepreneurial Studies: 225–37

Cooper A C, Schendel D (1976) Strategic responses to technological threats. *Business Horizons* 19(1): 61–9

Cooper A C, Willard G E, Woo C Y (1986) Strategies of high-performing new and small firms: a re-examination of the niche concept. *Journal of Business Venturing* 1: 247–60

Cooper A C, Woo C Y, Dunkelberg W C (1988) Entrepreneurs' perceived chances for success. *Journal of Business Venturing* 3: 97–108

Cooper J (1986) The civilian production of the Soviet defence industry. In Amann R, Cooper J M (eds) *Technical progress and Soviet economic development* Oxford, Basil Blackwell: 31–50

Cooper R G (1984a) How new product strategies impact on performance. *Journal of Product Innovation Management* 1: 5–18

Cooper R G (1984b) New product strategies: what distinguishes top performers? *Journal of Product Innovation Management* 1: 151–64

Cooper R G, Kleinschmidt E J (1987) New products: what separates winners from losers? *Journal of Product Innovation Management* 4: 169–84

Coraggio J L (1975) Polarization, development, and integration. In Kuklinski A (ed) *Regional development and regional planning: international perspectives* Leyden, Sijthoff: 353–74

Corbeau A B, Sheridan J B (1988) An analysis of entrepreneurial potential in a geographically isolated urban area. In Kirchhoff B A, Long W A, McMullan W E, Vesper K H, Wetzel W E (eds) *Frontiers of entrepreneurship research 1988* Wellesley, MA, Babson College, Center for Entrepreneurial Studies: 271–2

Corporation for Enterprise Development (1988) *Making the grade: the 1988 development report card for the states* Washington, Corporation for Enterprise Development

Corsten H (1987) Technology transfer from universities to small and medium-sized enterprises – an empirical survey from the standpoint of such enterprises. *Technovation* 6: 57–68

Courchene T J, Melvin J R (1988) A neoclassical approach to regional economics. In Higgins B, Savoie D J (eds) *Regional development: essays in honour of François Perroux* Boston, Unwin Hyman: 169–89

Cox R N (1985) Lessons from 30 years of science parks in the USA. In Gibb J M (ed) *Science parks and innovation centres: their economic and social impact* Amsterdam, Elsevier: 17–24

Crane D (1977) Technological innovation in developing countries: a review of the literature. *Research Policy* 6: 374–95

Crook C (1989) Poor man's burden: the Third World. *The Economist* September 23

Cross M (1981) *New firm formation and regional development,* Westmead, Gower

Croulet C R (1988) Indigenous entrepreneurship in Western Samoa. In Fairbairn T I J (ed) *Island entrepreneurs: problems and performances in the Pacific* Honolulu, East–West Center: 77–102

Crow M M (1988) Assessing government influence on industrial R&D. *Research-Technology Management* 31(5): 47–52

Cuadrado Roura J R (1982) Regional economic disparities: an approach and some reflections on the Spanish case. *Papers of the Regional Science Association* 49: 113–30

Cuadrado Roura J R, Granados V, Aurioles J (1983) Technological dependency in a Mediterranean economy: the case of Spain. In Gillespie A (ed) *Technological change and regional development* London, Pion: 118–31

Cumberland J H (1973) *Regional development: experiences and prospects in the United States of America* 2nd edn. The Hague, Mouton

CURDS (1979) *The mobilisation of indigenous potential in the U.K.* Report to the Regional Policy Directorate of the European Community, Newcastle upon Tyne, University of Newcastle upon Tyne, Centre for Urban and Regional Development Studies

Curien H (1987) The revival of Europe. In Pierre A J (ed) *A high technology gap? Europe, America and Japan* New York, Council on Foreign Relations: 44–66

Cusumano M A (1988) Manufacturing innovation: lessons from the Japanese auto industry. *Sloan Management Review* 30(1): 29–39

Cutler R S (1989) A comparison of Japanese and U.S. high-technology transfer practices. *IEEE Transactions on Engineering Management* 36: 17–24

Cyert R M, Mowery D C (1987) *Technology and employment: innovation and growth in the U.S. economy* Washington, National Academy Press

Czamanski D Z (1981) A contribution to industrial location theory. *Environment and Planning A* 13: 29–42

Czamanski D Z, Czamanski S (1977) Industrial complexes: their typology, structure and relation to economic development. *Papers of the Regional Science Association* 38: 93–111

Czamanski S (1974) *Study of clustering of industries* Halifax, Dalhousie University, Institute of Public Affairs

Czamanski S, Czamanski D Z (1976) *Study of spatial industrial complexes* Halifax, Dalhousie University, Institute of Public Affairs

Dahlman C J (1989) Technological change in industry in developing countries. *Finance and Development* 26(2): 13–15

Dahlman C J, Sercovich F C (1984) *Local development and exports of technology: the comparative advantage of Argentina, Brazil, India, the Republic of Korea, and Mexico* Staff working paper 667. Washington, World Bank

Dahlman C, Westphal L (1982) Technological effort in industrial development – an interpretative survey of recent research. In Stewart F, James J (eds) *The economics of new technology in developing countries* London, Frances Pinter: 105–37

Dalborg H (1974) *Research and development – organization and location* Stockholm, Stockholm School of Economics, Economic Research Institute

Damesick P J, Wood P A (eds) (1987) *Regional problems, problem regions, and public policy in the United Kingdom* Oxford, Clarendon Press

Dana L P (1987) Entrepreneurship and venture creation – an international comparison of five Commonwealth nations. In Churchill N C, Hornaday J A, Kirchhoff B A, Krasner O J, Vesper K H (eds) *Frontiers of entrepreneurship research 1987* Wellesley, MA, Babson College, Center for Entrepreneurial Studies: 573–83

Daniels P W (1982) *Service industries: growth and location* Cambridge, Cambridge University Press

Daniels P W (1983) Service industries: supporting role or centre stage? *Area* 15: 301–9

Daniels P W (1987) Producer-services research: a lengthening agenda. *Environment and Planning A* 19: 569–71

Daniels P W, Holly B P (1983) Office location in transition: observations on research in Britain and America. *Environment and Planning A* 15: 1293–98

Danilov V J (1972) Research parks and regional development. In Cooper A C, Komives J L (eds) *Technical entrepreneurship: a symposium* Milwaukee, Center for Venture Management: 96–107

Danson M W (1982) The industrial structure and labour market segmentation: urban and regional implications. *Regional Studies* 16: 255–65

Darwent D F (1969) Growth poles and growth centers in regional planning – a review. *Environment and Planning* 1: 5–32

Davelaar E J, Nijkamp P (1987) The urban incubator hypothesis: old wine in new bottles? In Fischer M, Sauberer M (eds) *Society–Economy–Space* Vienna, AMR-Info: 198–213

Davelaar E J, Nijkamp P (1988) The incubator hypothesis: revitalization of metro areas? *Annals of Regional Science* 22(3): 48–65

Davelaar E J, Nijkamp P (1989a) The role of the metropolitan milieu as an incubation centre for technological innovations: a Dutch case study. *Urban Studies* 26: 517–25

Davelaar E J, Nijkamp P (1989b) Spatial dispersion of technological innovation: a case study for the Netherlands by means of partial least squares. *Journal of Regional Science* 29: 325–46

David P (1975) *Technical choice, innovation, and economic growth* Cambridge, MA, Cambridge University Press

Davis W E (1988) The dynamics of competition in information technology. In H Schütte (ed) *Strategic issues in information technology* Maidenhead, Pergamon Infotech: 75–87

Day G S (1975) A strategic perspective on product planning. *Journal of Contemporary Business* 4: 1–34

Dean J W (1987) *Deciding to innovate: how firms justify advanced technology* Cambridge, MA, Ballinger

DeBresson C (1989) Breeding innovation clusters: a source of dynamic development. *World Development* 17: 1–16

DeBresson C, Lampel J (1985) Beyond the life cycle: organizational and technological design. I: an alternative perspective. *Journal of Product Innovation Management* 2: 170–87

de Jong M W (1987) *New economic activities and regional dynamics* Amsterdam, Netherlands Geographical Studies

Delapierre M (1988) Technology bunching and industrial strategies. In Urabe K,

Child J, Kagano T (eds) *Innovation and management: international comparisons* Berlin, Walter de Gruyter: 145–63

Delbecq A, Weiss J W (1988) The business culture of Silicon Valley: Is it a model for the future? In Weiss J W (ed) *Regional cultures, managerial behavior, and entrepreneurship: an international perspective* New York, Quorum Books: 23–41

de Meyer A C L (1985) The flow of technological information in an R&D department. *Research Policy* 14: 315–28

Denison E F (1967) *Why growth rates differ: postwar experience in nine western countries* Washington, Brookings Institution

Denison E F (1985) *Trends in American economic growth, 1929–1982* Washington, Brookings Institution

de Oliveira Campos R (1982) Take-off and breakdown: vicissitudes of the developing countries. In Kindleberger C P, di Tella G (eds) *Economics in the long view: essays in honour of W W Rostow* vol 1, New York, New York University Press: 116–40

Dertouzos M L, Lester R K, Solow R M (1989) *Made in America: regaining the productive edge* Cambridge, MA, MIT Press

Desai A V (1980) The origin and direction of industrial R&D in India. *Research Policy* 9: 74–96

Desai A V (1984) India's technological capability: an analysis of its achievements and limits. *Research Policy* 13: 303–10

Desai A V (1985) Market structure and technology: their interdependence in Indian industry. *Research Policy* 14: 161–70

de Soto H (1989) *The other path: the invisible revolution in the Third World* New York, Harper and Row

de Souza A R (1985) Scientific authorship and technological potential *Journal of Geography* 84(4): 138

Deutermann E P (1966) Seeding science-based industry. *Business Review, Federal Reserve Bank of Philadelphia* May: 3–10

Devereux E, Ferguson T, Fisher W, Magee S, McDonald S, Sharpe E, Smilor R, Smith J, Szygenda S, Williams F, Woodson H, Wilson M (1987) *Economic growth and investment in higher education* Austin, University of Texas, Bureau of Business Research

Devine M D, Ballard S C, White I L, Chartock M A, Brosz A R, Calzonetti F J, Eckert M S, Hall T A, Leonard R L, Malecki E J, Miller G D, Rappaport E B, Rycroft R W (1981) *Energy from the west* Norman, OK, University of Oklahoma Press

de Woot P (1990) *High technology Europe: strategic issues for global competitiveness* Oxford, Basil Blackwell

Diamond D, Spence N (1983) *Regional policy evaluation: a methodological review and the Scottish example* Aldershot, Gower

Dicken P (1986) *Global shift: industrial change in a turbulent world* London, Harper and Row

Dickson D (1984) *The new politics of science* New York, Pantheon

Dickson K (1983) The influence of Ministry of Defence funding on semiconductor research and development in the United Kingdom. *Research Policy* 12: 113–20

Dilley R S (1986) Tourist brochures and tourist images. *Canadian Geographer* 30: 59–65

Dixon R, Thirlwall A P (1975a) A model of regional growth-rate differences on Kaldorian lines. *Oxford Economic Papers* 27: 201–14

Dixon R, Thirlwall A P (1975b) *Regional growth and unemployment in the United Kingdom* London, Macmillan

Docter J, van der Horst R, Stokman C (1989) Innovation processes in small and medium size companies. *Entrepreneurship and Regional Development* 1: 33–52

Dodgson M, Rothwell R (1988) Small firm policy in the U.K. *Technovation* 7: 231–47

Doeringer P B (1984) Internal labor markets and paternalism in rural areas. In Osterman P (ed) *Internal labor markets* Cambridge, MA, MIT Press: 271–89

Doeringer P B, Terkla D G, Topakian G C (1987) *Invisible factors in local economic development* Oxford, Oxford University Press

Donckels R (1989) New entrepreneurship: lessons from the past, perspectives for the future. *Entrepreneurship and Regional Development* 1: 75–84

Dore R (1984) Technological self-reliance: sturdy ideal or self-serving rhetoric. In Fransman M, King K (eds) *Technological capability in the Third World* London, Macmillan: 65–80

Dorfman N S (1983) Route 128: the development of a regional high technology economy. *Research Policy* 12: 299–316

Dorfman N S (1987) *Innovation and market structure: lessons from the computer and semiconductor industries* Cambridge, MA, Ballinger

Dormard S (1988) New technology policies at the regional level in France: Nord–Pas-de-Calais and Provence–Alpes–Côte-d'Azur compared. In Dyson K (ed) *Local authorities and new technologies: the European dimension* London, Croom Helm: 95–114

Dosi G (1982) Technological paradigms and technological trajectories: a suggested interpretation of the determinants and directions of technical change. *Research Policy* 11: 147–62

Dosi G (1984) *Technical change and industrial transformation* London: Macmillan

Dosi G (1988) Sources, procedures, and microeconomic effects of innovation. *Journal of Economic Literature* 26: 1120–71

Dosi G, Orsenigo L (1988) Coordination and transformation: an overview of structures, behaviours and change in evolutionary environments. In Dosi G, Freeman C, Nelson R, Silverberg G, Soete L (eds) *Technical change and economic theory* London, Pinter: 13–37

Douglass M (1988) The transnationalization of urbanization in Japan. *International Journal of Urban and Regional Research* 12: 425–54

Doutriaux J (1984) Evolution of the characteristics of (high-tech) entrepreneurial firms. In Hornaday J A, Tarpley F A, Timmons J A, Vesper K H (eds) *Frontiers of entrepreneurship research 1984* Wellesley, MA, Babson College, Center for Entrepreneurial Studies: 368–86

Doz Y (1986) *Strategic management in multinational companies* Oxford, Pergamon

Doz Y (1989) Competence and strategy. In Punset E, Sweeney G (eds) *Information resources and corporate growth* London, Pinter: 47–55

Draheim K P (1972) Factors influencing the formation of technical companies. In Cooper A C, Komives J L (eds) *Technical entrepreneurship: a symposium* Milwaukee, Center for Venture Management: 3–27

Dreyfack K, Port O (1986) Even American knowhow is headed abroad. *Business Week* March 3: 60–3

Drew D E (1985) *Strengthening academic science* New York: Praeger

Drucker P F (1989) *The new realities* New York, Harper and Row

Dube S C (1988) *Modernization and development: the search for alternative paradigms* London, Zed

Dubini P (1989a) The influence of motivations and environment on business start-ups: some hints for public policies. *Journal of Business Venturing* 4: 11–26

Dubini P (1989b) Which venture capital backed entrepreneurs have the best chances of succeeding? *Journal of Business Venturing* 4: 123–32

Duffey J (1988) Competitiveness and human resources. *California Management Review* 30(3): 92–100

Dumbleton J H (1986) *Management of high-technology research and development* Amsterdam, Elsevier

Dunford M (1986) Integration and unequal development: the case of southern Italy. In Scott A J, Storper M (eds) *Production, work, territory* Boston, Allen and Unwin: 225–45

Dunford M F (1988) *Capital, the state, and regional development* London, Pion

Dunning J H (1979) Explaining changing patterns of international production: in defence of the eclectic theory. *Oxford Bulletin of Economics and Statistics* 41: 269–95

Dunning J H (1988) *Multinationals, technology and competitiveness* London, Unwin Hyman

Dunning J H, Norman G (1983) The theory of the multinational enterprise: an application to multinational office location. *Environment and Planning A* 15: 675–92

Dunning J H, Norman G (1987) The location choice of offices of international companies. *Environment and Planning A* 19: 613–31

Dutton J M, Thomas A (1985) Relating technological change and learning by doing. In Rosenbloom R S (ed) *Research on technological innovation, management and policy* vol 2, Greenwich, CT, JAI Press: 187–224

Dyckman J W, Swyngedouw E A (1988) Public and private technological innovation strategies in a spatial context: the case of France. *Environment and Planning C: Government and Policy* 6: 401–13

Easterlin R (1981) Why isn't the whole world developed? *Journal of Economic History* 61: 1–19

Edquist C, Jacobsson S (1988) *Flexible automation: the global diffusion of new technology in the engineering industry* Oxford, Basil Blackwell

Egge K A, Simer F J (1988) An analysis of the advice given by recent entrepreneurs to prospective entrepreneurs. In Kirchhoff B A, Long W A, McMullan W E, Vesper K H, Wetzel W E (eds) *Frontiers of entrepreneurship research 1988* Wellesley, MA, Babson College, Center for Entrepreneurial Studies: 119–33

Eisinger P K (1988) *The rise of the entrepreneurial state: state and local economic development policy in the United States* Madison, University of Wisconsin Press

Elder A H, Lind N S (1987) The implications of uncertainty in economic development: the case of Diamond Star Motors. *Economic Development Quarterly* 1(1): 30–40

Elkan W (1988) Entrepreneurs and entrepreneurship in Africa. *World Bank Research Observer* 3(2): 171–88

Ellis L W (1988) What we've learned: managing financial resources. *Research-Technology Management* 31(4): 21–38

El-Shakhs S (1982a) National and regional issues and policies in facing the challenges of the urban future. In Hauser P M, Gardner R W, Laquian A A, El-Shakhs S (eds) *Population and the urban future* Albany, State University of New York Press: 103–80

El-Shakhs S (1982b) Regional development and national integration: the Third World. In Fainstein N I, Fainstein S S (eds) *Urban policy under capitalism* Beverly Hills, CA, Sage: 137–58

Elzinga A (1987) Foresighting Canada's emerging science and technologies. In Brotchie J F, Hall P, Newton P W (eds) *The spatial impact of technological change* London, Croom Helm: 343–56

Emerson J (1971) An interindustry comparison of regional and national industrial structures. *Papers of the Regional Science Association* 26: 165–77

Emerson J (1976) Interregional trade effects in static and dynamic input–output models. In Polenske K R, Skolka J V (eds) *Advances in input–output analysis* Cambridge, MA, Ballinger: 263–77

Emerson J, Ringleb A (1977) A comparison of regional production structures. *Papers of the Regional Science Association* 39: 85–98

Emmerij L (1981) Basic needs and employment–oriented strategies reconsidered. In Misra R P, Honjo M (eds) *Changing perception of development problems* Singapore, Maruzen Asia: 177–93

Enos J L, Park W-H (1988) *The adoption and diffusion of imported technology: the case of Korea* London, Croom Helm

Ergas H (1987) Does technology policy matter? In Guile B R, Brooks H (eds) *Technology and global industry: companies and nations in the world economy* Washington, National Academy Press: 191–245

Erickson R A (1972) The 'lead firm' concept: an analysis of theoretical elements. *Tijdschrift voor Economische en Sociale Geografie* 63: 426–37

Erickson R A (1977) Sub-regional impact multipliers: income spread effects from a major defense installation. *Economic Geography* 53: 283–94

Erickson R A (1978) Purchasing patterns and the regional trade multiplier. *Growth and Change* 9(3): 49–51

Erickson R A (1987) Business climate studies: a critical evaluation. *Economic Development Quarterly* 1(1): 62–71

Erickson R A, Gavin N I, Cordes S M (1986) Service industries in interregional trade: the economic impacts of the hospital sector. *Growth and Change* 17(1): 17–27

Ernst D (ed) (1980) *The new international division of labour, technology and underdevelopment: consequences for the third world* New York, Campus Verlag

Ernst D, O'Connor D (1989) *Technology and global competition: the challenge for newly industrialising economies* Paris, Organisation for Economic Co-operation and Development

Etemad H, Séguin Dulude L (eds) (1986a) *Managing the multinational subsidiary: response to environmental changes and to host nation R&D policies* New York, St Martin's Press

Etemad H, Séguin Dulude L (1986b) Inventive activity in MNEs and their world product mandated subsidiaries. In Etemad H, Séguin Dulude L (eds) *Managing the multinational subsidiary: response to environmental changes and to host nation R&D policies* New York, St Martin's Press: 177–206

Etemad H, Séguin Dulude L (1987) Patenting patterns in 25 large multinational enterprises. *Technovation* 7: 1–15

Ettlie J E (1980) Manpower flows and the innovation process. *Management Science* 26: 1086–95

Ettlie J E, Rubenstein A H (1987) Firm size and product innovation. *Journal of Product Innovation Management* 4: 89–108

Ettlinger N (1981) Dependency and urban growth: a critical review and reformulation of the concepts of primacy and rank-size. *Environment and Planning A* 13: 1389–1400

Ettlinger N (1988) American fertility and industrial restructuring: a possible link? *Growth and Change* 19(3): 75–93

Evangelista M (1988) *Innovation and the arms race* Ithaca, NY, Cornell University Press

Evans A W (1986) Comparisons of agglomeration: or what Chinitz really said. *Urban Studies* 23: 387–9

Evans H E (1989) National development and rural–urban policy: past experience and new directions in Kenya. *Urban Studies* 26: 253–66

Ewers H-J (1986) Spatial dimensions of technological developments and employment effects. In Nijkamp P (ed) *Technological change, employment and spatial dynamics* Berlin, Springer-Verlag: 157–76

Ewers H-J, Wettmann R W (1980) Innovation-oriented regional policy. *Regional Studies* 14: 161–79

Fadem J A (1984) Automation and work design in the US: case studies of quality of working life impacts. In Warner M (ed) *Microprocessors, manpower and society* New York, St Martin's Press: 294–310

Fairbairn T I J (1988a) Finance for indigenous business development: an appraisal. In Fairbairn T I J (ed) *Island entrepreneurs: problems and performances in the Pacific* Honolulu, East–West Center: 209–26

Fairbairn T I J (1988b) Assessment and conclusions. In Fairbairn T I J (ed) *Island entrepreneurs: problems and performances in the Pacific* Honolulu, East–West Center: 269–76

Falemo B (1989) The firm's external persons: entrepreneurs or network actors? *Entrepreneurship and Regional Development* 1: 167–77

Falk W W, Lyson T A (1988) *High tech, low tech, no tech* Albany, State University of New York Press

Fall P N (1989) The informal sector in developing countries: understanding the organization and mechanisms for workable government policy encouragement. *Journal of Business Venturing* 4: 291–7

Farley J, Glickman N J (1986) R&D as an economic development strategy: the Microelectronics and Computer Technology Corporation comes to Austin, Texas. *Journal of the American Planning Association* 52: 407–18

Farrell K (1983) High tech highways. *Venture* 5(9): 38–50

Fawkes S D, Jacques J K (1987) Problems of adoption and adaptation of energy-conserving innovations in UK beverage and dairy industries *Research Policy* 16: 1–15

Feeser H R, Willard G E (1988) Incubators and performance: a comparison of high and low growth high tech firms. In Kirchhoff B A, Long W A, McMullan W E, Vesper K H, Wetzel W E (eds) *Frontiers of entrepreneurship research 1988* Wellesley, MA, Babson College, Center for Entrepreneurial Studies: 549–63

Feigenbaum E A, McCorduck P (1983) *The fifth generation: artificial intelligence and Japan's computer challenge to the world* Reading, MA, Addison-Wesley

Feller I (1988) Political and administrative aspects of state high-technology programmes. In Roessner J D (ed) *Government innovation policy: design, implementation, evaluation* New York, St Martin's Press: 75–89

Felsenstein D, Shachar A (1988) Locational and organizational determinants of R&D employment in high technology firms. *Regional Studies* 22: 477–86

Ferguson C H (1983) The microelectronics industry in distress. *Technology Review* 86(6): 24–37

Ferguson R F, Ladd H F (1988) Massachusetts. In Fosler R S (ed) *The new economic role of American states* New York, Oxford University Press: 19–87

Firestone O J (1974) Regional economic and social disparity. In Firestone O J (ed) *Regional economic development* Ottawa, University of Ottawa Press: 205–67

Firn J R (1975) External control and regional development: the case of Scotland. *Environment and Planning A* 7: 393–414

Fischer M M (1990) The micro-electronic revolution and its impact on labour and employment. In Cappellin R, Nijkamp P (eds) *The spatial context of technological development* Aldershot, Avebury: 43–74

Fischer M M, Nijkamp P (1988) The role of small firms for regional revitalization. *Annals of Regional Science* 22(1): 28–42

Fisher A G B (1933) Capital and the growth of knowledge. *Economic Journal* 43: 379–89

Fisher P S (1988) State venture capital funds as an economic development strategy. *Journal of the American Planning Association* 54: 166–77

Flamm K (1987) *Targeting the computer: government support and international competition* Washington, Brookings Institution

Flamm K (1988) *Creating the computer: government, industry, and high technology* Washington, Brookings Institution

Flammang R A (1979) Economic growth and economic development: counterparts or competitors? *Economic Development and Cultural Change* 28: 47–61

Florida R, Kenney M (1988a) Venture capital, high technology and regional development. *Regional Studies* 22: 33–48

Florida R, Kenney M (1988b) Venture capital-financed innnovation and technological change in the USA. *Research Policy* 17: 119–37

Florida R, Kenney M (1988c) Venture capital and high technology entrepreneurship. *Journal of Business Venturing* 3: 301–19

Flynn M S, Cole D E (1988) The US automobile industry: technology and competitiveness. In Hicks D A (ed) *Is new technology enough? Making and remaking US basic industries* Washington, American Enterprise Institute: 86–161

Flynn P M (1986) Technological change, the 'training cycle', and economic development. In Rees J (ed) *Technology, regions, and policy*. Totowa, NJ, Rowman and Littlefield: 282–308

Flynn P M (1988) *Facilitating technological change: the human resource challenge* Cambridge, MA, Ballinger

Folmer H, Nijkamp P (1985) Methodological aspects of impact analysis of regional economic policy. *Papers of the Regional Science Association* 57: 165–81

Fong C O (1986) *Technological leap: Malaysian industry in transition* Oxford, Oxford University Press

Foray D, Gibbons M, Ferné G (1989) *Major R&D programmes for information technology* Paris, Organisation for Economic Co-operation and Development

Forje J W (1988) In search of a strategy for a national science and technology policy in Africa. In Wad A (ed) *Science, technology, and development* Boulder, CO, Westview Press: 229–57

Fornengo G (1988) Manufacturing networks: telematics in the automotive industry. In Antonelli C (ed) *New information technology and industrial change: the Italian case* Dordrecht, Kluwer Academic: 33–56

Fortescue S (1985) Project planning in Soviet R&D. *Research Policy* 14: 267–82

Fosler F S (ed) 1988 *The new economic role of American States* New York, Oxford University Press

Foster R (1986) *Innovation: the attacker's advantage* New York, Simon and Schuster

Foxall G, Johnston B (1987) Strategies of user-initiated innovation. *Technovation* 6: 77–102

Fox Przeworski J (1986) Changing intergovernmental relations and urban economic development. *Environment and Planning C: Government and Policy* 4: 423–38

Frame J D (1983) *International business and global technology* Lexington, MA, Lexington Books

Fransman M (1984) Promoting technological capability in the capital goods sector: the case of Singapore. *Research Policy* 13: 33–54

Fransman M (1986a) *Technology and economic development* Boulder, CO, Westview Press

Fransman M (ed) (1986b) *Machinery and economic development* New York, St Martin's Press

Fransman M (1986c) Machinery in economic development In Fransman M (ed) *Machinery and economic development* New York, St Martin's Press: 1–53

Fransman M, King K (eds) (1984) *Technological capability in the Third World* New York, St Martin's Press

Fredriksson C G, Lindmark L G (1979) From firms to systems of firms: a study of interregional dependence in a dynamic society. In Hamilton F E I, Linge G J R (eds) *Spatial analysis, industry and the industrial environment*, vol. 1: *Industrial systems* London, John Wiley: 155–86

Freear J, Wetzel W E (1988) Equity financing for new technology-based firms. In Kirchhoff B A, Long W A, McMullan W E, Vesper K H, Wetzel W E (eds) *Frontiers of entrepreneurship research 1988* Wellesley, MA, Babson College, Center for Entrepreneurial Studies: 347–67

Freedman A E (1983) New technology-based firms: critical location factors. In Hornaday J A, Timmons J A, Vesper K H (eds) *Frontiers of entrepreneurship research 1983* Wellesley, MA, Babson College, Center for Entrepreneurial Studies: 478–94

Freeman C (1982) *The economics of innovation* 2nd edn. Cambridge, MA, MIT Press

Freeman C (1987) *Technology policy and economic performance: lessons from Japan* London, Pinter

Freeman C, Lundvall B-A (eds) (1988) *Small countries facing the technological revolution* London, Pinter

Freeman C, Perez C (1988) Structural crises of adjustment, business cycles and investment behaviour. In Dosi G, Freeman C, Nelson R, Silverberg G, Soete L (eds) *Technical change and economic theory* London, Pinter: 38–66

Freeman D B, Norcliffe G B (1985) *Rural enterprise in Kenya: development and spatial organization of the nonfarm sector* Research paper 214, Chicago, University of Chicago, Department of Geography

Freundlich N (1989) Spreading the risks of R&D. *Business Week* June 16: 60–4

Frey D N (1989) Junk your 'linear' R&D! *Research-Technology Management* 32(3): 7–8

Frey R S, Dietz T, Marte J (1986) The effect of economic dependence on urban primacy: a cross-national panel analysis. *Urban Affairs Quarterly* 21: 359–68

Friar J, Horwitch M (1986) The emergence of technology strategy: a new dimension of strategic management. In Horwitch M (ed) *Technology in the modern corporation: a strategic perspective* Oxford, Pergamon: 50–85

Friedman D (1988) *The misunderstood miracle: industrial development and political change in Japan* Ithaca, NY, Cornell University Press

Friedmann J (1966) *Regional development policy: a case study of Venezuela* Cambridge, MA, MIT Press

Friedmann J (1986a) The world city hypothesis. *Development and Change* 17: 69–83

Friedmann J (1986b) Regional development in industrialised countries:

endogenous or self-reliant? In Bassand M, Brugger E A, Bryden J M, Friedmann J, Stuckey B (eds) *Self-reliant development in Europe* Aldershot, Gower: 203–16

Friedmann J, Douglass M (1978) Agropolitan development: towards a new strategy for regional planning in Asia. In Lo F-C, Salih K (eds) *Growth pole strategy and regional development policy* Oxford, Pergamon: 163–92

Friedmann J, Weaver C (1979) *Territory and function: the evolution of regional planning* London, Edward Arnold

Friedmann J, Wolff G (1982) World city formation: an agenda for research and action. *International Journal of Urban and Regional Research* **6**: 309–43

Frischtak C (1986) Brazil. In Rushing F W, Brown C G (eds) *National policies for developing high technology industries: international comparisons* Boulder, CO, Westview Press: 31–69

Frisk T (1988) The future state of information technology: a technological assessment. In H Schütte (ed) *Strategic issues in information technology* Maidenhead, Pergamon Infotech: 15–26

Fröbel F, Heinrichs J, Kreye O (1980) *The new international division of labour* Cambridge, Cambridge University Press

Fuchs R J, Demko G J (1979) Geographic inequality under socialism. *Annals of the Association of American Geographers* **69**: 304–18

Fuentes A, Ehrenreich B (1987) Women in the global factory. In Peet R (ed) *International capitalism and industrial restructuring* Boston, Allen and Unwin: 201–15

Fuenzalida E E (1979) The problem of technological innovation in Latin America. In Villamil J J (ed) *Transnational capitalism and national development* Atlantic Highlands, NJ, Humanities Press: 115–27

Fuller R A (1983) Decentralized R&D organization. In Brown J K, Elvers L M (eds) *Research and development: key issues for management* Report 842, New York, The Conference Board: 34–7

Furtado C (1980) Development: theoretical and conceptual considerations. In Pajestka J, Feinstein C H (eds) *The relevance of economic theories* New York, St Martin's Press: 200–16

Fusfeld H I (1986) *The technical enterprise: present and future patterns* Cambridge, MA, Ballinger

Fusfeld H I, Haklisch C S (1987) Collaborative industrial research in the US. *Technovation* **5**: 305–15

Gaile G L (1980) The spread–backwash concept. *Regional Studies* **14**: 15–25

Galbraith C S (1985) High-technology location and development: the case of Orange County. *California Management Review* **28**(1): 98–109

Galbraith J R, Nathanson D A (1978) *Strategy implementation: the role of structure and process* St Paul, MN, West

Galuszka P, Brady R (1989) Gorbachev's reforms: will they work? *Business Week* June 5: 52–63

Galuszka P, King R (1989) Cold feet in Siberia. *Business Week* March 27: 48

Galuszka P, Marbach W D, Brady R (1988) What will they do when they get the right stuff? *Business Week* November 7: 82–6

Galuszka P, Marbach W D, Brady R, Javetski B, Schares G (1988) Soviet technology. *Business Week* November 7: 68–78

Gamser M S (1988) Innovation, technical assistance, and development: the importance of technology users. *World Development* 16: 711–21

Gana J A (1981) Spatial allocation of resources and regional development in Nigeria. In Mabogunje A L, Misra R P (eds) *Regional development alternatives: international perspectives* Singapore, Maruzen Asia: 197–220

Ganne B (1989) Regional dynamics of innovation: a look at the Rhône-Alpes region. *Entrepreneurship and Regional Development* 1: 147–54

Gansler J (1980) *The defense industry* Cambridge, MA, MIT Press

Ganz C (1980) Linkages between knowledge, diffusion, and utilization. *Knowledge: Creation, Diffusion, Utilization* 1: 591–612

Garvin D A (1983) Spin-offs and the new firm formation process. *California Management Review* 25(2): 3–20

Gaston R J, Bell S E (1987) Business angels: mobilizing private sector seed capital. *Economic Development Commentary* 11(1): 1–5

Geisler E, Rubenstein A H (1989) University–industry relations: a review of major issues. In Link A N, Tassey G (eds) *Cooperative research and development: the industry–university–government relationship* Boston, Kluwer Academic: 43–62

Gerking S D, Isserman A M (1981) Bifurcation and the time pattern of impacts in the economic base model. *Journal of Regional Science* 21: 451–67

Gerking S D, Weirick W N (1983) Compensating differences and interregional wage differentials. *Review of Economics and Statistics* 65: 483–7

Germidis D (ed) (1977) *Transfer of technology by multinational corporations* Paris: Organisation for Economic Co-operation and Development

Germidis D (ed) (1980) *International subcontracting: a new form of investment* Paris: Organisation for Economic Co-operation and Development

Gershuny J I, Miles I D (1983) *The new service economy* London, Frances Pinter

Gertler M S (1984) Regional capital theory. *Progress in Human Geography* 8: 50–81

Gertler M (1986a) Regional dynamics of manufacturing and non-manufacturing investment in Canada. *Regional Studies* 20: 523–34

Gertler M S (1986b) Discontinuities in regional development. *Environment and Planning D: Society and Space* 4: 71–84

Gertler M S (1988a) Some problems of time in economic geography. *Environment and Planning A* 20: 151–64

Gertler M S (1988b) The limits to flexibility: comments on the post-Fordist vision of production and its geography. *Transactions, Institute of British Geographers* NS 13: 419–32

Gertler M S (1989) Resurrecting flexibility? A reply to Schoenberger. *Transactions, Institute of British Geographers* NS 14: 109–12

Gerwin D (1989) Manufacturing flexibility in the CAM era. *Business Horizons* 32(1): 78–84

Ghali M, Akiyama M, Fujiwara J (1981) Models of regional growth: an empirical evaluation. *Regional Science and Urban Economics* 11: 175–90

Giaoutzi M (1985) Factors affecting the capacity of technological change: the case of less developed countries. *Papers of the Regional Science Association* 58: 73–82

Gibb J M (ed) (1985) *Science parks and innovation centres: their economic and social impact* Amsterdam, Elsevier

Gibbs D C (1987) Technology and the clothing industry. *Area* 19: 313–20

Gibbs D C (1988) Restructuring in the Manchester clothing industry: technical change and interrelationships between manufacturers and retailers. *Environment and Planning A* 20: 1219–33

Gibbs D C, Alderman N, Oakey R P, Thwaites A T (1985) *The location of research and development in Great Britain* Newcastle upon Tyne, University of Newcastle, Center for Urban and Regional Development Studies

Gibbs D C, Edwards A (1985) The diffusion of new production innovations in British industry. In Thwaites A T, Oakey R P (eds) *The regional impact of technological change* London, Frances Pinter: 132–63

Gibson L J, Worden M A (1981) Estimating the economic base multiplier: a test of alternative procedures. *Economic Geography* 57: 146–59

Giese E (1987) The demand for innovation-oriented regional policy in the Federal Republic of Germany: origins, aims, policy tools and prospects of realisation. In Brotchie J F, Hall P, Newton P W (eds) *The spatial impact of technological change* London, Croom Helm: 240–53

Gilbert A (1988) The new regional geography in English and French-speaking countries. *Progress in Human Geography* 12: 208–28

Gilbert A G (1975) A note on the incidence of development in the vicinity of a growth centre. *Regional Studies* 9: 325–33

Gilbert A G, Goodman D E (1976) Regional income disparities and economic development: a critique. In Gilbert A G (ed) *Development planning and spatial structure* New York, John Wiley: 113–41

Gillespie A E, Goddard J B, Hepworth M E, Williams H (1989) Information and communications technology and regional development: an information economy perspective. *Science Technology Industry Review* 5: 85–111

Gillespie A, Howells J, Williams H, Thwaites A (1987) Competition, internationalisation and the regions: the example of information technology production industries in Europe. In Breheny M J, McQuaid R W (eds) *The development of high technology industries: an international survey* London, Croom Helm: 113–42

Gillespie A, Williams H (1988) Telecommunications and the reconstruction of regional comparative advantage. *Environment and Planning A* 20: 1311–21

Gillis W R (1987) Can service-producing industries provide a catalyst for regional economic growth? *Economic Development Quarterly* 1: 249–56

Gilmer R W (1990) Identifying service-sector exports from major Texas cities. *Economic Review, Federal Reserve Bank of Dallas* July: 1–16

Gilmour J M (1975) The dynamics of spatial change in the export region. In Collins L, Walker D F (eds) *Locational dynamics of manufacturing activity* New York, John Wiley: 59–82

Gilpin R (1975) *Technology, economic growth, and international competitiveness* Washington, US Government Printing Office

Glasmeier A K (1986) High-tech industries and the regional division of labor. *Industrial Relations* 25: 197–211

Glasmeier A K (1988a) Factors governing the development of high tech industry agglomerations: a tale of three cities. *Regional Studies* 22: 287–301

Glasmeier A K (1988b) The Japanese Technopolis programme: high-tech development strategy or industrial policy in disguise? *International Journal of Urban and Regional Research* 12: 268–84

Glasmeier A K, McCluskey R E (1987) US auto parts production: an analysis of the organization and location of a changing industry. *Economic Geography* 63: 142–59

Glasson J, van der Wee D, Barrett B (1988) A local income and employment multiplier analysis of a proposed nuclear power station development at Hinkley Point in Somerset. *Urban Studies* 25: 248–61

Glazer S (1986) Businesses take root in university parks. *High Technology* 6(1): 40–7

Gleave D, Palmer D (1979) The relationship between geographic and occupational mobility in the context of regional economic growth. In Hobcraft J, Rees P (eds) *Regional demographic development* London, Croom Helm: 188–210

Glickman N J (1971) An econometric model of the Philadelphia region. *Journal of Regional Science* 11: 15–32

Glickman N J (1977) *Econometric analysis of regional systems* New York, Academic Press

Gober-Myers P (1978) Employment-motivated migration and economic growth in post-industrial market economies. *Progress in Human Geography* 2: 207–29

Goddard J B (1975) *Office location in urban and regional development* Oxford, Oxford University Press

Goddard J B (1978) The location of non-manufacturing activities within manufacturing industries. In Hamilton F E I (ed) *Contemporary industrialization* London, Longman: 62–85

Goddard J, Gillespie A (1988) Advanced telecommunications development and regional economic development. In Giaoutzi M, Nijkamp P (eds) *Informatics and regional development* Aldershot, Avebury: 121–46

Goddard J B, Thwaites A T (1986) New technology and regional development policy. In Nijkamp P (ed) *Technological change, employment and spatial dynamics* Berlin, Springer-Verlag: 91–114

Goddard J, Thwaites A, Gibbs D (1986) The regional dimension to technological change in Great Britain. In Amin A, Goddard J B (eds) *Technological change, industrial restructuring and regional development* London, Allen and Unwin: 140–56

Gold B (1979) *Productivity, technology, and capital* Lexington MA, Lexington Books

Goldfarb R S, Yezer A M J (1976) Evaluating alternative theories of intercity and interregional wage differentials. *Journal of Regional Science* 16: 345–63

Goldfarb R S, Yezer A M J (1987) Interregional wage differential dynamics. *Papers of the Regional Science Association* 62: 45–56

Goldhar J (1986) In the factory of the future, innovation *is* productivity. *Research Management* 29(2): 26–33

Goldhar J D, Jelinek A (1983) Plan for economies of scope. *Harvard Business Review* 61(6): 141–8

Goldman A (1982) Short product life cycles: implications for the marketing activities of small high-technology companies. *R&D Management* 12(2): 81–9

Goldman M I (1987) *Gorbachev's challenge: economic reform in the age of high technology* New York, W W Norton

Goldstein H A, Luger M I (1990) Science/technology parks and regional development theory. *Economic Development Quarterly* 4: 64–78

Goodman E, Bamford J, Saynor P (eds) (1989) *Small firms and industrial districts in Italy* London, Routledge

Goodman R (1979) *The last entrepreneurs* New York, Simon and Schuster

Gordon R, Kimball L (1987) The impact of industrial structure on global high technology location. In Brotchie J F, Hall P, Newton P W (eds) *The spatial impact of technological change* London, Croom Helm: 157–84

Gore C (1984) *Regions in question: space, development theory and regional policy* London, Methuen

Gorman M, Sahlman W A (1989) What do venture capitalists do? *Journal of Business Venturing* 4: 231–48

Goto A, Wakasugi R (1988) Technology policy. In Komiya R, Okuno M, Suzumura K (eds) *Industrial policy of Japan* Tokyo, Academic Press: 183–204

Gottmann J (1961) *Megalopolis* Cambridge, MA, MIT Press

Graham C P (1988) Technology and development: a case study of the Brazilian firm Engesa, S.A. Unpublished MA thesis, University of Florida

Graham J, Gibson K, Horvath R, Shakow D M (1988) Restructuring in U.S. manufacturing: the decline of monopoly capitalism. *Annals of the Association of American Geographers* 78: 473–90

Graham M B W (1985) Industrial research in the age of big science. In Rosenbloom R S (ed) *Research on technological innovation, management, and policy* vol 2, Greenwich, CT, JAI Press: 47–79

Graham M B W (1986a) *RCA and the videodisc: the business of research* Cambridge, Cambridge University Press

Graham M B W (1986b) Corporate research and development: the latest transformation. In Horwitch M (ed) *Technology in the modern corporation: a strategic perspective* Oxford, Pergamon: 86–102

Grant Thornton Inc (1989) *10th annual Grant Thornton manufacturing climates study* Chicago, Grant Thornton Inc

Gray H P (1970) *International travel – international trade* Lexington, MA, Lexington Books

Green A E (1988) The north–south divide: an examination of the evidence. *Transactions, Institute of British Geographers* NS 13: 179–98

Green M B (1989) Patterns of preference for venture capital investment in the United States of America, 1970–85. *Environment and Planning C: Government and Policy* 7: 205–22

Green M B, McNaughton R B (1989) Interurban variation in venture capital investment characteristics. *Urban Studies* 26: 199–213

Greenwood M J (1973) The influence of family and friends on geographic labor mobility in a less-developed country: the case of India. *Review of Regional Studies* 3(3): 27–36

Greenwood M J (1975) Research on internal migration in the United States: a survey. *Journal of Economic Literature* 13: 397–433

Greenwood M J (1985) Human migration: theory, models, and empirical studies. *Journal of Regional Science* 25: 521–44

Greenwood M J, Hunt G L, Pfalzgraff E L (1987) The economic effects of space science activities of Colorado and the western United States. *Annals of Regional Science* 21(2): 21–44

Griffin K (1978) *International inequality and national poverty* London, Macmillan

Griliches Z (ed) (1984) *R&D, patents, and productivity* Chicago, University of Chicago Press

Grime K (1987) The evolution of a naval shipbuilding firm in a small economy: Vickers at Barrow-in-Furness. In Bateman M, Riley R (eds) *The geography of defence* Totowa NJ, Barnes and Noble: 141–70

Gripaios P, Bishop P, Gripaios R, Herbert C (1989) High technology industry in a peripheral area: the case of Plymouth. *Regional Studies* 23: 151–7

Gripaios P, Herbert C (1987) The role of new firms in economic growth: some evidence from South West England. *Regional Studies* 21: 270–3

Groshen E L (1987) Can services be a source of export-led growth? Evidence from the Fourth District. *Federal Reserve Bank of Cleveland Economic Review* 3: 2–15

Gros-Pietro G M, Rolfo S (1989) Flexible automation and firm size: some empirical evidence on the Italian case. *Technovation* 9: 493–503

Gross N (1989) MITI: the sugar daddy to end all sugar daddies. *Business Week* October 23: 112

Grossberg A J (1982) Metropolitan industrial mix and cyclical employment stability. *Regional Science Perspectives* 12(2): 13–25

Gruenstein J M L (1984) Targeting high tech in the Delaware Valley. *Business Review, Federal Reserve Bank of Philadelphia* May–June: 3–14

Grunwald J, Flamm K (1985) *The global factory: foreign assembly and international trade* Washington, Brookings Institution

Gudgin G, Crum R, Bailey S (1979) White-collar employment in UK manufacturing industry. In Daniels P W (ed) *Spatial patterns of office growth and location* New York, Wiley: 127–57

Guile B R, Brooks H (eds) (1987) *Technology and global industry: companies and nations in the world economy* Washington, National Academy Press

Guile B R, Quinn J B (eds) (1988) *Technology in services* Washington, National Academy Press

Gulowsen J (1988) Skills, options and unions: united and strong or divided and weak? In Hyman R, Streeck W (eds) *New technology and industrial relations* Oxford, Basil Blackwell: 160–73

Gumbel P (1989) Western money, technology fall on infertile Soviet soil. *Wall Street Journal* December 1: A10

Gupta A (1986) India. In Rushing F W, Brown C G (eds) *National policies for developing high technology industries: international comparisons* Boulder, CO, Westview Press: 89–110

Gupta A K, Wilemon D (1988) Why R&D resists using marketing information. *Research-Technology Management* 31(6): 36–41

Haar N E, Starr J, MacMillan I C (1988) Informal risk capital investors: investment patterns on the east coast of the USA. *Journal of Business Venturing* 3: 11–29

Hadjimichalis C (1987) *Uneven development and regionalism: state, territory and class in southern Europe* London, Croom Helm

Hagen E E (1963) How economic growth begins: a theory of social change. *Journal of Social Issues* 19: 20–34

Hagey M J, Malecki E J (1986) Linkages in high technology industries: a Florida case study. *Environment and Planning A* 18: 1477–98

Hahn F H, Matthews R C O (1964) The theory of economic growth: a survey. *Economic Journal* 74: 779–902

Haider D (1986) Economic development: changing practices in a changing US economy. *Environment and Planning C: Government and Policy* 4: 451–69

Hakam A N, Chang Z-Y (1988) Patterns of technology transfer in Singapore: the case of the electronics and computer industry. *International Journal of Technology Management* 3: 181–8

Håkanson L (1979) Towards a theory of location and corporate growth. In Hamilton F E I, Linge G J R (eds) *Spatial analysis, industry and the industrial environment* vol 1: *Industrial systems* London, John Wiley: 115–38

Håkansson H (ed) (1987) *Industrial technological development: a network approach* London, Croom Helm

Håkansson H (1989) *Corporate technological behaviour: co-operation and networks* London, Routledge

Håkansson H, Laage-Hellman J (1984) Developing a network R&D strategy. *Journal of Product Innovation Management* 1: 224–37

Hall C M (1989) The definition and analysis of hallmark tourist events. *GeoJournal* 19: 263–8

Hall P (1982) The geography of innovation: planning for the fifth Kondratieff. *Environments* 14(2): 9–11

Hall P (1987) The anatomy of job creation: nations, regions and cities in the 1960s and 1970s. *Regional Studies* 21: 95–106

Hall P, Breheny M, McQuaid R, Hart D (1987) *Western sunrise: the genesis and growth of Britain's major high tech corridor* London, Allen and Unwin

Hall P, Markusen A R (eds) (1985) *Silicon landscapes* Boston, Allen and Unwin

Hall P, Markusen A R, Osburn R, Wachsman B (1983) The American computer software industry: economic development prospects. *Built Environment* 9(1): 29–39

Hall P, Preston P (1988) *The carrier wave: new information technology and the geography of innovation 1846–2003* London, Unwin Hyman

Hallwood C P (1988) Host regions and the globalization of the offshore oil supply industry: the case of Aberdeen. *International Regional Science Review* 11: 155–66

Hambrecht W R (1984) Venture capital and the growth of Silicon Valley. *California Management Review* 26(2): 74–82

Hamel G, Prahalad C K (1985) Do you really have a global strategy? *Harvard Business Review* 61(4): 139–48

Hamfelt C, Lindberg A K (1987) Technological development and the individual's contact network. In Håkansson H (ed) *Industrial technological development: a network approach* London, Croom Helm: 177–219

Hamilton F E I, Linge G J R (1983) Regional economies and industrial systems. In Hamilton F E I, Linge G J R (eds) *Spatial analysis, industry and the industrial environment* vol 3: *Regional industrial systems* Chichester, John Wiley: 1–57

Hamilton W F (1986) Corporate strategies for managing emerging technologies. In Horwitch M (ed) *Technology in the modern corporation: a strategic perspective* Oxford, Pergamon: 103–18

Hammer M, Mangurian G E (1987) The changing value of communications technology. *Sloan Management Review* 28(2): 65–71

Hammermesh R G, Silk S B (1979) How to compete in stagnant industries. *Harvard Business Review* 57(5): 161–8

Hammond G T (1990) *Countertrade, offsets and barter in international political economy* London, Pinter

Hampton W J (1988) How does Japan Inc. pick its American workers? *Business Week* October 3: 84–8

Hansen N M (1967) Development pole theory in a regional context. *Kyklos* 20: 709–27

Hansen N M (1971) *Intermediate-size cities as growth centers* New York, Praeger

Hansen N M (1973) *Location preferences, migration, and regional growth* New York, Praeger

Hansen N M (1980) Dualism, capital–labor ratios and the regions of the U.S.: a comment. *Journal of Regional Science* 20: 401–3

Hansen N (1987) The evolution of the French regional economy and French regional theory. *Review of Regional Studies* 17(3): 5–13

Hansen N M (1988) Economic development and regional heterogeneity: a reconsideration of regional policy for the United States. *Economic Development Quarterly* 2(2): 107–18

Hanson R (ed) (1983) *Rethinking urban policy* Washington, National Academy Press

Harber A D, Samson D A (1989) Japanese management practices: an integrative framework. *International Journal of Technology Management* 4: 283–303

Harding C F (1989) Location choices for research labs: a case study approach. *Economic Development Quarterly* 3: 223–34

Hargrave A (1985) *Silicon Glen: reality or illusion?* Edinburgh, Mainstream Publishing

Harner D, Haynes K E (1975) The impact of natural growth centers: an empirical investigation in west Texas. *Review of Regional Studies* 5(2): 84–97

Harper M (1984) *Small business in the Third World* Chichester, John Wiley

Harper M (1989) Partnerships of large and small firms: lessons for the Third

World. In Commission of the European Communities (ed) *Partnership between large and small firms* London, Graham and Trotman: 80–9

Harper R A (1982) Metropolitan areas as transactional centers. In Christian C, Harper R A (eds) *Modern metropolitan systems* Columbus, OH, Merrill: 87–109

Harrigan F, McGilvray J, McNicoll I (1980) A comparison of regional and national technical structures. *Economic Journal* 90: 795–810

Harrington J W (1987) Strategy formulation, organisational learning, and location. In van der Knaap B, Wever E (eds) *New technology and regional development* London, Croom Helm: 63–74

Harris C S (1986) Establishing high-technology enterprises in metropolitan areas. In Bergman E M (ed) *Local economies in transition* Durham, NC, Duke University Press: 165–84

Harris R I D (1987) The role of manufacturing in regional growth. *Regional Studies* 21: 301–12

Harris R I D (1988) Technological change and regional development in the UK: evidence from the SPRU database on innovations. *Regional Studies* 22: 361–74

Harrison B (1982) The tendency toward instability and inequality underlying the 'revival' of New England. *Papers of the Regional Science Association* 50: 41–65

Harrison B (1984) Regional restructuring and 'good business climates': the economic transformation of New England since World War II. In Sawers L, Tabb W K (eds) *Sunbelt/snowbelt: urban development and regional restructuring* New York, Oxford University Press: 48–96

Harrison B, Bluestone B (1988) *The great U-turn: corporate restructuring and the polarizing of America* New York, Basic Books

Harrison B, Sum A (1979) The theory of 'dual' or segmented labor markets. *Journal of Economic Issues* 13: 687–706

Harrison R T, Mason C M (1988) Risk finance, the equity gap and new venture formation in the United Kingdom: the impact of the Business Expansion Scheme. In Kirchhoff B A, Long W A, McMullan W E, Vesper K H, Wetzel W E (eds) *Frontiers of entrepreneurship research 1988* Wellesley, MA, Babson College, Center for Entrepreneurial Studies: 595–609

Hartmann G, Nicholas I, Sorge A, Warner M (1985) Computerized machine tools, manpower consequences and skill utilization: a study of British and West German manufacturing firms. In Rhodes E, Wield D (eds) *Implementing new technologies* Oxford, Basil Blackwell: 352–60

Harvey D (1988) The geographical and geopolitical consequences of the transition from Fordist to flexible accumulation. In Sternlieb G, Hughes J W (eds) *America's new market geography* New Brunswick, NJ, Center for Urban Policy Research: 101–34

Harvey D, Scott A (1989) The practice of human geography: theory and empirical specificity in the transition from Fordism to flexible accumulation. In Macmillan B (ed) *Remodelling geography* Oxford, Basil Blackwell: 217–29

Haslett B (1984) Venture capital and regional high-technology development. In Office of Technology Assessment *Technology, innovation, and regional economic development* Washington, US Government Printing Office: 41–50

Haug P (1986) US high technology multinationals and Silicon Glen. *Regional Studies* 20: 103–16

Haustein H-D, Maier H (1980) Basic improvement and pseudo-innovations and their impact on efficiency. *Technological Forecasting and Social Change* 16: 243–65

Hayes R H, Wheelwright S C (1984) *Restoring our competitive edge: competing through manufacturing* New York, John Wiley

Hayes R H, Wheelwright S C (1988) Matching process technology with product/market requirements. In Tushman M L, Moore W L (eds) *Readings in the management of innovation* 2nd edn Cambridge, MA, Ballinger: 417–43

Hayes R H, Wheelwright S C, Clark K B (1988) *Dynamic manufacturing* New York, Free Press

Hayter R (1982) Truncation, the international firm and regional policy. *Area* 14: 277–82

Hayter R (1986) Export performance and export potentials: western Canadian exports of manufactured end products. *Canadian Geographer* 30: 26–39

Hayter R, Watts H D (1983) The geography of enterprise: an appraisal. *Progress in Human Geography* 7: 157–81

Headrick D R (1988) *The tentacles of progress: technology transfer in the age of imperialism, 1850–1940* Oxford, Oxford University Press

Hébert R F, Link A N (1982) *The entrepreneur* New York, Praeger

Heilbroner R L (1963) *The great ascent: the struggle for economic development in our time* New York, Harper and Row

Hekman J S, Strong J S (1981) The evolution of New England industry. *New England Economic Review* March/April: 35–46

Helm L, Takahashi K, Arnold B (1985) Japan's secret weapon: exploited women. *Business Week* March 4: 54–5

Hemming R, Mansour A M (1988) Is privatization the answer? *Finance and Development* 25(3): 31–3

Henderson B D (1984) *The logic of business strategy* Cambridge, MA, Ballinger

Henderson J (1989) *The globalisation of high technology production* London, Routledge

Henderson J, Scott A J (1987) The growth and internationalisation of the American semiconductor industry: labour processes and the spatial organisation of production. In Breheny M J, McQuaid R W (eds) *The development of high technology industries: an international survey* London, Croom Helm: 37–79

Henderson Y K (1989) The emergence of the venture capital industry. *New England Economic Review* July/August: 64–79

Henry D L (1984) Searching for the right high-tech site. *Area Development* 19(9): 14–15, 92–9

Henry D L (1987) How Japanese executives select US sites. *Area Development* 22(8): 24–8, 120–46

Hepworth M (1986) The geography of technological change in the information economy. *Regional Studies* 20: 407–24

Hepworth M (1987) Information technology as spatial systems. *Progress in Human Geography* 11: 157–80

Hepworth M (1989) *Geography of the information economy* London, Belhaven Press

Hepworth M E, Green A E, Gillespie A E (1987) The spatial division of information labour in Great Britain. *Environment and Planning A* 19: 793–806

Hermansen T (1972) Development poles and development centres in national and regional development: elements of a theoretical framework. In Kuklinski A (ed) *Growth poles and growth centres in regional planning.* The Hague, Mouton 1–67

Hewings G J D (1977) *Regional industrial analysis and development* London, Methuen

Hewings G J D (1985) *Regional input–output analysis* Beverly Hills, CA, Sage

Hewings G J D, Jensen R J (1986) Regional, interregional and multiregional input–output analysis. In Nijkamp P (ed) *Handbook of regional and urban economics* vol 1: *Regional economics* Amsterdam, North–Holland: 295–355

Hewings G J D, Sonis M, Jensen R C (1988) Fields of influence of technological change in input–output models. *Papers of the Regional Science Association* 64: 25–36

Hicks D L (1986) *Automation technology and industrial renewal* Washington, American Enterprise Institute

Higgins B (1981) National development and regional policy. In Prantilla E B (ed) *National development and regional policy* Singapore, Maruzen Asia: 15–55

Higgins B (1983) From growth poles to systems of interactions in space. *Growth and Change* 14(4): 3–13

Higgins B (1988) Regional development and efficiency of the national economy. In Higgins B, Savoie D J (eds) *Regional development: essays in honour of François Perroux* Boston, Unwin Hyman: 193–224

Higgins B, Savoie D J (eds) (1988) *Regional economic development: essays in honour of François Perroux* Boston, Unwin Hyman

Hildebrand G H, Mace A (1950) The employment multiplier in an expanding industrial market: Los Angeles County, 1940–47. *Review of Economics and Statistics* 32: 241–49

Hill R C (1989) Comparing transnational production systems: the automobile industry in the USA and Japan. *International Journal of Urban and Regional Research* 13: 462–80

Hilling D, Hoyle B S (1984) Spatial approaches to port development. In Hoyle B S, Hilling D (eds) *Seaport systems and spatial change* Chichester, John Wiley: 1–19

Hiraoka L S (1989) Japanese automobile manufacturing in an American setting. *Technological Forecasting and Social Change* 35: 29–49

Hirschhorn L (1979) The urban crisis: a post-industrial perspective. *Journal of Regional Science* 19: 109–18

Hirschhorn L (1984) *Beyond mechanization* Cambridge, MA, MIT Press

Hirschman A (1958) *The strategy of economic development* New Haven, Yale University Press

Hisrich R D (1989) Women entrepreneurs: problems and prescriptions for success in the future. In Hagan O, Rivchun C, Sexton D (eds) *Women-owned businesses* New York, Praeger: 3–32

Hitomi K (1989) Non-mass, multi-product, small-sized production: the state of the art. *Technovation* 9: 357–69

Hitt M A, Ireland R D, Goryunov I Y (1988) The context of innovation: investment in R&D and firm performance. In Gattiker U E, Larwood L (eds) *Managing technological development: strategic and human resources issues* Berlin: Walter de Gruyter: 73–91

Hjalager A-M (1989) Why no entrepreneurs? Life modes, everyday life and unemployment strategies in an underdeveloped region. *Entrepreneurship and Regional Development* 1: 85–97

Hlavacek J D, Ames B C (1986) Segmenting industrial and high-tech markets. *Journal of Business Strategy* 7(2): 39–50

Hoare A G (1985) Industrial linkage studies. In Pacione M (ed) *Progress in industrial geography* London, Croom Helm: 40–81

Hobday M (1988) Evaluating collaborative R&D programmes in information technology: the case of the U.K. Alvey programme. *Technovation* 8: 271–98

Hobday M (1989) Corporate strategies in the international semiconductor industry. *Research Policy* 18: 225–38

Hoerr J (1989) The payoff from teamwork. *Business Week* July 10: 56–62

Hoffman C (1972) The role of the commercial loan officer in the formation and growth of new and young technical companies. In Cooper A C, Komives J L (eds) *Technical entrepreneurship: a symposium* Milwaukee, Center for Venture Management: 165–88

Hoffman K, Kaplinsky R (1988) *Driving force: the global restructuring of technology, labor, and investment in the automobile and components industries* Boulder, CO, Westview Press

Hoffman K, Rush H (1988) *Micro-electronics and clothing: the impact of technical change on a global industry* New York, Praeger

Hofstede G (1980) *Culture's consequences: international differences in work-related values* Beverley Hills, CA, Sage

Hollier G P (1988) Regional development. In Pacione M (ed) *The geography of the third world: progress and prospects* London, Routledge: 232–70

Holman M (1974) *The political economy of the space program* Palo Alto, CA, Pacific Books

Holmes J (1986) The organization and locational structure of production subcontracting. In Scott A J, Storper M (eds) *Production, work, territory* Boston, Allen and Unwin: 80–106

Holmes J (1987) Technical change and the restructuring of the North American automobile industry. In Chapman K, Humphrys G (eds) *Technical change and industrial policy* Oxford, Basil Blackwell: 121–56

Hoppes R B (1982) Disaggregate tertiary earnings per capita – a spatial analysis. *Regional Science Perspectives* 12(2): 36–51

Hornaday J A, Churchill N C (1987) Current trends in entrepreneurial research. In Churchill N C, Hornaday J A, Kirchhoff B A, Krasner O J, Vesper K H (eds) *Frontiers of entrepreneurship research 1987* Wellesley, MA, Babson College, Center for Entrepreneurial Studies: 1–21

Horwitch M (ed) (1986) *Technology in the modern corporation: a strategic perspective* Oxford, Pergamon

Horwitz P (1979) Direct government funding of research and development: intended and unintended consequences. In Hill C T, Utterback J M (eds) *Technological innovation for a dynamic economy* Oxford, Pergamon: 255–91

Hout T, Porter M E, Rudden E (1982) How global companies win out. *Harvard Business Review* 60(5): 98–108

Houttuin G (1985) Venture capitalism. In Sweeney G (ed) *Innovation policies* London, Frances Pinter: 189–96

Howells J R L (1984) The location of research and development: some observations and evidence from Britain. *Regional Studies* 18: 13–29

Howells J (1986) Industry–academic links in research and innovation: a national and regional development perspective. *Regional Studies* 20: 472–6

Howells J (1988) *Economic, technological and locational trends in European services* Aldershot, Avebury

Howells J, Charles D (1988) Research and technological development in the 'less-favoured' regions of the European Community: a UK dimension. In Dyson K (ed) *Local authorities and new technologies: the European dimension* London, Croom Helm: 24–48

Howland M I (1984a) Regional variations in cyclical employment. *Environment and Planning A* 16: 863–77

Howland M I (1984b) Age of capital and regional business cycles. *Growth and Change* 15(2): 29–37

Howland M I (1988a) *Plant closings and worker displacement: the regional issues* Kalamazoo, MI, W E Upjohn Institute for Employment Research

Howland M I (1988b) Plant closures and local economic conditions. *Regional Studies* 22: 193–207

Hoyt H (1954) Homer Hoyt on development of economic base concept. *Land Economics* 30: 182–91

Hudson R (1983) Regional labour reserves and industrialisation in the EEC. *Area* 15: 223–30

Hudson R (1989) Labour-market changes and new forms of work in old industrial regions: maybe flexibility for some but not flexible accumulation. *Environment and Planning D: Society and Space* 7: 5–30

Hudson R, Lewis J (eds) (1985) *Uneven development in southern Europe* London, Methuen

Hughes Aircraft (1978) *R&D productivity* 2nd edn Culver City, CA, Hughes Aircraft Company

Hughes H (ed) (1988) *Achieving industrialization in East Asia* Cambridge, Cambridge University Press

Hughes K (1988) The interpretation and measurement of R&D intensity – a note. *Research Policy* 17: 301–7

Hull F, Hage J, Azumi K (1984) Strategies for innovation in Japan and America. *Technovation* 2: 121–39

Hull F, Hage J, Azumi K (1985) R&D management strategies: America versus Japan. *IEEE Transactions on Engineering Management* 32: 78–83

Hull F M, Collins P D (1987) High-technology batch production systems: Woodward's missing type. *Academy of Management Journal* 30(4): 786–97

Humphrey C R, Erickson R A, McCluskey R E (1989) Industrial development groups, external connections, and job generation in local communities. *Economic Development Quarterly* 3: 32–45

Humphrey C R, Erickson R A, Ottensmeyer E J (1988) Industrial development groups, organizational resources, and the prospects for effecting growth in local economies. *Growth and Change* 19(3): 1–21

Hunt H A, Hunt T L (1983) *The human resource implications of robotics* Kalamazoo, MI, W E Upjohn Institute for Employment Research

Huskey L (1985) Import substitution: the hidden dynamic in the growth of frontier regions. *Growth and Change* 16(4): 43–55

Hyman R (1988) Flexible specialization: miracle or myth? In Hyman R, Streeck W (eds) *New technology and industrial relations* Oxford, Basil Blackwell: 48–60

Hyman R, Streeck W (eds) (1988) *New technology and industrial relations* Oxford, Basil Blackwell

Hymer S (1975) The multinational corporation and the law of uneven development. In Radice H (ed) *International firms and modern imperialism* Harmondsworth, Penguin: 37–62

Illeris S (1986) New firm formation in Denmark: the importance of the cultural background. In Keeble D, Wever E (eds) *New firms and regional development in Europe* London, Croom Helm: 141–50

Illeris S (1989) *Services and regions in Europe* Aldershot, Avebury

Imai K (1986) Japan's industrial policy for high technology industry. In Patrick H (ed) *Japan's high technology industries: lessons and limitations of industrial policy* Seattle, University of Washington Press: 137–69

Imai K (1988) Industrial policy and technological innovation. In Komiya R, Okuno M, Suzumura K (eds) *Industrial policy of Japan* Tokyo, Academic Press: 205–29

Ingalls G L, Martin W E (1988) Defining and identifying NICs. In Norwine J, Gonzales A (eds) *The Third World: states of mind and being* Boston, Unwin Hyman: 82–98

Ioannou L (1985) States move to stake entrepreneurs. *Venture* 7(7): 60–9

Ironside R G, Williams A G (1980) The spread effects of a spontaneous growth centre: commuter expenditure patterns in the Edmonton metropolitan region, Canada. *Regional Studies* 14: 313–32

Isard W (1956) *Location and space-economy* Cambridge MA, MIT Press

Isard W (1960) *Methods of regional analysis* New York, John Wiley

Ishitani H, Kaya Y (1989) Robotization in Japanese manufacturing industries. *Technological Forecasting and Social Change* 35: 97–131

ISNAR (1987) *Working to strengthen national agricultural research systems – ISNAR and its strategy* The Hague, International Service for National Agricultural Research Systems

Isserman A M (1977) The location quotient approach to measuring regional economic impacts. *Journal of the American Institute of Planners* 43: 33–41

Isserman A M (1980) Estimating export activity in a regional economy: a theoretical and empirical analysis of alternative methods. *International Regional Science Review* 5: 155–84

Isserman A M (1985) Economic–demographic modeling with endogenously determined birth and migration rates: theory and prospects. *Environment and Planning A* 17: 25–45

Isserman A, Taylor C, Gerking S, Schubert U (1986) Regional labor market analysis. In Nijkamp P (ed) *Handbook of regional and urban economics* vol 1: *Regional economics* Amsterdam, North-Holland: 543–80

Itami H (1989) Mobilising invisible assets: a key for successful corporate strategy. In Punset E, Sweeney G (eds) *Information resources and corporate growth* London, Pinter: 36–46

Jaikumar R (1986) Post-industrial manufacturing. *Harvard Business Review* 64(6): 69–76

James D D (1979) The economic case for more indigenous scientific and technological research and development in less developed countries. In Street J H, James D D (eds) *Technological progress in Latin America: the prospects for overcoming dependency* Boulder, CO, Westview Press: 83–108

James D D, Street J H, Jedlicka A D (1980) Issues in indigenous research and development in Third World countries. *Social Science Quarterly* 60: 588–603

James J (1989) *Improving traditional rural technologies* London, Macmillan

Janssen B, van Hoogstraten P (1989) The 'new infrastructure' and regional development. In Albrechts L, Moulaert F, Roberts P, Swyngedouw E (eds) *Regional policy at the crossroads: European perspectives* London, Jessica Kingsley: 52–66

Jarboe K P (1986) Location decisions of high-technology firms: a case study. *Technovation* 4: 117–29

Jarillo J C (1989) Entrepreneurship and growth: the strategic use of external resources. *Journal of Business Venturing* 4: 133–47

Jensen R C, Hewings G J D (1985) Shortcut 'input–output' multipliers: a requiem. *Environment and Planning A* 17: 747–59

Jensen R C, West G R, Hewings G J D (1988) The study of regional economic structure using input–output tables. *Regional Studies* 22: 209–20

Jéquier N (1979) Appropriate technology: some criteria. In Bhalla A S (ed) *Towards global action for appropriate technology* Oxford, Pergamon: 1–22

Jéquier N, Hu Y-S (1989) *Banking and the promotion of technological development* New York, St Martin's Press

Jewkes J, Sawers D, Stillerman R (1969) *The sources of invention* 2nd edn New York, W W Norton

Jin X-Y, Porter A L (1988) Technological innovation and development: prospects for China. *IEEE Transactions on Engineering Management* 35: 258–64

Johannisson B (1987) Entrepreneurship in a corporatist state: the case of Sweden. In Goffee R, Scase R (eds) *Entrepreneurship in Europe: the social processes* London, Croom Helm: 131–43

Johannisson B, Nilsson A (1989) Community entrepreneurs: networking for local development. *Entrepreneurship and Regional Development* 1: 3–19

Johansson B (1987) Information technology and the viability of spatial networks. *Papers of the Regional Science Association* 61: 51–64

Johansson B, Strömquist U (1981) Regional rigidities in the process of economic structural adjustment. *Regional Science and Urban Economics* 11: 363–75

Johansson B, Westin L (1987) Technical change, location, and trade. *Papers of the Regional Science Association* 62: 13–25

Johne F A (1986) The adoption of new technology in manufacturing firms. In Woodside A G (ed) *Advances in business marketing* Greenwich, CT, JAI Press: 141–62

Johne F A, Snelson P A (1988) Success factors in product innovation: a selective review of the literature. *Journal of Product Innovation Management* 5: 114–28

Johnson M L (1988) Labor environment and the location of electrical machinery employment in the US south. *Growth and Change* 19(2): 56–74

Johnson P S (1973) *Cooperative research in industry* New York, John Wiley

Johnson P S (1975) *The economics of invention and innovation* London, Martin Robertson

Johnson P (1986) *New firms: an economic perspective* London, Allen and Unwin

Johnson P S, Cathcart D G (1979) New manufacturing firms and regional development: some evidence from the Northern Region. *Regional Studies* 13: 269–80

Johnson W E, Lucas R D (1986) Strategic entrepreneurship in high-technology companies. In Gardner J R, Rachlin R, Sweeny H W A (eds) *Handbook of strategic planning* New York, Wiley: 20.1–20.78

Johnston R J (1986) The state, the region, and the division of labor. In Scott A J, Storper M (eds) *Production, work, territory* Boston, Allen and Unwin: 265–80

Johnston W B, Packer A E (1987) *Workforce 2000: work and workers for the twenty-first century* Indianapolis, Hudson Institute

Jones B (1982) *Sleepers, wake! Technology and the future of work* Melbourne, Oxford University Press

Jorgenson D W (1988) Productivity and economic growth in Japan and the United States. *American Economic Review* 78(2): 217–22

Joseph R A (1988) Technology parks and their contribution to the development of technology-oriented complexes in Australia. *Environment and Planning C: Government and Policy* 7: 173–92

Jowitt T (1988) Towards silicondale? Lessons from American experience for the Pennine towns and cities. In Dyson K (ed) *Local authorities and new technologies: the European dimension* London, Croom Helm: 124–39

Juma C (1989) *The gene hunters: biotechnology and the scramble for seeds* London, Zed

Juma C, Ojwang J B (1989) Innovation and sovereignty. In Juma C, Ojwang J B (eds) *Innovation and sovereignty* Nairobi, African Centre for Technology Studies: 7–27

Junkerman J (1987) Blue-sky management: the Kawasaki story. In Peet R (ed) *International capitalism and industrial restructuring* Boston, Allen and Unwin: 131–44

Jusenius C L, Ledebur L C (1977) *Where have all the firms gone? An analysis of the New England economy* Washington, US Economic Development Administration

Kagono T, Nonaka I, Sakakibara K, Okumura A (1985) *Strategic vs. evolutionary management: a U.S.–Japan comparison of strategy and organization* Amsterdam, North-Holland

Kaldor M (1980) Technical change in the defence industry. In Pavitt K (ed) *Technical innovation and British economic performance* London, Macmillan: 100–21

Kaldor M (1981) *The baroque arsenal* New York, Hill and Wang

Kaldor N (1970) The case for regional policies. *Scottish Journal of Political Economy* 17: 337–47

Kaneda M (1980) Position of Japanese industry. In Murata K (ed) *Industrial geography of Japan* London, Bell and Hyman: 28–32

Kanter R M (1984) Variations in managerial career structures in high-technology firms: the impact of organizational characteristics on internal labor market patterns. In Osterman P (ed) *Internal labor markets* Cambridge, MA, MIT Press: 109–31

Kantrow A (1980) The strategy–technology connection. *Harvard Business Review* 58(4): 6–21

Kaplinsky R (1984a) *Automation: the technology and society* London, Longman

Kaplinsky R (1984b) Trade in technology – who, what, where and when? In Fransman M, King K (eds) *Technological capability in the Third World* London, Macmillan: 139–60

Kapustin E (1980) Economic growth and labour productivity. In Matthews R C O (ed) *Economic growth and resources,* vol 2: *Trends and factors* New York, St Martin's Press: 130–44

Karaska G J, Linge G J R (1978) Applicability of the model of the territorial production complex outside the USSR. *Papers of the Regional Science Association* 40: 149–69

Kash D E (1989) *Perpetual innovation: the new world of competition* New York, Basic Books

Kassel S (1989) *Soviet advanced technologies in the era of restructuring* Santa Monica, CA, Rand Corporation

Katz J E (ed) (1986) *The implications of Third World military industrialization* Lexington, MA, Lexington Books

Katz J M (1982a) Technology and economic development: an overview of research findings. In Syrquin M, Teitel S (eds) *Trade, stability, technology, and equity in Latin America* London, Academic Press: 281–315

Katz J M (1982b) Technological change and development in Latin America. In French-Davis R, Tironi E (eds) *Latin America and the new international economic order* New York, St Martin's Press: 192–215

Katz R L (1988) *The information society: an international perspective* New York, Praeger

Kawashima T (1975) Urban agglomeration economies in manufacturing industries. *Papers of the Regional Science Association* 34: 157–75

Kawashima T, Stöhr W (1988) Decentralized technology policy: the case of Japan. *Environment and Planning C: Government and Policy* 6: 427–39

Kay N M (1979) *The innovating firm: a behavioural theory of corporate R&D* New York, St Martin's Press

Kay N M (1984) *The emergent firm: knowledge, ignorance and surprise in economic organisation* London, Macmillan

Keeble D (1967) Models of economic development. In Chorley R J, Haggett P (eds) *Models in geography* London, Methuen: 243–302

Keeble D (1988) High-technology industry and local environments in the United Kingdom. In Aydalot P, Keeble D (eds) *High technology industry and innovative environments: the European experience* London, Routledge: 65–98

Keeble D E (1989) High technology industry and regional development in Britain: the case of the Cambridge phenomenon. *Environment and Planning C: Government and Policy* 7: 153–72

Keeble D, Kelly T (1986) New firms and high-technology industry in the United Kingdom: the case of computer electronics. In Keeble D, Wever E (eds) *New firms and regional development in Europe* London, Croom Helm: 184–202

Keeble D, Wever E (eds) (1986) *New firms and regional development in Europe* London, Croom Helm

Keil S R, Mack R S (1986) Identifying export potential in the service sector. *Growth and Change* 17(2): 1–10

Kellerman A (1985) The evolution of service economies: a geographical perspective. *Professional Geographer* 37: 133–43

Kelley M R, Brooks H (1989) From breakthrough to follow-through. *Issues in Science and Technology* 5(3): 42–7

Kelly D L (1986) R&D and manufacturing operations: single or separate locations? *Area Development* 21(9): 10, 23, 32

Kendrick J W (1977) *Understanding productivity* Baltimore, Johns Hopkins University Press

Kennedy C, Thirlwall A P (1972) Technical progress: a survey. *Economic Journal* 82: 11–72

Kennedy P (1987) *The rise and fall of the great powers* New York, Random House

Kenney M (1986) Schumpeterian innovation and entrepreneurs in capitalism: a case study of the U.S. biotechnology industry. *Research Policy* 15: 21–31

Kenney M, Florida R (1988) Beyond mass production: production and the labor process in Japan. *Politics and Society* 16: 121–58

Kent C A (1984) The rediscovery of the entrepreneur. In Kent C A (ed) *The environment for entrepreneurship* Lexington, MA, Lexington Books: 1–19

Kierulff H E (1979) Finding – and keeping – corporate entrepreneurs. *Business Horizons* 22(1): 6–15

Kim L (1980) Stages of development of industrial technology in a developing country: a model. *Research Policy* 9(3): 254–77

Kim L, Lee H (1987) Patterns of technological change in a rapidly developing country: a synthesis. *Technovation* 6: 261–76

Kim L, Lee J, Lee J (1987) Korea's entry into the computer industry and its acquisition of technological capability. *Technovation* 6: 277–93

Kim S (1987) Diversity in urban labor markets and agglomeration economies. *Papers of the Regional Science Association* 62: 57–70

King A (1982) For better and for worse: the benefits and risks of information technology. In Bjorn-Andersen N, Earl M, Holst O, Mumford E (eds) *Information society: for richer, for poorer* Amsterdam, North-Holland: 35–56

King L J, Clark G L (1978) Government policy and regional development. *Progress in Human Geography* 2: 1–16

Kinsey J (1978) The application of growth pole theory in the Aire Métropolitaine Marseillaise. *Geoforum* 9: 245–67

Kipnis B A (1976) Local versus national coefficients in constructing regional input–output tables in small countries: a case study in northern Israel. *Journal of Regional Science* 16: 93–9

Kipnis B A, Swyngedouw E A (1988a) Manufacturing plant size – toward a regional strategy. A case study in Limburg, Belgium. *Urban Studies* 25: 43–52

Kipnis B A, Swyngedouw E A (1988b) Manufacturing research and development in a peripheral region: the case of Limburg, Belgium. *Professional Geographer* 40: 149–58

Kirby A (1985) Nine fallacies of local economic change. *Urban Affairs Quarterly* 21: 207–20

Kirchhoff B A, Phillips B D (1988) The effect of firm formation and growth on job creation in the United States. *Journal of Business Venturing* 3: 259–72

Kirk R (1987) Are business services immune to the business cycle? *Growth and Change* 18(2): 15–24

Kirn T J (1987) Growth and change in the service sector of the U.S.: a spatial perspective. *Annals of the Association of American Geographers* 77: 353–72

Kirzner I M (1984) The entrepreneurial process. In Kent C A (ed) *The environment for entrepreneurship* Lexington, MA, Lexington Books: 41–58

Kiser J W (1989) *Communist entrepreneurs: unknown innovators in the global economy* New York, Franklin Watts

Klaassen L (1987) The future of the larger European towns. *Urban Studies* 24: 251–7

Klaudt H (1987) Trends in small business start-ups in West Germany. In Goffee R, Scase R (eds) *Entrepreneurship in Europe: the social processes* London, Croom Helm: 26–38

Klein L R, Glickman N J (1977) Econometric model-building at regional level. *Regional Science and Urban Economics* 7: 3–23

Kleinknecht A (1987) *Innovation patterns in crisis and prosperity* New York, St. Martin's Press

Klimstra P D, Potts J (1988) What we've learned: managing R&D projects. *Research-Technology Management* 31(3): 23–39

Kline S J, Rosenberg N (1986) An overview of innovation. In Landau R, Rosenberg N (eds) *The positive sum strategy* Washington, National Academy Press: 275–305

Kloppenburg J, Kleiman D L (1987) Seeds of struggle: the geopolitics of genetic resources. *Technology Review* 90(2): 47–53

Knapp G, Huskey L (1988) Effects of transfers on remote regional economies: the transfer economy in rural Alaska. *Growth and Change* 19(2): 25–39

Knapp T A, Graves P E (1989) On the role of amenities in models of migration and economic development. *Journal of Regional Science* 29: 71–87

Knight R V (1986) The advanced industrial metropolis: a new type of world city. In Ewers H-J, Goddard J B, Matzerath H (eds) *The future of the metropolis* Berlin, Walter de Gruyter: 391–436

Knox P L, Agnew J A (1989) *The geography of the world economy* London, Edward Arnold

Kojima K, Ozawa T (1984) *Japan's general trading companies: merchants of economic development* Paris, Organisation for Economic Co-operation and Development

Kok J A A M, Pellenbarg P H (1987) Innovation decision-making in small and medium-sized firms: a behavioural approach concerning firms in the Dutch urban system. In van der Knaap B, Wever E (eds) *New technology and regional development* London, Croom Helm: 145–64

Kolm J E (1988) Regional and national consequences of globalizing industries of the Pacific rim. In Muroyama J H, Stever H G (eds) *Globalization of technology: international perspectives* Washington, National Academy Press: 106–40

Komiya R (1988) Introduction. In Komiya R, Okuno M, Suzumura K (eds) *Industrial policy of Japan* Tokyo, Academic Press: 1–22

Kono T (1984) *Strategy and structure of Japanese enterprises* London, Macmillan

Kono T (1988) Factors affecting the creativity of organizations – an approach from the analysis of new product development. In Urabe K, Child J, Kagano T (eds) *Innovation and management: international comparisons* Berlin, Walter de Gruyter: 105–44

Koshiro K (1987) Personnel planning, technological changes and outsourcing in the Japanese automobile industry: part 1. *International Journal of Technology Management* 2: 279–97

Kotler P, Fahey L, Jatusripitak S (1985) *The new competition* Englewood Cliffs, Prentice-Hall

Kowalski J S (1979) Development pole theory in the perspective of studies of the contacts of central, sectoral and regional systems. In Kuklinski A, Kultalahti O, Koskiaho B (eds) *Regional dynamics of socioeconomic change* Tampere, Finnpublishers: 209–18

Kowalski J S (1986) Regional conflicts in Poland: spatial polarization in a centrally planned economy. *Environment and Planning A* 18: 599–617

Kraar L (1989) Asia's rising export powers. *Fortune* 120 (13): 43–50

Krafcik J F (1988) Triumph of the lean production system. *Sloan Management Review* 30(1): 41–52

Krist H (1985) Innovation centres as an element of strategies for endogenous regional development. In Gibb J M (ed) *Science parks and innovation centres: their economic and social impact* Amsterdam, Elsevier: 178–88

Kristensen P H (1989) Denmark: an experimental laboratory for new industrial models. *Entrepreneurship and Regional Development* 1: 245–55

Krueger A O (1983) *Trade and employment in developing countries* vol 3: *Synthesis and conclusions* Chicago, University of Chicago Press

Krugman P (1979) A model of innovation, technology transfer, and the world distribution of income. *Journal of Political Economy* 87: 253–66

Krugman P R (ed) (1986) *Strategic trade policy and the new international economics* Cambridge, MA, MIT Press

Krumme G, Hayter R (1975) Implications of corporate strategies and product cycle adjustments for regional employment changes. In Collins L, Walker DF (eds) *Locational dynamics of manufacturing activity* New York, John Wiley: 325–56

Kuklinski A (ed) (1972) *Growth poles and growth centres in regional planning* The Hague, Mouton

Kuklinski A (ed) (1975a) *Regional disaggregation of national policies and plans* The Hague, Mouton

Kuklinski A (ed) (1975b) *Regional development and planning: international perspectives* Leyden, Sijthoff

Kuklinski A (ed) (1978) *Regional policies in Nigeria, India, and Brazil* The Hague, Mouton

Kuklinski A (ed) (1981) *Polarized development and regional policies: tribute to Jacques Boudeville* The Hague, Mouton

Kuklinski A, Petrella R (eds) (1972) *Growth poles and regional policies* The Hague, Mouton

Kulicke M, Krupp H (1987) The formation, relevance and public promotion of new technology-based firms. *Technovation* 6: 47–56

Kumar N (1987) Intangible assets, internalisation and foreign production: direct investments and licensing in Indian manufacturing. *Weltwirtschaftsliches Archiv* 123: 325–45

Kunzmann K R (1988) Military production and regional development in the Federal Republic of Germany. In Breheny M J (ed) *Defence expenditure and regional development* London, Mansell: 49–66

Kutay A (1989) Prospects for high technology based economic development in mature industrial regions: Pittsburgh as a case study. *Journal of Urban Affairs* 11: 361–79

Kutscher R E (1988) Structural changes of employment in the United States. In Candilis W O (ed) *United States service industries handbook* New York, Praeger: 23–44

Kutscher R E, Mark J A (1983) The service-producing sector: some common perceptions reviewed. *Monthly Labor Review* 106(4): 21–4

Kuttner R (1986) Party favors. *The New Republic* 195 (December 22): 21–2

Kuznets S (1972) Innovations and adjustments in economic growth. *Swedish Journal of Economics* 74: 431–51

Lacave M (1989) French urban research parks combine scientific research, economic development, and quality of life. *French Advances in Science and Technology* 3(4): 4–5

Lacroix R, Martin F (1988) Government and the decentralization of R&D. *Research Policy* 17: 363–73

Lakshmanan T R, Hua C-I (1987) Regional disparities in China. *International Regional Science Review* 11: 97–104

Lall S (1980) Developing countries as exporters of industrial technology. *Research Policy* 9: 24–52

Lall S (1985) Trade in technology by a slowly industrializing country: India. In Rosenberg N, Frischtak C (eds) *International technology transfer: concepts, measures, and comparisons* New York, Praeger: 45–76

Lande P S (1978) The interregional comparison of production functions. *Regional Science and Urban Economics* 8: 339–53

Lande P S (1982) The regional-development implications of industrial policy. In Bell M E, Lande P S (eds) *Regional dimensions of industrial policy* Lexington, MA, Lexington Books: 81–91

Landes D S (1980) The 'great drain' and industrialisation: commodity flows from periphery to centre in historical perspective. In Matthews R C O (ed) *Economic growth and resources* vol 2: *Trends and factors* New York, St Martin's Press: 294–327

Landes D S (1989) Rich country, poor country. *The New Republic* November 20: 23–7

Lane T (1966) The urban base multiplier: an evaluation of the state of the art. *Land Economics* 42: 339–47

Langdale J (1983) Competition in the United States' long-distance telecommunications industry. *Regional Studies* 17: 393–409

Langrish J, Gibbons M, Evans W G, Jevons F R (1972) *Wealth from knowledge* London, Macmillan

Lash S, Bagguley P (1988) Labour relations in disorganized capitalism: a five-nation comparison. *Environment and Planning D: Society and Space* 6: 321–38

Lasuen J R (1969) On growth poles. *Urban Studies* 6: 137–61

Lasuen J R (1973) Urbanisation and development – the temporal interaction between geographical and sectoral clusters. *Urban Studies* 10: 163–88

Law C M (1983) The defence sector in regional development. *Geoforum* 14: 169–84

Lawton-Smith H (1990) Innovation and technical links: the case of advanced technology industry in Oxfordshire. *Area* 22: 125–35

Leavitt F C (1968) Basic research at a distance. *Research Management* 11: 225–9

Leborgne D, Lipietz A (1988) New technologies, new modes of regulation: some spatial implications. *Environment and Planning D: Society and Space* 6: 263–80

Lee D, Brady R (1988) Long, hard days – at pennies an hour. *Business Week* October 31: 46–7

Lee D, Engardio P, Dunkin A (1989) U.S. importers aren't jumping ship – yet. Table title, Wages in China: still the lowest. *Business Week* June 26: 78

Lee E (1981) Basic-needs strategies: a frustrated response to development from below? In Söthr W B, Taylor D R F (eds) *Development from above or below?* New York, John Wiley: 107–22

Leff N H (1979a) Entrepreneurship and economic development: the problem revisited. *Journal of Economic Literature* 17: 46–64

Leff N H (1979b) 'Monopoly capitalism' and public policy in developing countries. *Kyklos* 32: 718–38

Le Heron R B (1973) Best-practice technology, technical leadership, and regional economic development. *Environment and Planning* 5: 735–49

Leibenstein H (1978) *General-efficiency theory and economic development* New York, Oxford University Press

Leinbach T R, Amrhein C (1987) A geography of the venture capital industry in the USA. *Professional Geographer* 39: 146–58

Lenz J E (1989) *Flexible manufacturing: benefits from the low-inventory factory* New York, Marcel Dekker

Leonard-Barton D (1984) Interpersonal communication patterns among Swedish and Boston-area entrepreneurs. *Research Policy* 13: 101–14

Leontief W (1983) Technological advance, economic growth, and the distribution of income. *Population and Development Review* 9: 403–10

Leslie S W (1980) Thomas Midgley and the politics of industrial research. *Business History Review* 54: 480–503

Leven C L (1964) Regional and interregional accounts in perspective. *Papers of the Regional Science Association* 13: 127–44

Leven C L (1985) Regional development analysis and policy. *Journal of Regional Science* 25: 569–92

Lever W F (1979) Industry and labour markets in Great Britain. In Hamilton F E I, Linge G J R (eds) *Spatial analysis, industry and the industrial environment* vol 1: *Industrial systems* Chichester, John Wiley: 89–114

Levin R (1982) The semiconductor industry. In Nelson R R (ed) *Government and technical progress: a cross-industry analysis* New York, Pergamon: 9–100

Levy J M (1981) *Economic development programs for cities, counties, and towns* New York, McGraw-Hill

Lewis G, Gross N, Levine J B, Verity J W, Therrien L, Cole P E (1989) Computers: Japan comes on strong. *Business Week* October 23: 104–12

Lewis H, Allison D (1982) *The real world war* New York, Coward, McCann and Geoghegan

Lewis J R, Williams A M (1987) Productive decentralization or indigenous growth? Small manufacturing enterprises and regional development in Central Portugal. *Regional Studies* 21: 343–61

Lewis P M (1986) The economic impact of the operation and closure of a nuclear power station. *Regional Studies* 20: 425–32

Lewis W C (1972) A critical examination of the export–base theory of urban–regional growth. *Annals of Regional Science* 6(3): 15–25

Lewis W C (1976) Export base theory and multiplier estimation: a critique. *Annals of Regional Science* 10(2): 58–70

Lifton D E, Lifton L R (1989) Applying the Japanese 'thin markets' strategy to industrial new product development. *International Journal of Technology Management* 4: 177–88

Lim L Y C (1980) Women workers in multinational corporations: the case of the electronics industry in Malaysia and Singapore. In Kumar K (ed) *Transnational enterprises: their impact on Third World societies and cultures* Boulder, CO, Westview Press: 109–36

Linge G J R, Hamilton F E I (1981) International industrial systems. In Hamilton F E I, Linge G J R (eds) *Spatial analysis, industry and the industrial environment* vol 2: *International industrial systems* London, John Wiley: 1–117

Link A N, Bauer L L (1989) *Cooperative research in US manufacturing* Lexington, MA, Lexington Books

Linn R A (1984) Product development in the chemical industry: a description of a maturing business. *Journal of Product Innovation Management* 1: 116–28

Lipietz A (1986) New tendencies in the international division of labor: regimes of accumulation and modes of regulation. In Scott A J, Storper M (eds) *Production, work, territory* Boston, Allen and Unwin: 16–40

Littler C R (1988) Technology, innovation and labour-management strategies. In Urabe K, Child J, Kagono T (eds) *Innovation and management: international comparisons* Berlin, Walter de Gruyter: 337–58

Lloyd P E (1989) Research and policy review 28. Fragmenting markets and the dynamic restructuring of production: issues for spatial policy. *Environment and Planning A* 21: 429–44

Lloyd P E, Reeve D E (1982) North-West England 1971–1977: a study in industrial decline and economic re-structuring. *Regional Studies* 16: 345–59

Lo F-C, Salih K (eds) (1978) *Growth pole strategy and regional development policy* Oxford, Pergamon

Lo F-C, Salih K, Douglass M (1981) Rural–urban transformation in Asia. In Lo F-C (ed) *Rural–urban relations and regional development* Singapore, Maruzen Asia: 7–43

Lock P (1986) Brazil: arms for export. In Brzoska M, Ohlson T (eds) *Arms production in the Third World* London, Taylor and Francis: 79–104

Lorenz E H (1989) The search for flexibility: subcontracting networks in British and French engineering. In Hirst P, Zeitlin J (eds) *Reversing industrial decline? Industrial structure and policy in Britain and her competitors* Oxford: Berg: 122–32

Lorenzoni G, Ornati O A (1988) Constellations of firms and new ventures. *Journal of Business Venturing* 3: 41–57

Lotchin R W (ed) (1984) *The martial metropolis* New York, Praeger

Louri H (1988) Urban growth and productivity: the case of Greece. *Urban Studies* 25: 433–8

Lovering J (1985) Regional intervention, defence industries, and the structuring of space in Britain: the case of Bristol and South Wales. *Environment and Planning D: Society and Space* 3: 85–107

Lovering J (1988) Islands of prosperity: the spatial impact of high-technology defence industry in Britain. In Breheny M J (ed) *Defence expenditure and regional development* London, Mansell: 29–48

Lovering J (1990) Fordism's unknown successor: a comment on Scott's theory of flexible accumulation and the re-emergence of regional economies. *International Journal of Urban and Regional Research* 14: 159–74

Lovering J, Boddy M (1988) The geography of military industry in Britain. *Area* 20: 41–51

Luger M I (1984) Does North Carolina's high-technology development policy work? *Journal of the American Planning Association* 50: 280–9

Luger M I (1985) The states and high-technology development: the case of North Carolina. In Whittington D (ed) *High hopes for high tech: microelectronics policy in North Carolina* Chapel Hill, University of North Carolina Press: 193–224

Luger M I, Evans W N (1988) Geographic differences in production technology. *Regional Science and Urban Economics* 18: 399–424

Lund L (1986) *Locating corporate R&D facilities* Research report number 892, New York, The Conference Board

Lundvall B-A (1988) Innovation as an interactive process: from user–producer interaction to the national system of innovation. In Dosi G, Freeman C, Nelson R, Silverberg G, Soete L (eds) *Technical change and economic theory* London, Pinter: 349–69

Luttrell W F (1972) Industrial complexes and regional economic development in Canada. In Kuklinski A (ed) *Growth poles and growth centres in regional planning* The Hague, Mouton: 243–62

Lutz N M (1988) Images of docility: Asian women and the world-economy. In Smith J, Collins J, Hopkins T K, Muhammad A (eds) *Racism, sexism, and the world-system* New York, Greenwood Press: 57–74

Lyson T A (1989) *Two sides to the sunbelt* New York, Praeger

McArthur R (1989) Locality and small firms: some reflections from the Franco-British project, 'Industrial systems, technical change and locality.' *Environment and Planning D: Society and Space* 7: 197–210

McArthur R (1990) Replacing the concept of high technology: towards a diffusion-based approach. *Environment and Planning A* 22: 811–28

Mabogunje A L (1978) Growth poles and growth centres in the regional development of Nigeria. In Kuklinski A (ed) *Regional policies in Nigeria, India, and Brazil* The Hague, Mouton: 1–93

MacCharles D C (1987) *Trade among multinationals: intra-industry trade and national competitiveness* London, Croom Helm

McCombie J S L (1988a) A synoptic view of regional growth and unemployment: I – the neoclassical theory. *Urban Studies* 25: 267–81

McCombie J S L (1988b) A synoptic view of regional growth and unemployment: II – the post-Keynesian theory. *Urban Studies* 25: 399–417

McCrackin B H (1984) Education's contribution to productivity and economic growth. *Economic Review, Federal Reserve Bank of Atlanta* 69(10): 8–23

McDermott P, Taylor M (1982) *Industrial organisation and location* Cambridge, Cambridge University Press

McDonald D W, Leahey H S (1985) Licensing has a role in technology strategic planning. *Research Management* 28(1): 35–40

Macdonald M, Perry J (1985) *Pension funds and venture capital: the critical links between savings, investment, technology, and jobs* Ottawa, Science Council of Canada

Macdonald R J (1985) Strategic alternatives in emerging industries. *Journal of Product Innovation Management* 2: 158–69

Macdonald S (1986) Headhunting in high technology. *Technovation* 4: 233–45

Macdonald S (1987) Towards higher high technology policy. In Brotchie J F, Hall P, Newton P W (eds) *The spatial impact of technological change* London, Croom Helm: 357–74

McGee T G (1971) *The urbanization process in the Third World* London, Bell

Macgregor B D, Langridge R J, Adley J, Chapman J (1986) The development of high technology industry in Newbury district. *Regional Studies* 20: 433–47

McGuinness N W, Conway H A (1986) World product mandates: the need for directed search strategies. In Etemad H, Séguin Dulude L (eds) *Managing the multinational subsidiary: response to environmental changes and to host nation R&D policies* New York, St Martin's Press: 136–58

McGuire S (1989) White sand, blue seas – and big dreams. *Newsweek* January 9: 37

McHenry W K (1987) The application of computer aided design at Soviet enterprises: an overview. In Baranson J, Dolan J M, Heslin M, Leneman B, McHenry W (eds) *Soviet automation: perspectives and prospects* Mt Airy, MD, Lomond Books: 57–76

Machlup F (1962) *The production and distribution of knowledge in the United States* Princeton, NJ, Princeton University Press

McHone W W (1987) Factors in the adoption of industrial development incentives by states. *Applied Economics* 19: 17–29

McIntyre J R, Papp D S (1986) *The political economy of international technology transfer* Westport, CT, Quorum Books

McIntyre S H (1988) Market adaptation as a process in the product life cycle of radical innovations and high technology products. *Journal of Product Innovation Management* 5: 140–9

McKay J, Whitelaw J S (1977) The role of large private and government organizations in generating flows of inter-regional migrants: the case of Australia. *Economic Geography* 53: 28–44

McKee D L (1974) Some reflections on urban dualism in mature economies. *Review of Regional Studies* 4(2): 74–8

McKee D L (1988) *Growth, development, and the service economy in the Third World* New York, Praeger

MacMillan I C, Kulow D M, Khoylian R (1989) Venture capitalists' involvement in their investments: extent and performance. *Journal of Business Venturing* 4: 27–47

McNaughton R B, Green M B (1989) Spatial patterns of Canadian venture capital investment. *Regional Studies* 23: 9–18

McNulty J E (1977) A test of the time dimension in economic base analysis. *Land Economics* 53: 359–68

MacPherson A (1988a) Industrial innovation in the small business sector: empirical evidence from metropolitan Toronto. *Environment and Planning A* 20: 953–71

MacPherson A (1988b) New product development among small Toronto manufacturers: empirical evidence on the role of technical service linkages. *Economic Geography* 64: 62–75

McQueen D H, Wallmark J T (1985) Support for new ventures at Chalmers University of Technology. In Gibb J M (ed) *Science parks and innovation centres: their economic and social impact* Amsterdam, Elsevier: 153–61

McUsic M (1987) US manufacturing: any cause for alarm? *New England Economic Review* January/February: 3–17

Magaziner I C, Hout T M (1980) *Japanese industrial policy* London, Policy Studies Institute

Mager N H (1987) *The Kondratieff waves* New York, Praeger

Maggi R, Haeni P K (1986) Spatial concentration, location and competitiveness: the case of Switzerland. *Regional Studies* 20: 141–9

Maidique M A (1980) Entrepreneurs, champions, and technological innovation. *Sloan Management Review* 21(2): 59–76

Maidique M A, Hayes R H (1984) The art of high-technology management. *Sloan Management Review* 25(2): 17–31

Maidique M A, Patch P (1988) Corporate strategy and technological policy. In Tushman M L, Moore W L (eds) *Readings in the management of innovation* 2nd edn Cambridge, MA, Ballinger: 236–48

Maidique M A, Zirger B J (1985) The new product learning cycle. *Research Policy* 14: 299–313

Maier H (1982) Innovation, efficiency, and the quantitative and qualitative demand for human resources. *Technological Forecasting and Social Change* 21: 15–31

Maillat D (1984) Conditions d'une stratégie de développement par le bas: le cas de la région horlogère Suisse. *Revue d'Economie Régionale et Urbaine*: 257–73

Maillat D (1990) SMEs, innovation and territorial development. In Cappellin R, Nijkamp P (eds) *The spatial context of technological development* Aldershot, Avebury: 331–51

Maillat D, Vasserot J-Y (1988) Economic and territorial conditions for indigenous revival in Europe's industrial regions. In Aydalot P, Keeble D (eds) *High technology industry and innovative environments: the European experience* London, Routledge: 163–83

Mair A, Florida R, Kenney M (1988) The new geography of automobile production: Japanese transplants in North America. *Economic Geography* 64: 352–73

Majumdar B A (1988) Industrial policy in action: the case of the electronics industry in Japan. *Columbia Journal of World Business* 23(3): 25–34

Malecki E J (1977) Firms and innovation diffusion: examples from banking. *Environment and Planning A* 9: 1291–305

Malecki E J (1979) Locational trends in R&D by large US corporations, 1965–1977. *Economic Geography* 55: 309–23

Malecki E J (1980a) Corporate organization of R and D and the location of technological activities. *Regional Studies* 14: 219–34

Malecki E J (1980b) Dimensions of R&D location in the United States. *Research Policy* 9: 2–22

Malecki E J (1981a) Science, technology, and regional economic development: review and prospects. *Research Policy* 10: 312–34

Malecki E J (1981b) Product cycles, innovation cycles, and regional economic change. *Technological Forecasting and Social Change* 19: 291–306

Malecki E J (1982) Federal R and D spending in the United States of America: some impacts on metropolitan economies. *Regional Studies* 16: 19–35

Malecki E J (1983) Technology and regional development: a survey. *International Regional Science Review* 8: 89–125

Malecki E J (1984a) High technology and local economic development. *Journal of the American Planning Association* 50: 260–9

Malecki E J (1984b) Military spending and the US defense industry: regional patterns of contracts and subcontracts. *Environment and Planning C: Government and Policy* 2: 31–44

Malecki E J (1985a) Industrial location and corporate organization in high technology industries. *Economic Geography* 61: 345–69

Malecki E J (1985b) Public sector research and development and regional economic performance. In Thwaites A T, Oakey R P (eds) *The regional impact of technological change* London, Frances Pinter: 115–31

Malecki E J (1986a) Word games and fuzzy thinking. *Environment and Planning A* 18: 289–90

Malecki E J (1986b) Research and development and the geography of high-technology complexes. In Rees J (ed) *Technology, regions, and policy* Totowa, NJ, Rowman and Littlefield: 51–74

Malecki E J (1986c) Technological imperatives and modern corporate strategy. In Scott A J, Storper M (eds) *Production, work, territory* Boston, Allen and Unwin: 67–79

Malecki E J (1987a) Hope or hyperbole? High tech and economic development. *Technology Review* 90(7): 44–52

Malecki E J (1987b) The R&D location decision of the firm and 'creative' regions. *Technovation* 6: 205–22

Malecki E J (1988) New firm startups: key to rural growth. *Rural Development Perspectives* 4(2): 18–23

Malecki E J (1989) What about people in high technology? Some research and policy considerations. *Growth and Change* 20(1): 67–79

Malecki E J (1990) Technological innovation and paths to regional economic growth. In Schmandt J, Wilson R W (eds) *Growth policy in an age of high technology* Boston, Unwin Hyman: 97–126

Malecki E J, Nijkamp P (1988) Technology and regional development: some thoughts on policy. *Environment and Planning C: Government and Policy* 6: 383–99

Malecki E J, Stark L (1988) Regional and industrial variation in defence spending: some American evidence. In Breheny M J (ed) *Defence expenditure and regional development* London, Mansell: 67–101

Malecki E J, Varaiya P (1986) Innovation and changes in regional structure. In Nijkamp P (ed) *Handbook of regional and urban economics* vol 1: *Regional economics* Amsterdam, North-Holland: 629–45

Malerba F (1985) Demand structure and technological change: the case of the European semiconductor industry. *Research Policy* 14: 283–97

Malizia E E (1985) The locational attractiveness of the Southeast to high-technology manufacturers. In Whittington D (ed) *High hopes for high tech: microelectronics policy in North Carolina* Chapel Hill, University of North Carolina Press: 173–90

Mandel E (1975) *Late capitalism* London, New Left Books

Mandel E (1980) *Long waves of capitalist development* Cambridge, Cambridge University Press

Mandelbaum K (1945) *The industrialization of backward areas* Institute of Statistics Monograph No. 2, Oxford, Basil Blackwell

Mangum G, Mayall D, Nelson K (1985) The temporary help industry: a response to the dual internal labor market. *Industrial and Labor Relations Review* 38: 599–611

Mansell R (1988) The role of information and telecommunication technologies in regional development. *Science Technology Industry Review* 3: 135–73

Mansfield E (1988a) Industrial innovation in Japan and the United States. *Science* 241: 1769–74

Mansfield E (1988b) The speed and cost of industrial innovation in Japan and the United States: external vs. internal technology. *Management Science* 34: 1157–68

Mansfield E (1989) The diffusion of industrial robots in Japan and the United States. *Research Policy* 18: 183–92

Mansfield E, Rapaport J, Romeo A, Villani E, Wagner S, Husic F (1977) *The production and application of new industrial technology* New York, Norton

Mansfield E, Romeo A, Schwartz M, Teece D, Wagner S, Brach P (1982) *Technology transfer, productivity, and economic policy* New York, Norton

Mansfield E, Teece D, Romeo R (1979) Overseas research and development by US-based firms. *Economica* 46: 187–96

Mantel S J, Rosegger G (1987) The role of third-parties in the diffusion of innovations: a survey. In Rothwell R, Bessant J (eds) *Innovation: adaptation and growth* Amsterdam, Elsevier: 123–34

Marchand C (1986) The transmission of fluctuations in a central place system. *Canadian Geographer* 30: 249–54

Marcum J M (1986) The technology gap: Europe at a crossroads. *Issues in Science and Technology* 2(4): 28–37

Marelli E (1983) Empirical estimation of intersectoral and interregional transfers of surplus value: the case of Italy. *Journal of Regional Science* 23: 49–70

Marelli E (1985) Economic policies and their effects upon regional economies. *Papers of the Regional Science Association* 58: 127–39

Maren M (1988) Textiles Togo. *Venture* 10 (12): 30–5

Marer P (1986) *East–west technology transfer: study of Hungary 1968–1984* Paris, Organisation for Economic Co-operation and Development

Mariani L (1989) Collaboration between large and small and medium-sized companies in the experience of the ENI Group. In Commission of the European Communities (ed) *Partnership between small and large firms* London, Graham and Trotman: 113–19

Markusen A R (1985) *Profit cycles, oligopoly, and regional development* Cambridge, MA, MIT Press

Markusen A R (1986a) Defence spending: a successful industrial policy? *International Journal of Urban and Regional Research* 10: 105–22

Markusen A R (1986b) Defense spending and the geography of high-tech industries. In Rees J (ed) *Technology, regions, and policy* Totowa, NJ, Rowman and Littlefield: 94–119

Markusen A (1986c) Neither ore, nor coal, nor markets: a policy-oriented view of steel sites in the USA. *Regional Studies* 20: 449–61

Markusen A R (1987) *Regions: the economics and politics of territory* Totowa NJ, Rowman and Allanheld

Markusen A R (1988) The military remapping of the United States. In Breheny M J (ed) *Defence expenditure and regional development* London, Mansell: 17–28

Markusen A R, Bloch R (1985) Defensive cities: military spending, high technology, and human settlements. In Castells M (ed) *High technology, space, and society* Beverly Hills, CA, Sage: 106–20

Markusen A, Hall P, Glasmeier A (1986) *High tech America* Boston, Allen and Unwin

Markusen A R, Teitz M B (1985) The world of small business: turbulence and survival. In Storey D J (ed) *Small firms in regional economic development* Cambridge, Cambridge University Press: 193–217

Marquis D G (1988) The anatomy of successful innovations. In Tushman M L, Moore W L (eds) *Readings in the management of innovation* 2nd edn Cambridge, MA, Ballinger: 79–87

Marshall J N, Wood P, Daniels P W, McKinnon A, Bachtler J, Damesick P, Thrift N, Gillespie A, Green A, Leyshon A (1988) *Services and uneven development* Oxford, Oxford University Press

Marshall M (1987) *Long waves of regional development* New York, St Martin's Press

Martin B R, Irvine J (1984) CERN: past performance and future prospects III. CERN and the future of world high-energy physics. *Research Policy* 13: 311–42

Martin F (1986) L'entrepreneurship et le développement local: une évaluation. *Canadian Journal of Regional Science* 9: 1–23.

Martin M J C (1984) *Managing technological innovation and entrepreneurship* Reston, VA, Reston Publishing

Martin R (1988) The political economy of Britain's north–south divide. *Transactions, Institute of British Geographers* NS 13: 389–418

Martin R (1989) The growth and geographical anatomy of venture capitalism in the United Kingdom. *Regional Studies* 23: 389–403

Martin R C (1979) Federal regional development programs and US problem regions. *Journal of Regional Science* 19: 157–70

Martin R C, Miley H W (1983) The stability of economic base multipliers: some empirical evidence. *Review of Regional Studies* 13(3): 18–25

Martin R L, Hodge J S C (1983) The reconstruction of British regional policy: 1. The crisis of conventional practice. *Environment and Planning C: Government and Policy* 1: 133–52

Martinelli F (1985) Public policy and industrial development in southern Italy: anatomy of a dependent industry. *International Journal of Urban and Regional Research* 9: 47–81

Martinez N, Nueño P (1988) Catalan regional business culture, entrepreneurship, and management behavior: an exploratory study. In Weiss J W (ed) *Regional cultures, managerial behavior, and entrepreneurship: an international perspective* New York, Quorum Books: 61–75

Marton K (1986) *Multinationals, technology, and industrialization: implications and impact in Third World countries* Lexington, MA, Lexington Books

Mason C M (1985) The geography of 'successful' small firms in the United Kingdom. *Environment and Planning A* 17: 1499–513

Mason C, Harrison R T (1985) The geography of small firms in the UK: towards a research agenda. *Progress in Human Geography* 9: 1–37

Mason E S (1982) Stages of economic growth revisited. In Kindleberger C P, di Tella G (eds) *Economics in the long view: essays in honour of W W Rostow* vol 1, New York, New York University Press: 116–40

Massey D (1973) The basic : service categorization in planning. *Regional Studies* 7: 1–15

Massey D (1979a) A critical evaluation of industrial-location theory. In Hamilton F E I, Linge G J R (eds) *Spatial analysis, industry and the industrial environment* vol 1: *Industrial systems* London, John Wiley: 57–72

Massey D (1979b) In what sense a regional problem? *Regional Studies* 13: 233–43

Massey D (1984) *Spatial divisions of labour* London, Macmillan

Massey D, Meegan R (1982) *The anatomy of job loss* London, Methuen

Mayer W, Pleeter S (1975) A theoretical justification for the use of location quotients. *Regional Science and Urban Economics* 5: 343–55

Meller P, Marfán M (1981) Small and large industry: employment generation, linkages, and key sectors. *Economic Development and Cultural Change* 29: 263–74

Mellor I, Ironside R G (1978) The incidence multiplier of a regional development programme. *Canadian Geographer* 22: 225–51

Mensch G (1979) *Stalemate in technology: innovations overcome the depression* Cambridge, MA, Ballinger

Mera K (1974) Trade-off between aggregate efficiency and interregional equity: the case of Japan. *Regional and Urban Economics* 4: 273–300

Meyer D R (1977) Agglomeration economies and urban–industrial growth: a clarification and review of concepts. *Regional Science Perspectives* 7(1): 80–91

Meyer D R (1983) Emergence of the American manufacturing belt: an interpretation. *Journal of Historical Geography* 9: 145–74

Meyer D R (1986) System of cities' dynamics in newly industrializing nations. *Studies in Comparative International Development* 21: 3–22

Meyer J R (1963) Regional economics: a survey. *American Economic Review* 53: 19–54

Meyer-Krahmer F (1985) Innovation behaviour and regional indigenous potential. *Regional Studies* 19: 523–34

Miernyk W H (1980) The tools of regional development policy: an evaluation. *Growth and Change* 11(2): 2–6

Miernyk W H, Shellhammer K L, Brown D M, Cochari R L, Gallagher C J, Wineman W H (1970) *Simulating regional economic development* Lexington, MA, Lexington Books

Miles I (1988) *Home informatics: information technology and the transformation of everyday life* London, Frances Pinter

Miller D L (1986) Mexico. In Rushing F W, Brown C G (eds) *National policies for developing high technology industries: international comparisons* Boulder, CO, Westview Press: 173–99

Miller H M, Brown E E, Centner T J (1986) Southern Appalachian handicrafts industry: implications for regional economic development. *Review of Regional Studies* 16(3): 50–8

Miller J P (1985) Rethinking small businesses as the best way to create rural jobs. *Rural Development Perspectives* 1(2): 9–12

Miller R, Coté M (1985) Growing the next silicon valley. *Harvard Business Review* 63(4): 114–23

Miller R, Coté M (1987) *Growing the next silicon valley* Lexington, MA, Lexington Books

Miller R E, Blair P D (1985) *Input–output analysis: foundations and extensions* Englewood Cliffs, NJ, Prentice-Hall

Mishan E J (1976) *The economic growth debate: an assessment* London, George Allen and Unwin

Mitchell G R, Hamilton W F (1988) Managing R&D as a strategic option. *Research-Technology Management* 31(3): 15–22

Mitchell R (1989) Nurturing those ideas. *Business Week* June 16: 106–18

MITI (1990) *Outline and present status of the Technopolis project* Japanese Ministry of International Trade and Industry, Industrial Location and Environmental Protection Bureau

Mittelman J H (1988) *Out from underdevelopment* New York, St Martin's Press

Moffett M (1990) Battling the odds against high-tech success in Mexico. *Wall Street Journal* February 9: B2

Mogee M E (1980a) The process of technological innovation in industry: a state-of-knowledge review for Congress. In *Special study on economic change* vol 3: *Research and innovation: developing a dynamic nation* Washington, US Government Printing Office: 171–256

Mogee M E (1980b) The relationship of federal support of basic research in universities to industrial innovation and productivity. In *Special study on economic change* vol 3: *Research and innovation: developing a dynamic nation* Washington, US Government Printing Office: 257–79

Mokry B W (1988) *Entrepreneurship and public policy: can government stimulate business startups?* New York, Quorum Books

Molero J (1983) Foreign technology in the Spanish economy: an analysis of the recent evolution. *Research Policy* 12: 269–86

Molina A H (1989) *The social basis of the microelectronics revolution* Edinburgh, Edinburgh University Press

Molle W, Beumer L, Boeckhout I (1989) The location of information intensive activities in the European Community. In Punset E, Sweeney G (eds) *Information resources and corporate growth* London, Pinter: 161–72

Monck C S P, Porter R B, Quintas P, Storey D J, Wynarczyk P (1988) *Science parks and the growth of high technology firms* London, Croom Helm

Monkiewicz J (1989) *International technology flows and the technology gap: the experience of Eastern European socialist countries in international perspective* Boulder, CO, Westview Press

Monkiewicz J, Maciejewicz J (1986) *Technology exports from the socialist countries* Boulder, CO, Westview Press

Moody H T, Puffer F W (1970) The empirical verification of the urban base multiplier: traditional and adjustment process models. *Land Economics* 46: 91–8

Moomaw R L (1981) Productivity and city size: a critique of the evidence. *Quarterly Journal of Economics* 94: 675–88

Moomaw R L (1983) Spatial productivity variations in manufacturing: a critical survey of cross-sectional analyses. *International Regional Science Review* 8: 1–22

Moomaw R L (1988) Agglomeration economies: localization or urbanization? *Urban Studies* 25: 150–61

Moore B, Rhodes J (1976) Regional economic policy and the movement of manufacturing firms to development areas. *Economica* 43: 17–31

Moore C L (1975) A new look at the minimum requirements approach to regional economic analysis. *Economic Geography* 51: 350–6

Moore C L, Jacobsen M (1984) Minimum requirements and regional economics, 1980. *Economic Geography* 60: 217–24

Morbey G K (1989) R&D expenditures and profit growth. *Research-Technology Management* 32(3): 20–3

More R (1985) Barriers to innovation: intraorganizational dislocations. *Journal of Product Innovation Management* 2: 205–7

Morehouse W, Gupta B (1987) India: success and failure. In Segal A (ed) *Learning by doing: science and technology in the developing world* Boulder, CO, Westview Press: 189–212

Morgan K, Sayer A (1985) A 'modern' industry in a 'mature' region: the remaking of management–labour relations. *International Journal of Urban and Regional Research* 9: 383–404

Morgan K, Sayer A (1988) *Microcircuits of capital: 'sunrise' industry and uneven development* Cambridge, Polity Press

Moriarty B M (1980) *Industrial location and community development* Chapel Hill, University of North Carolina Press

Morita K, Hiraoka H (1988) Technopolis Osaka: integrating urban functions and science. In Smilor R W, Kozmetsky G, Gibson D V (eds) *Creating the technopolis* Cambridge, MA, Ballinger: 23–49

Morrill R L (1973) On the size and spacing of growth centers. *Growth and Change* 4(2): 21–4

Morris J L (1988a) Producer services and the regions: the case of large accountancy firms. *Environment and Planning A* 20: 741–59

Morris J L (1988b) New technologies, flexible work practices, and regional sociospatial differentiation: some observations from the United Kingdom. *Environment and Planning D: Society and Space* 6: 301–19

Moseley M J (1974) *Growth centres in spatial planning* Oxford, Pergamon

Moss M L (1986) Telecommunications and the future of cities. *Land Development Studies* 3: 33–44

Moss M L (1987) Telecommunications, world cities, and urban policy. *Urban Studies* 24: 534–46

Moss M L (1988) Telecommunications: shaping the future. In Sternlieb G, Hughes J W (eds) *America's new market geography* New Brunswick, NJ, Center for Urban Policy Research: 255–75

Moudoud E (1989) *Modernization, the state, and regional disparity in developing countries* Boulder, CO, Westview Press

Moulaert F, Salinas P W (eds) (1982) *Regional analysis and the new international division of labor* Boston, Kluwer-Nijhoff

Moulaert F, Swyngedouw E (1989) Survey 15. A regulation approach to the geography of flexible production systems. *Environment and Planning D: Society and Space* 7(3): 336–7

Moulaert F, Swyngedouw E, Wilson P (1988) Spatial responses to Fordist and post-Fordist accumulation and regulation. *Papers of the Regional Science Association* 64: 11–23

Mowery D C (1983a) The relationship between intrafirm and contractual forms of industrial research in American manufacturing, 1900–1940. *Explorations in Economic History* 20: 351–74

Mowery D C (1983b) Economic theory and government technology policy. *Policy Sciences* 16: 27–43

Mowery D C (ed) (1988a) *International collaborative ventures in US manufacturing* Cambridge, MA, Ballinger

Mowery D C (1988b) The diffusion of new manufacturing technologies. In Cyert R M, Mowery D C (eds) *The impact of technological change on employment and economic growth* Cambridge, MA, Ballinger: 481–509

Mowery D C (1989) Collaborative ventures between US and foreign manufacturing firms. *Research Policy* 18: 19–32

Mowery D C, Rosenberg N (1979) The influence of market demand upon innovation: a critical review of some recent empirical studies. *Research Policy* 8: 102–53

Moyes A, Westhead P (1990) Environments for new firm formation in Great Britain. *Regional Studies* 24: 123–36

Mufson S (1985) Don't thank us, Pepsi; we just talked with a few bean buyers. *Wall Street Journal* July 17: 31

Mulkey D, Dillman B D (1976) Location effects of state and local industrial development subsidies. *Growth and Change* 7(2): 37–43

Müller J (1984) Facilitating an indigenous social organisation of production in Tanzania. In Fransman M, King K (eds) *Technological capability in the Third World* London, Macmillan: 375–90

Müller R (1979) The multinational corporation and the underdevelopment of the Third World. In Wilber C K (ed) *The political economy of development and underdevelopment* 2nd edn. New York, Random House: 151–78

Mulligan G (1987) Employment multipliers and functional types of communities: effects of public transfer payments. *Growth and Change* 18 (3): 1–11

Mulligan G F, Gibson L J (1984) Regression estimates of economic base multipliers for small communities. *Economic Geography* 60: 225–37

Myers D D, Hobbs D J (1985) Profile of location preferences for non-metropolitan high tech firms. In Hornaday J A, Shils E B, Timmons J A, Vesper K H (eds) *Frontiers of entrepreneurship research 1985* Wellesley, MA, Babson College, Center for Entrepreneurial Studies: 358–82

Myrdal G (1957) *Economic theory and underdeveloped regions* London, Duckworth

Mytelka L K (1985) Stimulating effective technology transfer: the case of textiles in Africa. In Rosenberg N, Frischtak C (eds) *International technology transfer: concepts, measures, and comparisons* New York, Praeger: 77–126

Nagamine H (ed) (1981) *Human needs and regional development* Singapore, Maruzen Asia

Nakarmi L, Shao M, Griffiths D (1989) South Korea's new destination: the wild blue yonder. *Business Week* September 11: 50

Narin F, Frame J D (1989) The growth of Japanese science and technology. *Science* 245: 600–5

Nash M (1985) *Making people productive* San Francisco, Jossey-Bass

National Academy of Sciences and National Academy of Engineering (1969) *The impact of science and technology on regional economic development* Washington, National Academy of Sciences

National Science Foundation (1988) Economic growth and corporate mergers dampen growth in company R&D. *Science Resources Studies Highlights* March 11

Nau H (1986) National policies for high technology development and trade: an international and comparative assessment. In Rushing F W, Brown C G (eds) *National policies for developing high technology industries* Boulder, CO, Westview Press: 9–29

Naylor P (1985) High and low technology businesses. In Gibb J M (ed) *Science parks and innovation centres: their economic and social impact* Amsterdam, Elsevier: 246–9

Near J P, Olshavsky R W (1985) Japan's success: luck or skill? *Business Horizons* 28(6): 15–22

Neff R, Magnusson P, Holstein W J (1989) Rethinking Japan. *Business Week* August 7: 44–52

Nelson K (1986) Labor demand, labor supply and the suburbanization of low-wage office work. In Scott A J, Storper M (eds) *Work, production, territory* Boston, Allen and Unwin: 149–71

Nelson R R (1981) Research on productivity growth and productivity differences: dead ends and new departures. *Journal of Economic Literature* 19: 1029–64

Nelson R R (1982) Government stimulus of technological progress: lessons from American history. In Nelson R R (ed) *Government and technical progress: a cross-industry study* New York, Pergamon: 451–82

Nelson R R (1984) *High-technology policies: a five-nation comparison* Washington, American Enterprise Institute

Nelson R R (1986) Incentives for entrepreneurship and supporting institutions. In Balassa B, Giersch H (eds) *Economic incentives* New York, St Martin's Press: 173–87

Nelson R R (1987) *Understanding technical change as an evolutionary process* Amsterdam, Elsevier

Nelson R R, Norman V D (1977) Technological change and factor mix over the product cycle. *Journal of Development Economics* 4: 3–24

Nelson R R, Winter S G (1977) In search of useful theory of innovation. *Research Policy* 6: 36–76

Nelson R R, Winter S G (1982) *An evolutionary theory of economic change* Cambridge, MA, Harvard University Press

Newman R J (1984) *Growth in the American South: changing regional employment and wage patterns in the 1960s and 1970s* New York, New York University Press

Nijkamp P (1982) Long waves or catastrophes in regional development. *Socio-Economic Planning Sciences* 16: 261–71

Nijkamp P, Mouwen A (1987) Knowledge centres, information diffusion and regional development. In Brotchie J F, Hall P, Newton P W (eds) *The spatial impact of technological change* London, Croom Helm: 254–70

Nijkamp P, Rietveld P (1986) Multiple objective decision analysis in regional economics. In Nijkamp P (ed) *Handbook of regional and urban economics* vol 1: *Regional economics* Amsterdam, North-Holland: 493–541

Nijkamp P, Rietveld P, Snickars F (1986) Regional and multiregional economic models: a survey. In Nijkamp P (ed) *Handbook of regional and urban economics* vol 1: *Regional economics* Amsterdam, North-Holland: 257–94

Nijkamp P, Schubert U (1985) Structural change in urban systems. *Sistemi Urbani* 7: 155–76

Nijkamp P, van der Mark R, Alsters T (1988) Evaluation of regional incubator profiles for small and medium sized enterprises. *Regional Studies* 22: 95–105

Nishioka H, Takeuchi A (1987) The development of high technology industry in Japan. In Breheny M J, McQuaid R W (eds) *The development of high technology industries: an international survey* London, Croom Helm: 262–95

Noble D F (1977) *America by design: science, technology and the rise of corporate capitalism* New York, Alfred A Knopf

Noble D F (1984) *Forces of production: a social history of industrial automation* New York, Alfred A Knopf

Nolan J E (1986) South Korea: an ambitious client of the United States. In Brzoska M, Ohlson T (eds) *Arms production in the Third World* London, Taylor and Francis: 215–31

Norcliffe G B (1983) Using location quotients to estimate the economic base and trade flows. *Regional Studies* 17: 161–8

Norcliffe G B (1985) The industrial geography of the Third World. In Pacione M (ed) *Progress in industrial geography* London, Croom Helm: 249–83

Norcliffe G B, Kotseff L E (1980) Local industrial complexes in Ontario. *Annals of the Association of American Geographers* 70: 68–79

Norman C (1989) HDTV: the technology *du jour*. *Science* 244: 761–4

Norris W C (1985) Cooperative R & D: a regional strategy. *Issues in Science and Technology* 1(2): 91–102

Norton R D, Rees J (1979) The product cycle and the decentralization of American manufacturing. *Regional Studies* 13: 141–51

Noyce R N (1982) Competition and cooperation – a prescription for the eighties. *Research Management* 25(2): 13–17

Noyelle T J (1982) The implications of industry restructuring for spatial organization in the United States. In Moulaert F, Salinas P W (eds) *Regional analysis and the new international division of labor.* Boston, Kluwer–Nijhoff: 113–33

Noyelle T J, Stanback T M (1983) *The economic transformation of American cities* Totowa NJ, Rowman and Allenheld

Nueño P, Oosterveld J P (1986) The status of technology strategy in Europe. In Horwitch M (ed) *Technology in the modern corporation: a strategic perspective* New York, Pergamon: 145–66

Nusbaumer J (1987a) *The services economy: lever to growth* Boston, Kluwer Academic

Nusbaumer J (1987b) *Services in the global market* Boston, Kluwer Academic

Nussbaum B, Bernstein A, Ehrlich E, Garland S B, Therrien L, Hammonds K H, Pennar K (1988) Needed: human capital. *Business Week* September 19: 100–41

Oakey R P (1981) *High technology industry and industrial location* Aldershot, Gower

Oakey R P (1983) New technology, government policy and regional manufacturing employment. *Area* 15: 61–5

Oakey R P (1984) *High technology small firms* New York, St. Martin's Press

Oakey R (1985) High-technology industry and agglomeration economies. In Hall P, Markusen A (eds) *Silicon landscapes* Boston, Allen and Unwin: 94–117

Oakey R P, Cooper S Y (1989) High technology industry, agglomeration and the potential for peripherally sited small firms. *Regional Studies* 23: 347–60

Oakey R, Rothwell R, Cooper S (1988) *The management of innovation in high-technology small firms* London, Pinter

Oakey R P, Thwaites A T, Nash P A (1980) The regional distribution of innovative manufacturing establishments in Britain. *Regional Studies* 14: 235–53

Oakey R P, Thwaites A T, Nash P A (1982) Technological change and regional development: some evidence on regional variations in product and process innovation. *Environment and Planning A* 14: 1073–86

Ochel W, Wegner M (1987) *Service economies in Europe: opportunities for growth* London, Pinter

O'Connor J (1985) Can entrepreneurship be found, trained or fostered in peripheral regions of the Community? Practice and issues in the context of a technological park. In Gibb J M (ed) *Science parks and innovation centres: their economic and social impact* Amsterdam, Elsevier: 208–17

Odagiri H (1985) Research activity, output growth, and productivity increase in Japanese manufacturing industries. *Research Policy* 14: 117–30

Odle M A (1979) Technology leasing as the latest imperialist phase: a case study of Guyana and Trinidad. *Social and Economic Studies* 28: 189–233

OECD (1977) *Restrictive regional policy measures* Paris, Organisation for Economic Co-operation and Development

OECD (1980a) *Regional policies in the United States* Paris, Organisation for Economic Co-operation and Development

OECD (1980b) *Technical change and economic policy* Paris, Organisation for Economic Co-operation and Development

OECD (1981a) *North/south technology transfer: the adjustments ahead* Paris, Organisation for Economic Co-operation and Development

OECD (1981b) *The future of university research* Paris, Organisation for Economic Co-operation and Development

OECD (1981c) *The measurement of scientific and technical activities: proposed standard practice for surveys of research and experimental development ('Frascati manual' 1980)* Paris, Organisation for Economic Co-operation and Development

OECD (1984) *OECD science and technology indicators: resources devoted to R and D* Paris, Organisation for Economic Co-operation and Development

OECD (1985a) *Venture capital in information technology* Paris, Organisation for Economic Co-operation and Development

OECD (1985b) *Countertrade: developing country practices* Paris, Organisation for Economic Co-operation and Development

OECD (1986a) *OECD science and technology indicators, No 2: R&D, invention and competitiveness* Paris, Organisation for Economic Co-operation and Development

OECD (1986b) *Venture capital: context, development and policies* Paris, Organisation for Economic Co-operation and Development

OECD (1987) *New roles for cities and towns: local initiatives for employment creation* Paris, Organisation for Economic Co-operation and Development

OECD (1988a) *The newly industrialising countries* Paris, Organisation for Economic Co-operation and Development

OECD (1988b) *Industrial revival through technology* Paris, Organisation for Economic Co-operation and Development

OECD (1988c) *New technologies in the 1990s: a socio-economic strategy* Paris, Organisation for Economic Co-operation and Development

OECD (1989a) *OECD science and technology indicators report number 3: R&D, production and diffusion of technology* Paris, Organisation for Economic Co-operation and Development

OECD (1989b) *Information technology and new growth opportunities* Paris, Organisation for Economic Co-operation and Development

OECD (1989c) *Mechanisms for job creation: lessons from the United States* Paris, Organisation for Economic Co-operation and Development

O'Farrell P N (1986a) Entrepreneurship and regional development: some conceptual issues. *Regional Studies* 20: 565–74

O'Farrell P N (1986b) The nature of new firms in Ireland: empirical evidence and policy implications. In Keeble D, Wever E (eds) *New firms and regional development in Europe* London, Croom Helm: 151–83

O'Farrell P N, Crouchley R (1983) Industrial closures in Ireland 1973–1981: analysis and implications. *Regional Studies* 17: 411–27

O'Farrell P N, Crouchley R (1987) Manufacturing-plant closures: a dynamic survival model. *Environment and Planning A* 19: 313–29

O'Farrell P N, Hitchens D N M W (1988a) The relative competitiveness and performance of small manufacturing firms in Scotland and the Mid-West of Ireland: an analysis of matched pairs. *Regional Studies* 22: 399–416

O'Farrell P N, Hitchens D N M W (1988b) Alternative theories of small-firm growth: a critical review. *Environment and Planning A* 20: 1365–83

O'Farrell P N, Pickles A R (1989) Entrepreneurial behaviour within male work histories: a sector-specific analysis. *Environment and Planning A* 21: 311–31

Office of Technology Assessment (1984) *Technology, innovation, and regional economic development* Washington, US Government Printing Office

Office of Technology Assessment (1989) *Holding the edge: maintaining the defense technology base* Washington, US Government Printing Office

Ohlson T (1986) The ASEAN countries: low-cost latecomers. In Brzoska M, Ohlson T (eds) *Arms production in the Third World* London, Taylor and Francis: 55–77

Ohmae K (1985) *Triad power: the coming shape of global competition* New York, Basic Books

O'hUallachain B (1984a) The identification of industrial complexes. *Annals of the Association of American Geographers* 74: 420–36

O'hUallachain B (1984b) Linkages and foreign direct investment in the United States. *Economic Geography* 60: 238–53

O'hUallachain B (1987) Regional and industrial implications of the recent buildup in American defense spending. *Annals of the Association of American Geographers* 77: 208–23

O'hUallachain B (1989) Agglomeration of services in American metropolitan areas. *Growth and Change* 20(3): 34–49

Okimoto D I (1986) Regime characteristics of Japanese industrial policy. In Patrick H (ed) *Japan's high technology industries: lessons and limitations of industrial policy* Seattle, University of Washington Press: 35–95

Okimoto D I (1989) *Between MITI and the market: Japanese industrial policy for high technology* Stanford, CA, Stanford University Press

Olleros F-J (1986) Emerging industries and the burnout of pioneers. *Journal of Product Innovation Management* 3: 5–18

Olleros F-J, Macdonald R (1988) Strategic alliances: managing complementarity to capitalize on emerging technologies. *Technovation* 7: 155–76

Olofsson C, Wahlbin C (1984) Technology-based new ventures from technical universities: a Swedish case. In Hornaday J A, Tarpley F A, Timmons J A, Vesper K H (eds) *Frontiers of entrepreneurship research 1984* Wellesley, MA, Babson College, Center for Entrepreneurial Studies: 192–211

Onda M (1988) Tsukuba science city complex and the Japanese technopolis strategy. In Smilor R W, Kozmetsky G, Gibson D V (eds) *Creating the technopolis: linking technology commercialization and economic development* Cambridge, MA, Ballinger: 51–68

Ondrack D A (1983) Responses to government industrial research policy: a comparison of foreign-owned and Canadian-owned firms. In Goldberg W H, Negandhi A R (eds) *Governments and multinationals: the policy of control versus autonomy* Cambridge, MA, Oelgeschlager, Gunn and Hain: 177–200

Onida F, Malerba F (1989) R&D cooperation between industry, universities and research organizations in Europe. *Technovation* 9: 131–95

Onkvisit S, Shaw J J (1986) Competition and product management: can the product life cycle help? *Business Horizons* 29(4): 51–62

Ooghe H, Bekaert A, van den Bossche P (1989) Venture capital in the U.S.A., Europe and Japan. *Management International Review* 29(1): 29–45

Oppenheim N (1980) *Applied models in urban and regional analysis* Englewood Cliffs, NJ, Prentice-Hall

Osborne D (1988) *Laboratories of democracy* Boston, Harvard Business School Press

Oshima K (1984) Technological innovation and industrial research in Japan. *Research Policy* 13: 285–301

Oshima K (1987) The high technology gap: a view from Japan. In Pierre A J (ed) *A high technology gap? Europe, America and Japan* New York, Council on Foreign Relations: 88–114

Osman-Rani H, Woon T K, Ali A (1986) *Effective mechanisms for the enhancement of technology and skills in Malaysia* Singapore, Institute of Southeast Asian Studies

Oster S (1979) Industrial search for new locations: an empirical analysis. *Review of Economics and Statistics* 61: 288–92

Osterman P (1988) *Employment futures: reorganization, dislocation, and public policy* New York, Oxford University Press

Otsuka K, Ranis G, Saxonhouse G (1988) *Comparative technology choice in development: the Indian and Japanese textile industries* New York, St Martin's Press

Pacione M (1982) Space preferences, locational decisions, and the dispersal of civil servants from London. *Urban Studies* 14: 323–33

Pack H (1982) The capital goods sector in LDCs: economic and technical development. In Syrquin M, Teitel S (eds) *Trade, stability, technology, and equity in Latin America* New York, Academic Press: 349–69

Pajestka J (1979) Factors of economic development and the new international economic order. In Adelman I (ed) *Economic growth and resources* vol 4: *National and international policies* New York, St Martin's Press: 83–91

Palda K S, Pazderka B (1982) International comparisons of R&D effort: the case of the Canadian pharmaceutical industry. *Research Policy* 11: 247–59

Pandit K, Casetti E (1989) The shifting pattern of sectoral labor allocation during development: developed versus developing countries. *Annals of the Association of American Geographers* 79: 329–44

Parisi A J (1989) How R&D spending pays off. *Business Week* June 16: 177–9

Park S-H, Chan K S (1989) A cross-country input–output analysis of intersectoral relationships between manufacturing and services and their employment implications. *World Development* 17: 199–212

Park S O, Wheeler J O (1983) The filtering-down process in Georgia: the third stage of the product life cycle. *Professional Geographer* 35: 18–31

Parsons D (1984) Employment stimulation and the local labour market: a case study of airport growth. *Regional Studies* 18: 423–8

Pascal A H, McCall J J (1980) Agglomeration economies, search costs, and industrial location. *Journal of Urban Economics* 8: 383–8

Patel P, Pavitt K (1987) Is Western Europe losing the technological race? *Research Policy* 16: 59–99

Patrick H (ed) (1986) *Japan's high technology industries: lessons and limitations of industrial policy* Seattle, University of Washington Press

Patton S G (1985) Tourism and local economic development: factory outlets and the Reading SMSA. *Growth and Change* 16(3): 64–74

Pavitt K (ed) (1980) *Technical innovation and British economic performance* London, Macmillan

Pavitt K (1984) Sectoral patterns of technical change: towards a taxonomy and a theory. *Research Policy* 13(6): 343–73

Pavitt K (1986a) 'Chips' and 'trajectories': how does the semiconductor influence the sources and directions of technical change? In MacLeod R M (ed) *Technology and the human prospect* London, Frances Pinter: 31–54

Pavitt K (1986b) Technology, innovation, and strategic management. In McGee J, Thomas H (eds) *Strategic management research* New York, John Wiley: 171–90

Pavitt K, Robson M, Townsend J (1989) Technological accumulation, diversification and organisation in UK companies, 1945–83. *Management Science* 35: 81–99

Pavitt K, Walker W (1976) Governmental policies towards industrial innovation: a review. *Research Policy* 5: 11–97

Pearce D (1987) *Tourism today: a geographical analysis* London, Longman

Peattie L R (1980) Anthropological perspectives on the concepts of dualism, the informal sector, and marginality in developing urban economies. *International Regional Science Review* 5: 1–32

Peattie L (1981) *Thinking about development* New York, Plenum

Peck F W, Townsend A R (1984) Contrasting experience of recession and spatial restructuring: British Shipbuilding, Plessey and Metal Box. *Regional Studies* 18: 319–38

Peck F, Townsend A (1987) The impact of technological change upon the spatial pattern of UK employment within major corporations. *Regional Studies* 21: 225–39

Peck J A (1989) Reconceptualising the local labour market: space, segmentation and the state. *Progress in Human Geography* 13: 42–61

Peck M J (1986) Joint R&D: the case of Microelectronics and Computer Technology Corporation. *Research Policy* 15: 219–31

Peck M J, Goto A (1981) Technology and economic growth: the case of Japan. *Research Policy* 10: 222–43

Pedersen P O (1978) Interaction between short- and long-run development in regions – the case of Denmark. *Regional Studies* 12: 683–700

Peet R (1987) The new international division of labor and debt crisis in the Third World. *Professional Geographer* 39: 172–8

Pelissero J P, Fasenfest D (1989) Suburban economic development policy. *Economic Development Quarterly* 3: 301–11

Pellenbarg P H, Kok J A A M (1985) Small and medium-sized firms in the

Netherlands' urban and rural regions. *Tijdschrift voor Economische en Sociale Geografie* 76: 242–52

Pennings J M (1982) The urban quality of life and entrepreneurship. *Academy of Management Journal* 25(1): 63–79

Perez C (1983) Structural change and the assimilation of new technologies in the economic and social system. *Futures* 15: 357–75

Perez C, Soete L (1988) Catching up in technology: entry barriers and windows of opportunity. In Dosi G, Freeman C, Nelson R, Silverberg G, Soete L (eds) *Technical change and economic theory* London, Pinter: 458–79

Perloff H, Dunn E, Lampard E, Muth R (1960) *Regions, resources, and economic growth* Baltimore, Johns Hopkins University Press

Perrin J-C (1974) *Le développement régional* Paris, Presses Universitaires de France

Perrin J-C (1988a) A deconcentrated technology policy – lessons from the Sophia-Antipolis experience. *Environment and Planning C: Government and Policy* 6: 415–25

Perrin J-C (1988b) New technologies, local synergies and regional policies in Europe. In Aydalot P, Keeble D (eds) *High technology industry and innovative environments: the European experience* London, Routledge: 139–62

Perrin J-C (1989) Action by local authorities and partnership schemes between small and large businesses: an evaluation of the main examples in France. In Commission of the European Communities (ed) *Partnership between small and large firms* London, Graham and Trotman: 188–93

Perrino A C, Tipping J W (1989) Global management of technology. *Research-Technology Management* 32(3): 12–19

Perrolle J A (1986) Intellectual assembly lines: the rationalization of managerial, professional, and technical work. *Computers and the Social Sciences* 2: 111–21

Perroux F (1955) Note sur la notion de pôle de croissance. *Economie Appliquée* 8: 307–20; translation in Livingstone I (ed) (1971) *Economic policy for development* Harmondsworth, Penguin: 278–89

Perroux F (1983) *A new concept of development* London, Croom Helm

Persky J (1978) Dualism, capital–labor ratios and the regions of the U.S. *Journal of Regional Science* 18: 373–81

Persky J, Klein W (1975) Regional capital growth and some of those other things we never talk about. *Papers of the Regional Science Association* 35: 181–90

Peters L (1989) Academic crossroads – the US experience. *Science Technology Industry Review* 5: 163–93

Peterson T (1989) Can Europe catch up in the high-tech race? *Business Week* October 23: 142–54

Peterson T, Maremont M (1989) Adding hustle to Europe's muscle. *Business Week* June 16: 32–4

Peterson T, Maremont M (1990) Suddenly, high tech is a three-way race. *Business Week* June 15: 118–23

Peterson T, Schares G (1988) Europe's high-tech titans put their chips on JESSI. *Business Week* November 14: 62

Pezzini M (1989) The small-firm economy's odd man out: the case of Ravenna. In Goodman E, Bamford J, Saynor P (eds) *Small firms and industrial districts in Italy* London, Routledge: 223–38

Pfeffermann G, Weigel D R (1988) The private sector and the policy environment. *Finance and Development* 25(4): 25–7

Pfister R L (1976) On improving export base studies. *Regional Science Perspectives* 6(1): 104–16

Piatier A (1984) *Barriers to innovation* London, Frances Pinter

Pickles A R, O'Farrell P N (1987) An analysis of entrepreneurial behaviour from male work histories. *Regional Studies* 21: 425–44

Pinstrup-Andersen P (1982) *Agricultural research and technology in economic development* London, Longman

Piore M J, Sabel C F (1983) Italian small business development: lessons for U.S. industrial policy. In Zysman J, Tyson L (eds) *American industry in international competition* Ithaca, NY, Cornell University Press: 391–421

Piore M J, Sabel C F (1984) *The second industrial divide: possibilities for prosperity* New York, Basic Books

Pisano G P, Russo M V, Teece D J (1988) Joint ventures and collaborative arrangements in the telecommunications equipment industry. In Mowery D C (ed) *International collaborative ventures in U.S. manufacturing* Cambridge, MA, Ballinger: 23–70

Pisano G P, Shan W, Teece D J (1988) Joint ventures and collaboration in the biotechnology industry. In Mowery D C (ed) *International collaborative ventures in U.S. manufacturing* Cambridge, MA, Ballinger: 183–222

Pistella F (1989) From Madrid to Copenhagen: the Milan conference within EUREKA developments. *Technovation* 9: 205–8

Plastrik P (1989) The Michigan strategic fund. In OECD. *Mechanisms for job creation: lessons from the United States* Paris, Organisation for Economic Co-operation and Development: 143–9

Pleeter S (1980) Methodologies of economic impact analysis: an overview. In Pleeter S (ed) *Economic impact analysis: methodologies and applications* Boston, Martinus Nijhoff: 7–31

Plucknett D L, Smith N J H, Ozgediz S (1990) *Networking in international agricultural research* Ithaca, NY, Cornell University Press

Polenske K R (1988) Growth pole theory and strategy reconsidered: domination, linkages, and distribution. In Higgins B, Savoie D J (eds) *Regional economic development: essays in honour of François Perroux* Boston, Unwin Hyman: 91–111

Pollock M A, Bernstein A (1986) The disposable employee is becoming a fact of corporate life. *Business Week* December 15: 52–6

Popovich M G, Buss T F (1989) Entrepreneurs find niche even in rural economies. *Rural Development Perspectives* 5(3): 11–14

Porat M U (1977) *The information economy* Washington, US Department of Commerce, Office of Telecommunications

Port O (1989) Back to basics. *Business Week* June 16: 14–18

Port O, King R, Hampton W J (1988) How the new math of productivity adds up. *Business Week* June 6: 103–14

Porter M E (1980) *Competitive strategy* New York, Free Press

Porter M E (1983a) The technological dimension of competitive strategy. In Rosenbloom R S (ed) *Research on technological innovation, management and policy* vol 1, Greenwich, CT, JAI Press: 1–33

Porter M E (1983b) *Cases in competitive strategy* New York, Free Press

Porter M E (1985) *Competitive advantage* New York, Free Press

Porter M E (ed) (1986) *Competition in global industries* Boston, Harvard Business School Press

Porter M E (1987) Changing patterns of international competition. In Teece D J (ed) *The competitive challenge* Cambridge, MA, Ballinger: 27–57

Porter M E (1990) *The competitive advantage of nations* New York, Free Press

Portes A, Castells M, Benton L A (eds) (1989) *The informal economy: studies in advanced and informal economies* Baltimore, Johns Hopkins University Press

Posner B G (1989) Inside out: outside contractors can provide a flexible alternative to hiring full-time employees. *Inc* 11(8): 120–2

Pottier C (1985) The adaptation of regional industrial structures to technical changes. *Papers of the Regional Science Association* 58: 59–72

Pottier C (1987) The location of high technology industries in France. In Breheny M J, McQuaid R W (eds) *The development of high technology industries: an international survey* London, Croom Helm: 192–222

Pottier C (1988) Local innovation and large firm strategies in Europe. In Aydalot P, Keeble D (eds) *High technology industry and innovative environments: the European experience* London, Routledge: 99–120

Power T M (1988) *The economic pursuit of quality* New York, M E Sharpe

Poznanski K (1980) A study of technical innovation in Polish industry. *Research Policy* 9: 232–53

Prahalad C K, Doz Y L (1981) Strategic control – the dilemma in headquarters–subsidiary relationship. In Otterbeck L (ed) *The management of headquarters–subsidiary relationships in multinational corporations* New York, St Martin's Press: 187–203

Prais S J (1988) Qualified manpower in engineering: Britain and other industrially advanced countries. *National Institute Economic Review* 127: 76–83

Prasad A J (1981) Licensing as an alternative to foreign investment for technology transfer. In Hawkins R G, Prasad A J (eds) *Technology transfer and economic development* Research in international business and finance, vol 2, Greenwich, CT, JAI Press: 193–218

Pratt R (1968) An appraisal of the minimum requirements technique. *Economic Geography* 44: 117–24

Pred A R (1966) *The spatial dynamics of U.S. urban-industrial growth 1800–1914* Cambridge, MA, MIT Press

Pred A R (1976) The interurban transmission of growth in advanced economies: empirical findings versus regional planning assumptions. *Regional Studies* 10: 151–71

Pred A R (1977) *City-systems in advanced economies* London, Hutchinson

Premus R (1982) *Location of high technology firms and regional economic development* Washington, US Government Printing Office

Premus R (1985) *Venture capital and innovation* Washington, US Government Printing Office

Premus R (1986) High technology and state economic development strategies. In Redburn R S, Buss T F, Ledebur L C (eds) *Revitalizing the US economy* New York, Praeger: 99–113

Premus R (1988) US technology policies and their regional effects. *Environment and Planning C: Government and Policy* 6: 441–8

Press F (1987) Technological competition and the Western alliance. In Pierre A J (ed) *A high technology gap? Europe, America and Japan* New York, Council on Foreign Relations: 11–43

Prestowitz C L (1988) *Trading places: how we allowed Japan to take the lead* New York, Basic Books

Prud'homme R (1974) Regional economic policy in France, 1962–1972. In Hansen N M (ed) *Public policy and regional economic development: the experience of nine western countries* Cambridge, MA, Ballinger: 33–63

Pudup M B (1987) From farm to factory: structuring and location of the US farm machinery industry. *Economic Geography* 63: 203–22

Pullen M J, Proops J L R (1983) The North Staffordshire regional economy: an input–output assessment. *Regional Studies* 17: 191–200

Pulver G C, Hustedde R J (1988) Regional variables that influence the allocation of venture capital: the role of banks. *Review of Regional Studies* 18(2): 1–9

Quinn J B (1979) Technological innovation, entrepreneurship, and strategy. *Sloan Management Review* 20(3): 19–30

Quinn J B, Baruch J J, Paquette P C (1987) Technology in services. *Scientific American* 257(6): 50–8

Rada J F (1984) Advanced technologies and development: are conventional ideas about comparative advantage obsolete? *Trade and Development* 5: 275–98

Radnor M, Kaufman S (1988) Facing the future: the need for international technology intelligence and sourcing. In Wad A (ed) *Science, technology, and development* Boulder, CO, Westview Press: 305–12

Ragas W R, Miestchovich I J, Nebel E C, Ryan T P, Lacho K J (1987) Louisiana superdome: public costs and benefits 1975–1984. *Economic Development Quarterly* 1: 226–39

Rahman A (1979) Science, technology and development in a new social order. In Goldsmith M E, King A (eds) *Issues of development: towards a new role for science and technology* New York, Pergamon: 57–70

Rajan A (1987) *Services – the second industrial revolution?* London, Butterworths

Randall J N (1987) Scotland. In Damesick P J, Wood P A (eds) *Regional problems, problem regions, and public policy in the United Kingdom* Oxford, Clarendon Press: 218–37

Ray D M, Brewis T N (1976) The geography of income and its correlates. *Canadian Geographer* 20: 41–71

Ray D M, Turpin D V (1987) Factors influencing entrepreneurial events in Japanese high technology venture business. In Churchill N C, Hornaday J A, Kirchhoff B A, Krasner O J, Vesper K H (eds) *Frontiers of entrepreneurship*

research 1987 Wellesley, MA, Babson College, Center for Entrepreneurial Studies: 557–72

Ray G F (1969) The diffusion of new technology. *National Institute Economic Review* 48: 40–83

Ray G F (1980) Innovation in the long cycle. *Lloyds Bank Review* 135: 14–28

Ray G F (1984) *The diffusion of mature technologies* Cambridge, Cambridge University Press

Ray G F (1989) Full circle: the diffusion of technology. *Research Policy* 18: 1–18

Raynauld A (1988) Regional development in a federal state. In Higgins B, Savoie D J (eds) *Regional economic development: essays in honour of François Perroux* Boston, Unwin Hyman: 225–43

Reddy A K N (1979) National and regional technology groups and institutions: an assessment. In Bhalla A S (ed) *Towards global action for appropriate technology* Oxford, Pergamon: 63–137

Rees J (1979) Technological change and regional shifts in American manufacturing. *Professional Geographer* 31: 45–54

Rees J (1980) Government policy and industrial location in the United States. In *Special study on economic change* vol 7: *State and local finance: adjustments in a changing economy* Washington, US Government Printing Office: 128–79

Rees J (1987) What happened to macroeconomics? *Environment and Planning A* 19: 139–41

Rees J (1989) Regional development and policy. *Progress in Human Geography* 13: 576–88

Rees J, Briggs R, Oakey R (1984) The adoption of new technology in the American machinery industry. *Regional Studies* 18: 489–504

Rees J, Briggs R, Oakey R (1986) The adoption of new technology in the American machinery industry. In Rees J (ed) *Technology, regions, and policy* Totowa, NJ, Rowman and Littlefield: 187–217

Rees J, Stafford H A (1984) High-technology location and regional development: the theoretical base. In Office of Technology Assessment *Technology, innovation, and regional economic development* Washington, US Government Printing Office: 97–107

Rees J, Weinstein B L, Gross H T (1988) *Regional patterns of military procurement and their implications* Washington, Sunbelt Institute

Regional Studies Association (1983) *Report of an inquiry into regional problems in the United Kingdom* Norwich, Geo Books

Reich R B (1984) Collusion course. *The New Republic* February 27: 18–21

Reich R (1987) The rise of techno-nationalism. *Atlantic* 259(5): 63–9

Reid G C, Jacobsen L R (1988) *The small entrepreneurial firm* Aberdeen, Aberdeen University Press

Reid J N (1988) Entrepreneurship as a community development strategy for the rural south. In Beaulieu L J (ed) *The rural South in crisis: challenges for the future* Boulder, CO: Westview Press: 325–43

Reid S D, Reid N (1988) Public policy and promoting manufacturing under licensing. *Technovation* 7: 401–14

Reiner T A, Wolpert J (1981) The nonprofit sector in the metropolitan economy. *Economic Geography* 57: 23–33

Renaud B M (1973) Conflicts between national growth and regional income equality in a rapidly growing economy: the case of Korea. *Economic Development and Cultural Change* 21: 429–45

Reynolds L G (1983) The spread of economic growth to the third world. *Journal of Economic Literature* 21: 941–80

Rhee Y W (1990) The catalyst model of development: lessons from Bangladesh's success with garment exports. *World Development* 18: 333–46

Richardson H W (1971) *Urban economics* London, Penguin

Richardson H W (1972) *Input–output and regional economics* New York, John Wiley

Richardson H W (1973) *Regional growth theory* London, Macmillan

Richardson H W (1976) Growth pole spillovers: the dynamics of backwash and spread. *Regional Studies* 10: 1–9

Richardson H W (1978) Growth centers, rural development, and national urban policy: a defense. *International Regional Science Review* 3: 133–52

Richardson H W (1979) Aggregate efficiency and interregional equity. In Folmer H, Oosterhaven J (ed) *Spatial inequalities and regional development* Boston, Martinus Nijhoff: 161–83

Richardson H W (1985) Input–output and economic base multipliers: looking backward and forward. *Journal of Regional Science* 25: 607–61

Richardson H W (1988) A review of techniques for regional policy analysis. In Higgins B, Savoie D J (eds) *Regional economic development: essays in honour of François Perroux* Boston, Unwin Hyman: 142–68

Richardson H W, Townroe P M (1986) Regional policies in developing countries. In Nijkamp P (ed) *Handbook of regional and urban economics* vol 1: *Regional economics* Amsterdam, North-Holland: 647–78

Riche R W, Hecker D E, Burgan J U (1983) High technology today and tomorrow: a small slice of the employment pie. *Monthly Labor Review* 106(11): 50–8

Richter L K (1989) *The politics of tourism in Asia* Honolulu, University of Hawaii Press

Riddell J B (1970) *The spatial dynamics of modernization in Sierra Leone* Evanston, IL, Northwestern University Press

Riddell J B (1981a) The geography of modernization in Africa: a re-examination. *Canadian Geographer* 25: 290–9

Riddell J B (1981b) Beyond the description of spatial pattern: the process of proletarianization as a factor in population migration in West Africa. *Progress in Human Geography* 5: 370–92

Riddell J B (1987) Geography and the study of Third World underdevelopment. *Progress in Human Geography* 11: 264–74

Riddell R (1985) *Regional development policy* Aldershot, Gower

Riddle D I (1986) *Service-led growth: the role of the service sector in world development* New York, Praeger

Riedel J (1988) Economic development in East Asia: doing what comes naturally? In Hughes H (ed) *Achieving industrialization in East Asia* Cambridge, Cambridge University Press: 1–38

Riedle K (1989) Demand for R&D activities and the trade off between in-house and external research: a viewpoint from industry with reference to large companies and small and medium-sized enterprises. *Technovation* 9: 213–25

Riggs H E (1983) *Managing high-technology companies* Belmont, CA, Wadsworth

Rink D R, J E Swan (1979) **Product life cycle research: a literature review.** *Journal of Business Research* 7: 219–42

Rip A, Nederhof A J (1986) Between dirigism and laissez-faire: effects of implementing the science policy priority for biotechnology in the Netherlands. *Research Policy* 15: 253–68

Ritterbush S D (1988) Entrepreneurship in an ascribed status society: the Kingdom of Tonga. In Fairbairn T I J (ed) *Island entrepreneurs: problems and performances in the Pacific* Honolulu, East–West Center: 137–63

Ritterbush S D, Pearson J (1988) Pacific women in business: constraints and opportunities. In Fairbairn T I J (ed) *Island entrepreneurs: problems and performances in the Pacific* Honolulu, East–West Center: 195–207

Roberts E B (ed) (1987) *Generating technological innovation* Oxford, Oxford University Press

Roberts E B (1988) What we've learned: managing invention and innovation. *Research-Technology Management* 31(1): 11–29

Roberts E B, Fusfeld H I (1981) Staffing the innovative technology-based organization. *Sloan Management Review* 22(3): 19–34

Roberts E B, Peters D H (1983) Commercial innovations from university faculty. *Research Policy* 10: 108–26

Robertson G E, Allen D N (1986) From kites to computers: Pennsylvania's Ben Franklin Partnership. *Technovation* 4: 29–43

Robinson F D (1985) University and industry cooperation in microelectronics research. In Whittington D (ed) *High hopes for high tech: microelectronics policy in North Carolina* Chapel Hill, University of North Carolina Press: 113–44

Robinson R B (1987) Emerging strategies in the venture capital industry. *Journal of Business Venturing* 2: 53–77

Robinson R D (1988) *The international transfer of technology: theory, issues, and practice* Cambridge, MA, Ballinger

Robson M, Townsend J, Pavitt K (1988) Sectoral patterns of production and use of innovations in the UK: 1945–1983. *Research Policy* 17: 1–14

Rock A (1987) Strategy vs. tactics from a venture capitalist. *Harvard Business Review* 65(6): 63–7

Rodan G (1989) *The political economy of Singapore's industrialization: national state and international capital* New York, St Martin's Press

Rodgers A (1979) *Economic development in retrospect* New York, Winston/Wiley

Rodriguez L J, Krienke A (1982) The economic effect of a military base on a small metropolitan area. *Texas Business Review* 56: 138–40

Roessner J D (1987) Technology policy in the United States: structures and limitations. *Technovation* 5: 229–45

Roessner J D (1989) Evaluation of government innovation programs: introduction. *Research Policy* 18: 309–12

Rogers E M (1976) Communication and development: the passing of the dominant paradigm. *Communication Research* 3: 121–48

Rogers E M (1982) Information exchange and technological innovation. In Sahal D (ed) *The transfer and utilization of technical knowledge* Lexington, MA, Lexington Books: 105–23

Rogers E M (1986) The role of the research university in the spin-off of high-technology companies. *Technovation* 4: 169–81

Rogers E M, Larsen J (1984) *Silicon Valley fever* New York, Basic Books

Rogers G, Shaffer R, Pulver G (1988) Identification of local capital markets for policy research. *Review of Regional Studies* 18(1): 55–66

Röling N G, Ascroft J, Chege F W (1976) The diffusion of innovations and the issue of equity in rural development. In Rogers E M (ed) *Communication and development* Beverly Hills, CA, Sage: 63–78

Romsa G, Blenman M, Nipper J (1989) From the economic to the political: regional planning in West Germany. *Canadian Geographer* 33: 47–57

Ronayne J (1984) *Science in government* Oxford, Basil Blackwell

Rondinelli D A (1983) *Secondary cities in developing countries* Beverly Hills, CA, Sage

Ronstadt R (1984) R&D abroad by U.S. multinationals. In Stobaugh R, Wells L T (eds) *Technology crossing borders* Boston, Harvard Business School Press: 241–64

Roobeek A J M (1990) *Beyond the technology race: an analysis of technology policy in seven industrial countries* Amsterdam, Elsevier

Rose A (1984) Technological change and input–output analysis: an appraisal. *Socio-Economic Planning Sciences* 18: 305–18

Rosenberg N (1976) On technological expectations. *Economic Journal* 86: 523–35

Rosenberg N (1982) *Inside the black box: technology and economics* Cambridge, MA, Cambridge University Press

Rosenberg R (1985) What companies look for. *High Technology* 5(1): 30–7

Rosenbloom R S, Abernathy W J (1982) The climate for innovation in industry: the role of management attitudes and practices in consumer electronics. *Research Policy* 11: 209–25

Rosenbloom R S, Cusumano M A (1987) Technological pioneering and competitive advantage: the birth of the VCR industry. *California Management Review* 29(4): 51–76

Rosenfeld S A (1989–90) Regional development European style. *Issues in Science and Technology* 6(2): 63–70

Rosenfeld S A, Bergman E M (1989) *Making connections: After the factories revisited* Research Triangle Park, NC, Southern Growth Policies Board

Rosenfeld S A, Bergman E M, Rubin S (1985) *After the factories: changing employment patterns in the rural south* Research Triangle Park, NC, Southern Growth Policies Board

Rosenstein J (1988) The board and strategy: venture capital and high technology. *Journal of Business Venturing* 3: 159–70

Rossant J, Galuszka P, Reed S (1989) Hey! Want a hot little alloy from Shemyakin? *Business Week* June 26: 82

Rossant J, Reed S, Griffiths D (1989) Israel has everything it needs – except peace. *Business Week* December 4: 54–8

Rossini F A, Porter A L (1987) Who's using computers in industrial R&D – and for what? *Research Management* 29(3): 39–44

Rostow W W (1960) *The stages of economic growth: a non-communist manifesto* Cambridge, Cambridge University Press

Rothblatt D N (1971) *Regional planning: the Appalachian experience* Lexington, MA, Heath Lexington

Rothwell R (1980) The impact of regulation on innovation: some U.S. data. *Technological Forecasting and Social Change* 17: 7–34

Rothwell R (1985) Venture finance, small firms and public policy in the UK. *Research Policy* 14: 253–65

Rothwell R, Freeman C, Horlsey A, Jervis V T P, Robertson A B, Townsend J (1974) SAPPHO updated – Project SAPPHO phase II. *Research Policy* 3: 258–91

Rothwell R, Zegveld W (1982) *Innovation and the small and medium sized firm* Hingham, MA, Kluwer Nijhoff

Rothwell R, Zegveld W (1985) *Reindustrialization and technology* London, Longman

Rothwell R, Zegveld W (1988) An assessment of government innovation policies. In Roessner J D (ed) *Government innovation policy: design, implementation, evaluation* New York, St Martin's Press: 19–35

Round J I (1983) Nonsurvey techniques: a critical review of the theory and the evidence. *International Regional Science Review* 8: 189–212

Rubenstein A H (1964) Organization factors affecting research and development decision-making in large decentralized companies. *Management Science* 10: 618–33

Rubenstein A H (1980) The role of imbedded technology in the industrial innovation process. In *Special study on economic change* vol 3: *Research and innovation: developing a dynamic nation* Washington, US Government Printing Office: 380–414

Rubenstein A H (1989) *Managing technology in the decentralized firm* New York, John Wiley

Rubenstein A H, Ginn M E (1985) Project management at significant interfaces in the R&D/innovation process. In Dean B V (ed) *Project management: methods and studies* Amsterdam, North-Holland: 173–99

Rubenstein J M (1986) Changing distribution of the American automobile industry. *Geographical Review* 76: 288–300

Rubenstein J M (1988) Changing distribution of American motor-vehicle-parts suppliers. *Geographical Review* 78: 288–98

Rubin B M, Zorn C (1985) Sensible state and local economic development. *Public Administration Review* 45: 333–9

Rullani E, Zanfei A (1988a) Networks between manufacturing and demand – cases from textile and clothing industries. In Antonelli C (ed) *New information technology and industrial change: the Italian case* Dordrecht, Kluwer Academic: 57–95

Rullani E, Zanfei A (1988b) Area networks: telematic connections in a traditional textile district. In Antonelli C (ed) *New information technology and industrial change: the Italian case* Dordrecht, Kluwer Academic: 97–113

Russo M (1985) Technical change and the industrial district: the role of interfirm relations in the growth and transformation of ceramic tile production in Italy. *Research Policy* 14: 329–43

Ruttan V W (1982) *Agricultural research policy* Minneapolis, University of Minnesota Press

Ruttan V W (1986) Toward a global agricultural research system: a personal view. *Research Policy* 15: 307–27

Ryans J K, Shanklin W L (1986) *Guide to marketing for economic development: competing in America's second civil war* Columbus, OH, Publishing Horizons

Sabel C (1982) *Work and politics: the division of labor in industry* Cambridge, Cambridge University Press

Sabel C (1989) Flexible specialization and the re-emergence of regional economies. In Hirst P, Zeitlin J (eds) *Reversing industrial decline? Industrial structure and policy in Britain and her competitors* Oxford: Berg: 17–70

Sabel C, Herrigel G, Kazis R, Deeg R (1987) How to keep mature industries innovative. *Technology Review* 90(3): 27–35

Sachs J D (ed) (1989) *Developing country debt and the world economy* Chicago, University of Chicago Press

Sagasti F R (1979) *Technology, planning, and self-reliant development: a Latin American view* New York, Praeger

Sagasti F R (1988) Market structure and technological behavior in developing countries. In Wad A (ed) *Science, technology, and development* Boulder, CO, Westview Press: 149–68

Sagdeev R Z (1988) Science and *perestroika*: a long way to go. *Issues in Science and Technology* 4(4): 48–52

Salih K, Young M L, Rasiah R (1988) The changing face of the electronics industry in the periphery: the case of Malaysia. *International Journal of Urban and Regional Research* 12: 375–403

Salita D C, Juanico M B (1983) Export processing zones: new catalysts for economic development. In Hamilton F E I, Linge G J R (eds) *Spatial analysis, industry and the industrial environment* vol 2: *International industrial systems* London, John Wiley: 441–61

Salter W E G (1966) *Productivity and technical change* 2nd edn Cambridge, Cambridge University Press

Samuels R J, Whipple B C (1989) The FSX and Japan's strategy for aerospace. *Technology Review* 92(7): 43–51

Sanchez P (1988) Technology exports by Spanish companies. *Science Technology Industry Review* 4: 167–89

Sanderson S W, Williams G, Ballenger T, Berry B J L (1987) Impacts of computer-aided manufacturing on offshore assembly and future manufacturing locations. *Regional Studies* 21: 131–41

Santos M (1979) *The shared space: the two circuits of the urban economy in underdeveloped countries* London, Methuen

Sassen-Koob S (1989) New York City's informal economy. In Portes A, Castells

M, Benton L A (eds) *The informal economy: studies in advanced and less developed countries* Baltimore, Johns Hopkins University Press: 60–77

Sato K (1978) Did technological progress accelerate in Japan? In Tsuru S (ed) *Economic growth and resources* vol 5: *Growth and resources problems related to Japan* New York, St Martin's Press 153–81

Sauer W J, Currid M, Schwab A J (1988) The emergence of the high tech industry in southwestern Pennsylvania: a case study of Pittsburgh. In Kirchhoff B A, Long W A, McMullan W E, Vesper K H, Wetzel W E (eds) *Frontiers of entrepreneurship research 1988* Wellesley, MA, Babson College, Center for Entrepreneurial Studies: 415–29

Savey S (1983) Organization of production and the new spatial division of labour in France. In Hamilton F E I, Linge G J R (eds) *Spatial analysis, industry and the industrial environment* vol 3: *Regional industrial systems* London, John Wiley: 103–20

Savoie D J (1986) *Regional economic development: Canada's search for solutions* Toronto, University of Toronto Press

Saxenian A (1983a) The genesis of Silicon Valley. *Built Environment* 9(1): 7–17

Saxenian A (1983b) The urban contradictions of Silicon Valley: regional growth and the restructuring of the semiconductor industry. *International Journal of Urban and Regional Research* 7: 237–62

Saxenian A (1988) The Cheshire cat's grin: innovation and regional development in England. *Technology Review* 91(2): 66–75

Sayer R A (1985) Industry and space: a sympathetic critique of radical research. *Environment and Planning D: Society and Space* 3: 3–29

Sayer A (1986) Industrial location on a world scale: the case of the semiconductor industry. In Scott A J, Storper M (eds) *Production, work, territory* Boston, Allen and Unwin: 107–23

Sayer A (1989) Postfordism in question. *International Journal of Urban and Regional Research* 13: 666–95

Sayer A, Morgan K (1986) The electronics industry and regional development in Britain. In Amin A, Goddard J B (eds) *Technological change, industrial restructuring and regional development* London, Allen and Unwin: 157–87

Sayer A, Morgan K (1987) High technology industry and the international division of labour: the case of electronics. In Breheny M J, McQuaid R W (eds) *The development of high technology industries: an international survey* London, Croom Helm: 10–36

Schaffer W A (1981) Determinants of regional decline. In Buhr W, Friedrich P (eds) *Lectures on regional stagnation* Baden Baden, Nomos: 11–27

Schamp E W (1987) Technology parks and interregional competition in the Federal Republic of Germany. In van der Knaap B, Wever E (eds) *New technology and regional development* London, Croom Helm: 119–35

Schares G E (1989) Reunification? East German industry says 'Not so fast'. *Business Week* December 25: 66–7

Schares G E, Brady R, Maremont M, Peterson T, Keller J (1989) The East bloc's $100 billion phone bill. *Business Week* November 20: 139–42

Schaub J D (1984) New and expanding firms provide new jobs in rural Georgia. *Rural Development Perspectives* 1(1): 26–9

Schell D W (1983) Entrepreneurial activity: a comparison of three North Carolina communities. In Hornaday J A, Timmons J A, Vesper K H (eds) *Frontiers of entrepreneurship research 1983* Wellesley, MA, Babson College Center for Entrepreneurial Studies: 495–518

Schell D W (1984) The development of the venture capital industry in North Carolina: a new approach. In Hornaday J A, Tarpley F A, Timmons J A, Vesper K H (eds) *Frontiers of entrepreneurship research 1984* Wellesley, MA, Babson College, Center for Entrepreneurial Studies: 55–72

Scherer F M (1982) Inter-industry technology flows in the United States. *Research Policy* 11: 227–45

Schiffman J R, Shao M (1986) South Korea and Taiwan: two strategies. *Wall Street Journal* May 1: 36

Schlesinger J M (1990) A new Nippon? MITI 'vision' for 1990s seeks a mellow Japan others shouldn't fear. *Wall Street Journal* July 3: A1, A4

Schlieper A (1988) Case studies of local technology and innovation policies in West Germany: success and failure. In Dyson K (ed) *Local authorities and new technologies: the European dimension* London, Croom Helm: 115–23

Schmandt J, Wilson R (eds) (1987) *Promoting high–technology industry: initiatives and policies for state government* Boulder, CO, Westview Press

Schmenner R W (1982) *Making business location decisions* Englewood Cliffs, NJ, Prentice-Hall

Schmenner R W (1988) The merit of making things fast. *Sloan Management Review* 30(1): 11–17

Schmenner R W, Huber J C, Cook R L (1987) Geographic differences and the location of new manufacturing facilities. *Journal of Urban Economics* 21: 83–104

Schmookler J (1966) *Invention and economic growth* Cambridge, MA, Harvard University Press

Schoenberger E (1986) Competition, competitive strategy, and industrial change: the case of electronic components. *Economic Geography* 62: 321–33

Schoenberger E (1987) Technological and organizational change in automobile production: spatial implications. *Regional Studies* 21: 199–214

Schoenberger E (1988) From Fordism to flexible accumulation: technology, competitive strategies, and international location. *Environment and Planning D: Society and Space* 6: 245–62

Schoenberger E (1989) Thinking about flexibility: a response to Gertler. *Transactions, Institute of British Geographers* NS 14: 98–108

Schoenberger R J (1986) *World class manufacturing* New York, Free Press

Scholing E, Timmermann V (1988) Why LDC growth rates differ: measuring 'unmeasurable' influences. *World Development* 16: 1271–84

Schumpeter J A (1934) *The theory of economic development* Oxford, Oxford University Press

Sciberras E (1985) Technical innovation and international competitiveness in the television industry. In Rhodes E, Wield D (eds) *Implementing new technologies* Oxford, Basil Blackwell: 177–90

Science Council of Canada (1979) *Forging the links: a technology policy for Canada* Ottawa, Science Council of Canada

Science Council of Canada (1980) *Multinationals and industrial strategy: the role of world product mandates* Ottawa, Science Council of Canada

Science Council of Canada (1984) *Canadian industrial development: some policy directions* Ottawa, Science Council of Canada

Scott A J (1983a) Industrial organization and the logic of intra-metropolitan location, II: a case study of the printed circuits industry in the greater Los Angeles region. *Economic Geography* 59: 343–67

Scott A J (1983b) Location and linkage systems: a survey and reassessment. *Annals of Regional Science* 17(1): 1–39

Scott A J (1984) Industrial organization and the logic of intra-metropolitan location, III: a case study of the women's dress industry in the greater Los Angeles region. *Economic Geography* 60: 3–27

Scott A J (1985) Location processes, urbanization, and territorial development: an exploratory essay. *Environment and Planning A* 17: 479–501

Scott A J (1986) High technology industry and territorial development: the rise of the Orange County complex. *Urban Geography.* 7: 3–45

Scott A J (1987) The semiconductor industry in South East Asia: organization, location and the international division of labour. *Regional Studies* 21: 143–60

Scott A J (1988a) Flexible production systems and regional development: the rise of new industrial spaces in North America and western Europe. *International Journal of Urban and Regional Research* 12: 171–86

Scott A J (1988b) *Metropolis: from the division of labor to urban form* Berkeley, University of California Press

Scott A J (1988c) *New industrial spaces* London, Pion

Scott A J (1988d) Flexible production systems and regional development: the rise of new industrial spaces in North America and western Europe. *International Journal of Urban and Regional Research* 12: 171–86

Scott A J, Angel D P (1987) The US semiconductor industry: a locational analysis. *Environment and Planning A* 19: 875–912

Scott A J, Angel D P (1988) The global assembly-operations of US semiconductor firms: a geographical analysis. *Environment and Planning A* 20: 1047–67

Scott A J, Kwok E C (1989) Inter-firm subcontracting and locational agglomeration: a case study of the printed circuits industry in Southern California. *Regional Studies* 23: 405–16

Scott A J, Storper M (1987) High technology industry and regional development: a theoretical critique and reconstruction. *International Social Science Journal* 34: 215–32

Scott M (1982) Mythology and misplaced pessimism: the real failure record of new, small businesses. In Watkins D, Stanworth J, Westrip A (eds) *Stimulating new firms* Aldershot, Gower: 220–44

Scott M F G (1981) The contribution of investment to growth. *Scottish Journal of Political Economy* 28: 211–26

Scott-Stevens S (1987) *Foreign consultants and counterparts: problems in technology transfer* Boulder, CO, Westview Press

Seers D, Schaffer B, Kiljunen M L (eds) (1979) *Underdeveloped Europe: studies in core–periphery relations* Brighton, Harvester

Segal A (ed) (1987) *Learning by doing: science and technology in the developing world* Boulder, CO, Westview Press

Segal N S (1986) Universities and technological entrepreneurship in Britain: some implications of the Cambridge phenomenon. *Technovation* 4: 189–204

Sen A (1979) Followers' strategy for technological development. *The Developing Economies* 17: 506–28

Sen F (1988) The dilemma of managing R&D in India. In Wad A (ed) *Science, technology, and development* Boulder, CO, Westview Press: 279–91

Seninger S F (1985) Employment cycles and process innovation in regional structural change. *Journal of Regional Science* 25: 259–72

Seninger S F (1989) *Labor force policies for regional economic development: the role of employment and training programs* New York, Praeger

Senker J (1985) Small high technology firms: some regional implications. *Technovation* 3: 243–62

Sforzi F (1989) The geography of industrial districts in Italy. In Goodman E, Bamford J, Saynor P (eds) *Small firms and industrial districts in Italy* London, Routledge: 153–73

Shaffer R E, Pulver G C (1985) Regional variations in capital structure of new small businesses: the Wisconsin case. In Storey D J (ed) *Small firms in regional economic development* Cambridge, Cambridge University Press: 166–92

Shahidullah M (1985) Institutionalization of modern science and technology in non-western societies: lessons from Japan and India. *Knowledge: Creation, Diffusion, Implementation* 6: 437–60

Shaiken H (1984) *Work transformed: automation and labor in the computer age* New York, Holt, Rinehart and Winston

Shanklin W L, Ryans J K (1984) *Marketing high technology* Lexington, MA, Lexington Books

Shapero A (1971) *An action program for entrepreneurship: the design of action experiments to elicit technical company formations in the Ozarks region* Report prepared for the Ozarks Regional Commission, Austin, TX, Multi-Disciplinary Research Inc

Shapero A (1972) The process of technical company formation in a local area. In Cooper A C, Komives J L (eds) *Technical entrepreneurship: a symposium* Milwaukee, Center for Venture Management: 63–95

Shapero A (1984) The entrepreneurial event. In Kent C A (ed) *The environment for entrepreneurship* Lexington, MA, Lexington Books: 21–40

Shapero A (1985) *Managing professional people* New York, Free Press

Sharif M N (1986) Measurement of technology for national development. *Technological Forecasting and Social Change* 29: 119–72

Sharp M (1987) National policies towards biotechnology. *Technovation* 5: 281–304

Sharp M, Shearman C (1987) *European technological collaboration* London, Routledge and Kegan Paul

Sharpe B (1988) Informal work and development in the west. *Progress in Human Geography* 12: 315–36

Shaw G, Williams A (1988) Tourism and employment: reflections on a pilot study of Looe, Cornwall. *Area* 20: 23–34

Shearman C, Burrell G (1987) The structures of industrial development. *Journal of Management Studies* 24: 325–45

Sheppard E (1982) City size distributions and spatial economic change. *International Regional Science Review* 7: 127–51

Shinjo K (1988) The computer industry. In Komiya R, Okuno M, Suzumura K (eds) *Industrial policy of Japan* Tokyo, Academic Press: 333–65

Shmelev N, Popov V (1989) *The turning point: revitalizing the Soviet economy* New York, Doubleday

Shrivastava P, Sonder W E (1987) The strategic management of technological innovations: a review and a model. *Journal of Management Studies* 24: 25–41

Siebert H (1969) *Regional economic growth: theory and policy* Scranton, PA, International Textbook

Silveira M P W (ed) (1985) *Research and development: linkages to production in developing countries* Boulder, CO, Westview Press

Silver A D (1980) Life without venture capitalists. *Venture* 2 (March): 26–7

Simon D F, Rehn D (1987) Innovation in China's semiconductor industry: the case of Shanghai. *Research Policy* 16: 259–77

Simon D F, Rehn D (1988) *Technological innovation in China* Cambridge, MA, Ballinger

Simon D F, Schive C (1986) Taiwan. In Rushing F W, Brown C G (eds) *National policies for developing high technology industries* Boulder, CO, Westview Press: 201–26

Simpson E S (1987) *The developing world: an introduction* London, Longman

Sinclair T, Sutcliffe C (1988) The economic effects on destination areas of foreign involvement in the tourism industry: a Spanish application. In Goodall B, Ashworth G (eds) *Marketing in the tourism industry: the promotion of destination regions* London, Croom Helm: 111–32

Singelmann J (1978) *From agriculture to services: the transformation of industrial employment* Beverly Hills, CA, Sage

Singelmann J (1981) Southern industrialization. In Poston D L, Weller R H (eds) *The population of the South* Austin, University of Texas Press: 175–97

Singerman P A (1989) Centers for advanced technology transfer: the Ben Franklin Partnership. In OECD *Mechanisms for job creation: lessons from the United States* Paris, Organisation for Economic Co-operation and Development: 123–7

Singh M S (1988) The changing role of the periphery in the international industrial arena exemplified by Malaysia and Singapore. In Linge G J R (ed) *Peripheralisation and industrial change* London, Routledge: 72–93

Singh M S, Choo C S (1981) Spatial dynamics in the growth and development of multinational corporations in Malaysia. In Hamilton F E I, Linge G J R (eds) *Spatial analysis, industry and the industrial environment* vol 2: *International industrial systems* London, John Wiley: 481–507

Skoro C L (1988) Rankings of state business climates. *Economic Development Quarterly* 2: 138–52

Skowronski S (1987) Transfer of technology and industrial cooperation. *Technovation* 7: 17–22

Sloan C (1981) A good business climate: what it really means. *The New Republic* January 3 and 10: 12–15

Smilor R W, Gill M D (1986) *The new business incubator* Lexington, MA, Lexington Books

Smilor R W, Kozmetsky G, Gibson D V (eds) (1988a) *Creating the technopolis: linking technology commercialization and economic development* Cambridge, MA, Ballinger

Smilor R W, Kozmetsky G, Gibson D V (1988b) The Austin/San Antonio corridor: the dynamics of a developing technopolis. In Smilor R W, Kozmetsky G, Gibson D V (eds) *Creating the technopolis: linking technology commercialization and economic development* Cambridge, MA, Ballinger: 145–83

Smith C (1986) Class relations, diversity and location – technical workers. In Armstrong P, Carter B, Smith C, Nichols T *White collar workers, trade unions, and class* London, Croom Helm: 76–106

Smith C, Ayukawa Y (1989) Japan venture capital: a different game than US, *Venture Japan* 1(4): 60–6

Smith D M (1981) *Industrial location: an economic geographical analysis* 2nd edn New York, John Wiley

Smith N (1984) *Uneven development* Oxford, Basil Blackwell

Smith N, Dennis W (1987) The restructuring of geographical scale: coalescence and fragmentation of the northern core region. *Economic Geography* 63: 160–82

Soete L (1986) Technological innovation and long waves: an inquiry into the nature and wealth of Christopher Freeman's thinking. In MacLeod R M (ed) *Technology and the human prospect* London, Frances Pinter: 214–38

Soete L, Turner R (1984) Technology diffusion and the rate of technical change. *Economic Journal* 94: 612–23

Soja E W (1968) *The geography of modernization in Kenya* Syracuse, Syracuse University Press

Soja E W (1979) The geography of modernization – a radical reappraisal. In Obudho R A, Taylor D R F (eds) *The spatial structure of development: a study of Kenya* Boulder, CO, Westview Press: 28–45

Soja E W (1984) A materialist interpretation of spatiality. In Forbes D K, Rimmer P J (eds) *Uneven development and the geographical transfer of value* Canberra, Australian National University, Department of Human Geography: 43–77

Solow R S (1957) Technical change and the aggregate production function. *Review of Economics and Statistics* 39: 312–20

Souder W E (1977) *An exploratory study of the coordinating mechanisms between R and D and marketing as an influence on the innovation process: final report* Pittsburgh, University of Pittsburgh, Department of Industrial Engineering

Souder W E (1983) Organizing for modern technology and innovation: a review and synthesis. *Technovation* 2: 27–44

Soukup W R, Cooper A C (1983) Strategic response to technological change in the electronic components industry. *R&D Management* **13**: 219–30

Spekman R E (1988) Strategic supplier selection: understanding long-term buyer relationships. *Business Horizons* **31**(4): 75–81

Spence A M, Hazard H A (eds) (1988) *International competitiveness* Cambridge, MA, Ballinger

Spenner K I (1988) Technological change, skill requirements, and education: the case for uncertainty. In Cyert R M, Mowery D C (eds) *The impact of technological change on employment and growth* Cambridge, MA, Ballinger: 131–84

Spital F C (1983) Gaining market share advantage in the semiconductor industry by lead time in innovation. In Rosenbloom R S (ed) *Research on technological innovation, management and policy* vol 1, Greenwich, CT, JAI Press: 55–67

Spring W J (1989) New England's training systems and regional economic competitiveness. *New England Economic Indicators* second quarter: iv–ix

Stabler J C, Howe E C (1988) Service exports and regional growth in the post-industrial era. *Journal of Regional Science* **28**: 303–15

Stabler M J (1988) The image of destination regions: theoretical and empirical aspects. In Goodall B, Ashworth G (eds) *Marketing in the tourism industry: the promotion of destination regions* London, Croom Helm: 133–61

Stalk G, Hout T M (1990) *Competing against time: how time-based competition is reshaping global markets* New York, Free Press

Stamas G D (1981) The puzzling lag in southern earnings. *Monthly Labor Review* **104**(6): 27–36

Stanback T M (1987) *Computerization and the transformation of employment: government, hospitals, and universities* Boulder, CO, Westview Press

Stanback T M, Bearse P J, Noyelle T J, Karasek R A (1981) *Services: the new economy* Totowa, NJ, Allanheld, Osmun

Stanback T M, Noyelle T (1982) *Cities in transition* Totowa, NJ, Allanheld, Osmun

Standing G (1989) Global feminization through flexible labor. *World Development* **17**: 1077–95

Stankiewicz R (1985) A new role for universities in technological innovation? In Sweeney G (ed) *Innovation policies: an international perspective* London, Frances Pinter: 114–51

Starr M K, Biloski A J (1985) The decision to adopt new technology – effects on organizational size. In Rhodes E, Wield D (eds) *Implementing new technologies* Oxford, Basil Blackwell: 303–15

Steed G P F (1978) Product differentiation, locational protection and economic integration: Western Europe's clothing industries. *Geoforum* **9**: 307–18

Steed G P F (1981) International location and comparative advantage: the clothing industry and developing countries. In Hamilton F E I, Linge G J R (eds) *Spatial analysis, industry and the industrial environment* vol 2: *International industrial systems* London, John Wiley: 265–303

Steed G P F (1982) *Threshold firms: backing Canada's winners* Ottawa, Science Council of Canada

Steed G P F (1989) *Not a long shot: Canadian industrial science and technology policy* Ottawa, Science Council of Canada

Steedman H, Wagner K (1989) Productivity, machinery and skills: clothing manufacture in Britain and Germany. *National Institute Economic Review* **128**: 40–57

Steele D B (1972) A numbers game (or the return of regional multipliers). *Regional Studies* **6**: 115–30

Steele L W (1975) *Innovation in big business* New York, Elsevier

Steele L W (1988) What we've learned: selecting R&D programs and objectives. *Research-Technology Management* **31**(2): 17–36

Steele L W (1989) *Managing technology: the strategic view* New York, McGraw-Hill

Steiner P (1987) Contrasts in regional potentials: some aspects of regional economic development. *Papers of the Regional Science Association* **61**: 79–92

Stevens B, Lahr M L (1988) Regional economic multipliers: definition, measurement, and application. *Economic Development Quarterly* **2**: 88–96

Stevens C (1990) 1992: the European technology challenge. *Research-Technology Management* **33**(1): 17–23

Stewart C T, Nihei Y (1987) *Technology transfer and human factors* Lexington, MA, Lexington Books

Stewart F (1978) *Technology and underdevelopment* 2nd edn London, Macmillan

Stewart F (1981) International technology transfer: issues and policy options. In Streeten P, Jolly R (eds) *Recent issues in world development* Oxford, Pergamon: 67–110

Stewart F (1987) The case for appropriate technology: a reply to R S Eckaus. *Issues in Science and Technology* **3**(4): 101–9

Stiglitz J E (1987) Learning to learn, localized learning and technological progress. In Dasgupta P, Stoneman P (eds) *Economic policy and industrial performance* Cambridge, Cambridge University Press: 125–53

Stobaugh R (1985) Creating a monopoly: product innovation in petrochemicals. In Rosenbloom R S (ed) *Research on technological innovation, management and policy* vol 2, Greenwich, CT, JAI Press: 81–112

Stöhr W B (1981) Development from below: the bottom-up and periphery-inward development paradigm. In Stöhr W B, Taylor D R F (eds) *Development from above or below?* New York, John Wiley: 39–72

Stöhr W (1982) Structural characteristics of peripheral areas: the relevance of the stock-in-trade variables of regional science. *Papers of the Regional Science Association* **49**: 71–84

Stöhr W B (1986a) Regional innovation complexes. *Papers of the Regional Science Association* **59**: 29–44

Stöhr W (1986b) Changing external conditions and a paradigm shift in regional development strategies? In Bassand M, Brugger E A, Bryden J M, Friedmann J, Stuckey B (eds) *Self-reliant development in Europe* Aldershot, Gower: 59–73

Stöhr W (1988) Regional policy, technology complexes and research/science parks. In Giaoutzi M, Nijkamp P (eds) *Informatics and regional development* Aldershot, Avebury: 201–14

Stöhr W, Taylor D R F (1981) Development from above or below? The dialectics of regional planning in developing countries. In Mabogunje A L, Misra R P (eds) *Regional development alternatives: international perspectives* Singapore, Maruzen Asia: 9–26

Stöhr W, Tödtling F (1977) Spatial equity – some antitheses to current regional development doctrine. *Papers of the Regional Science Association* 38: 33–53

Stokman C, Docter J (1987) Innovation in manufacturing medium-size and small enterprise: knowledge breeds prospects. In Rothwell R, Bessant J (eds) *Innovation: adaptation and growth* Amsterdam, Elsevier: 213–26

Storey D J (1982) *Entrepreneurship and the new firm* London, Croom Helm

Storey D J, Johnson S (1987) Regional variations in entrepreneurship in the UK. *Scottish Journal of Political Economy* 34: 161–73

Storper M (1988) Big structures, small events, and large processes in economic geography. *Environment and Planning A* 20: 165–85

Storper M (1989) Industrial policy at the crossroads: production flexibility, the region, and the state. *Environment and Planning D: Society and Space* 7: 235–43

Storper M, Scott A J (1989) The geographical foundations and social regulation of flexible production complexes. In Wolch J, Dear M (eds) *The power of geography: how territory shapes social life* Boston, Unwin Hyman: 21–40

Storper M, Walker R (1984) The spatial division of labor: labor and the location of industries. In Sawers L, Tabb W K (eds) *Sunbelt/snowbelt* New York, Oxford University Press: 19–47

Storper M, Walker R (1989) *The capitalist imperative: territory, technology, and industrial growth* Oxford, Basil Blackwell

Strassoldo R (1981) Center and periphery: socio-ecological perspectives. In Kuklinski A (ed) *Polarized development and regional policies: tribute to Jacques Boudeville* The Hague, Mouton: 71–102

Streeten P (1979a) From growth to basic needs. *Finance and Development* 16 (3): 28–31

Streeten P (1979b) Development ideas in historical perspective. In *Toward a new strategy for development. A Rothko Chapel colloquium* New York, Pergamon: 21–52

Stretton H (1978) *Urban planning in rich and poor countries* Oxford, Oxford University Press

Su S (1988) Information technology as a threat and an opportunity. In Schütte H (ed) *Strategic issues in information technology* Maidenhead, Pergamon Infotech: 103–13

Suarez–Villa L (1984) Industrial export enclaves and manufacturing change. *Papers of the Regional Science Association* 54: 89–111

Summers G F, Evans S D, Clemente F, Beck E M, Minkoff J (1976) *Industrial invasion of nonmetropolitan America* New York, Praeger

Sun M (1989) Investors' yen for U.S. technology. *Science* 246: 1238–41

Sundquist J L (1975) *Dispersing population: what America can learn from Europe* Washington, Brookings Institution

Sveikauskas L (1979) Interurban differences in the innovative nature of production. *Journal of Urban Economics* 6: 211–27

Sveikauskas L, Gowdy J, Funk M (1988) Urban productivity: city size or industry size? *Journal of Regional Science* 28: 185–202

Swales J K (1983) A Kaldorian model of cumulative causation: regional growth with induced technical change. In Gillespie A (ed) *Technological change and regional development* London, Pion: 68–87

Sweeney G P (1985) Innovation is entrepreneur-led. In Sweeney G (ed) *Innovation policies: an international perspective* London, Frances Pinter: 80–113

Sweeney G P (1987) *Innovation, entrepreneurs and regional development* New York, St Martin's Press

Syme G J, Shaw B J, Fenton D M, Mueller W S (eds) (1989) *The planning and evaluation of hallmark events* Aldershot, Avebury

Szentes T (1971) *The political economy of underdevelopment* Budapest, Akadémia Kiadó

Taaffe E J, Morrill R L, Gould P R (1963) Transport expansion in underdeveloped countries: a comparative analysis. *Geographical Review* 53: 503–29

Takeuchi A (1987) Two elements supporting high position of Tokyo region in the national system of Japanese machinery industry. *Geographical Reports of Tokyo Metropolitan University* 22: 129–38

Takeuchi A, Mori H (1987) Spontaneous technological center of Japanese machinery industry in provincial area. *Report of Researches, Nippon Institute of Technology* 17(3): 265–81

Tam M-Y, Persky J (1982) Regional convergence and national inequality. *Review of Economics and Statistics* 64: 161–4

Tanaka M (1989) Japanese-style evaluation systems for R&D projects: the MITI experience. *Research Policy* 18: 361–78

Tani A (1989) International comparisons of industrial robot populations. *Technological Forecasting and Social Change* 34: 191–210

Tata R J, Schultz R R (1988) World variation in human welfare: a new index of development status. *Annals of the Association of American Geographers* 78: 580–93

Tatsuno S (1986) *The technopolis strategy* New York, Prentice Hall Press

Taylor A (1989) Why U.S. carmakers are losing ground. *Fortune* 120(9): 96–116

Taylor C A (1982) Econometric modeling of urban and other substate areas: an analysis of alternative methodologies. *Regional Science and Urban Economics* 12: 425–48

Taylor M J (1986) The product cycle model: a critique. *Environment and Planning A* 18: 751–61

Taylor M J (1987) Enterprise and the product-cycle model: conceptual ambiguities. In van der Knaap B, Wever E (eds) *New technology and regional development* London, Croom Helm: 75–93

Taylor M J, Thrift N J (1982a) Industrial linkage and the segmented economy: 1. some theoretical proposals. *Environment and Planning A* 14: 1601–13

Taylor M, Thrift N (eds) (1982b) *The geography of multinationals* London, Croom Helm

Taylor M J, Thrift N J (1983) Business organisation, segmentation and location. *Regional Studies* 17: 445–66

Taylor M, Thrift N (eds) (1986) *Multinationals and the restructuring of the world economy* The geography of multinationals, vol 2, London, Croom Helm

Taylor P J (1985) *Political geography: world-economy, nation-state and locality* London, Longman

Tchijov I, Sheinin R (1989) Flexible manufacturing systems (FMS): current diffusion and main advantages. *Technological Forecasting and Social Change* 35: 277–93

Teece D J (1980) Economies of scope and the scope of the enterprise. *Journal of Economic Behavior and Organization* 1: 223–47

Teece D J (1981) Technology transfer and R&D activities of multinational firms: some theory and evidence. In Hawkins R G, Prasad A J (eds) *Technology transfer and economic development* Research in international business and finance, vol 2. Greenwich, CT, JAI Press: 39–74

Teece D J (1982) Towards an economic theory of the multiproduct firm. *Journal of Economic Bahavior and Organization* 3: 39–63

Teece D J (1986a) Firm boundaries, technological innovation, and strategic management. In Thomas L G (ed) *The economics of strategic planning* Lexington, MA, Lexington Books: 187–99

Teece D J (1986b) Profiting from technological innovation: implications for integration, collaboration, licensing and public policy. *Research Policy* 15: 285–305

Teece D J (ed) (1987) *The competitive challenge* Cambridge, MA, Ballinger

Teece D J (1988) Technological change and the nature of the firm. In Dosi G, Freeman C, Nelson R, Silverberg G, Soete L (eds) *Technical change and economic theory* London, Pinter: 256–81

Teitel S (1981) Towards an understanding of technical change in semi-industrialized countries. *Research Policy* 10: 127–47

Telesio P (1990) Are East Europe's executives prepared for the 1990s? *Wall Street Journal* January 15: A10

Templeman J, Peterson T, Schares G E, Kapstein J (1989) The shape of Europe to come. *Business Week* 60–4

Teubal M (1986) *Innovation performance, learning, and government policy* Madison, University of Wisconsin Press

Therkildsen O (1981) The relationship between economic growth and regional inequality: a critical re-appraisal. In Buhr W, Friedrich P (eds) *Regional development under stagnation* Baden–Baden, Nomos: 91–105

Thirlwall A P (1972) *Growth and development* Cambridge MA, Schenkman

Thirlwall A P (1974) Regional economic disparities and regional policy in the Common Market. *Urban Studies* 11: 1–12

Thirlwall A P (1980) Regional problems are 'balance-of-payments' problems. *Regional Studies* 14: 419–25

Thirlwall A P (1983) A plain man's guide to Kaldor's growth laws. *Journal of Post Keynesian Economics* 5: 345–58

Thomas D B (1979) Building scientific and technological capabilities in LDCs –

a survey of some economic development issues. In Thomas D B, Wionczek M S (eds) *Integration of science and technology with development* New York, Pergamon: 3–16

Thomas L J (1983) The centralized research organization. In Brown J K, Elvers L M (eds) *Research and development: key issues for management* Report 842, New York, The Conference Board: 30–3

Thomas M D (1972a) Growth pole theory: an examination of some of its basic concepts. In Hansen N M (ed) *Growth centers in regional economic development* New York, Free Press: 50–81

Thomas M D (1972b) The regional problem, structural change and growth pole theory. In Kuklinski A R (ed) *Growth poles and growth centres in regional planning* The Hague, Mouton: 69–102

Thomas M D (1975) Growth pole theory, technological change, and regional economic growth. *Papers of the Regional Science Association* **34**: 3–25

Thomas M D (1985) Regional economic development and the role of innovation and technological change. In Thwaites A T, Oakey R P (eds) *The regional economic impact of technological change* London, Frances Pinter: 13–35

Thomas M D (1986) Growth and structural change: the role of technical innovation. In Amin A, Goddard J B (eds) *Technological change, industrial restructuring and regional development* London, Allen and Unwin: 115–39

Thomas M D (1987a) Schumpeterian perspectives on entrepreneurship in economic development: a commentary. *Geoforum* **18**: 173–86

Thomas M D (1987b) The innovation factor in the process of microeconomic industrial change: conceptual explorations. In van der Knaap B, Wever E (eds) *New technology and regional development* London, Croom Helm: 21–44

Thomas M D (1988) Technical entrepreneurship in high technology industries: exploratory conceptualizations of firm formation and location processes. Paper presented at the North American Meetings of the Regional Science Association, Toronto, November

Thomas S (1988) The development and appraisal of nuclear power. Part II. The role of technical change. *Technovation* **7**: 305–39

Thompson C (1987) Definitions of 'high technology' used by state programs in the USA: a study in variations in industrial policy under a federal system. *Environment and Planning C: Government and Policy* **5**: 417–31

Thompson C (1988a) High-technology development and recession: the local experience in the United States, 1980–1982. *Economic Development Quarterly* **2**: 153–67

Thompson C (1988b) Some problems with R&D/SE&T-based definitions of high technology industry. *Area* **20**: 265–77

Thompson C (1989a) High-technology theories and public policy. *Environment and Planning C: Government and Policy* **7**: 121–52

Thompson C (1989b) The geography of venture capital. *Progress in Human Geography* **13**: 62–98

Thompson W R (1965) *A preface to urban economics* Baltimore, Johns Hopkins University Press

Thompson W R (1968) Internal and external factors in the development of urban economies. In Perloff H S, Wingo L (eds) *Issues in urban economics* Baltimore, Johns Hopkins University Press: 43–62

Thompson W R (1987) Policy-based analysis for local economic development. *Economic Development Quarterly* 1(3): 203–13

Thompson W R, Thompson P R (1987) National industries and local occupational strengths: the cross-hairs of targeting. *Urban Studies* 24: 547–60

Thomson R (1989) *The path to mechanized shoe production in the United States* Chapel Hill, University of North Carolina Press

Thorelli H B (1986) Networks: between markets and hierarchies. *Strategic Manangement Journal* 7: 37–51

Thrift N J (1989) New times and spaces? The perils of transition models. *Environment and Planning D: Society and Space* 7: 127–9

Thrift N, Taylor M (1989) Battleships and cruisers: the new geography of multinational corporations. In Gregory D, Walford R (eds) *Horizons in human geography* London, Macmillan: 279–97

Thwaites A T (1978) Technological change, mobile plants and regional development. *Regional Studies* 12: 445–61

Thwaites A T (1982) Some evidence of regional variations in the introduction and diffusion of industrial products and processes within British manufacturing industry. *Regional Studies* 16: 371–81

Thwaites A T (1983) The employment implications of technological change. In Gillespie A (ed) *Technological change and regional development* London, Pion: 36–53

Thwaites A, Alderman N (1990) The location of R&D: retrospect and prospect. In Cappellin R, Nijkamp P (eds) *The spatial context of technological development* Aldershot, Avebury: 17–42

Tiebout C M (1956) Exports and regional growth. *Journal of Political Economy* 64: 160–9

Tiebout C M (1962) *The community economic base study* New York, Committee for Economic Development

Time Inc (1989) *Corporate site selection for new facilities: a study conducted among the largest US companies* New York, Time Inc

Timmons J A, Broehl W G, Frye J M (1980) Developing Appalachian entrepreneurs. *Growth and Change* 11(3): 44–50

Timmons J A, Muzyka D F, Stevenson H H, Bygrave W D (1987) Opportunity recognition: the core of entrepreneurship. In Churchill N C, Hornaday J A, Kirchhoff B A, Krasner O J, Vesper K H (eds) *Frontiers of entrepreneurship research 1987* Wellesley, MA, Babson College, Center for Entrepreneurial Studies: 109–23

Tirman J (ed) (1984) *The militarization of high technology* Cambridge, MA, Ballinger

Tisdell C A (1981) Science and technology policy: priorities of governments. London, Chapman and Hall

Toda T (1987) The location of high-technology industry in Japan and the technopolis plan. In Brotchie J, Hall P, Newton P W (eds) *The spatial impact of technological change* London, Croom Helm: 271–83

Todd D (1974) An appraisal of the development pole concept in regional analysis. *Environment and Planning A* 6: 291–306

Todd D (1980a) The defence sector in regional development. *Area* 12: 115–21

Todd D (1980b) Welfare or efficiency: can the growth center offer a compromise? *Growth and Change* 11(3): 39–43

Todd D (1988) *Defence industries: a global perspective* London, Routledge

Todd D, Simpson J (1985) Aerospace, the state and the regions: a Canadian perspective. *Political Geography Quarterly* 4: 111–30

Tolosa H C (1981) Key issues in Latin American spatial development. In Prantilla E B (ed) *National development and regional policy* Singapore, Maruzen Asia: 261–76

Tomlin B (1981) Inter-location technical communication in a geographically-dispersed research organization. *R&D Management* 11(1): 19–23

Törnqvist G (1983) Creativity and the renewal of regional life. In Buttimer A (ed) *Creativity and context* Lund Studies in Geography, series B, number 50, Lund, Gleerup: 91–112

Townsend A (1986) Spatial aspects of the growth of part-time employment in Britain. *Regional Studies* 20(4): 313–30

Toye J (1987) *Dilemmas of development* Oxford, Basil Blackwell

Tsuruta T (1988) The rapid growth era. In Komiya R, Okuno M, Suzumura K (eds) I*ndustrial policy of Japan* Tokyo, Academic Press: 49–87

Tucker J B (1985) Managing the industrial miracle. *High Technology* 5(8): 22–30

Tushman M L (1979) Managing communication networks in R&D laboratories. *Sloan Management Review* 20: 37–49

Twiss B (1980) *Managing technological innovation* 2nd edn London, Longman

Tyebjee T T (1988) A typology of joint ventures: Japanese strategies in the United States. *California Management Review* 31(1): 75–86

Tyebjee T T, Bruno A V (1984a) Venture capital: investor and investee perspectives. *Technovation* 2: 185–208

Tyebjee T T, Bruno A V (1984b) A model of venture capitalist investment activity. *Management Science* 30: 1051–66

Tyebjee T T, Bruno A V (1986) Negotiating venture capital financing. *California Management Review* 29(1): 45–59

Tyebjee T, Vickery L (1988) Venture capital in Western Europe. *Journal of Business Venturing* 3: 123–36

Tymes E R, Krasner O J (1983) Informal risk capital in California. In Hornaday J A, Timmons J A, Vesper K H (eds) *Frontiers of entrepreneurship research 1983* Wellesley, MA, Babson College, Center for Entrepreneurial Studies: 347–68

Udell G G, Potter T A (1989) Pricing new technology. *Research-Technology Management* 32(4): 14–18

Uekusa M (1988) The oil crisis and after. In Komiya R, Okuno M, Suzumura K (eds) *Industrial policy of Japan* Tokyo, Academic Press: 89–117

Ullman E L, Dacey M F (1960) The minimum requirements approach to the urban economic base. *Papers and Proceedings of the Regional Science Association* 6: 175–94

Ullmann J E (1988) *The anatomy of industrial decline: productivity, investment, and location in US manufacturing* New York, Quorum Books

UN Centre on Transnational Corporations (1982) *Transnational corporations in international tourism* New York, United Nations

UN Conference on Trade and Development (UNCTAD) (1984) *Multimodal transport and containerization* New York, United Nations

Unesco (1988) *Statistical yearbook* Paris, United Nations Educational, Scientific and Cultural Organisation

UN Office for Science and Technology (1980) Mobilizing science and technology for increasing the endogenous capabilities in developing countries. In Standke K-H, Anandakrishnan M (eds) *Science, technology and society: needs, challenges and limitations* New York, Pergamon: 476–502

Urquidi V L (1986) Science, technology, and endogenous development: some notes on the objectives and the possibilities. In Smith K R, Fesharaki F, Holdren J P (eds) *Earth and the human future: essays in honor of Harrison Brown* Boulder, CO, Westview Press: 208–27

Urrutia M, Krueger A O (1989) Remarks on country studies. In Sachs J D (ed) *Developing country debt and the world economy* Chicago, University of Chicago Press: 212–22

Urry J (1986) Capitalist production, scientific management and the service class. In Scott A J, Storper M (eds) *Production, work, territory* Boston, Allen and Unwin: 43–66

Urry J (1987) Some social and spatial aspects of services. *Environment and Planning D: Society and Space* 5: 5–26

US Bureau of the Census (1981) *Shipments to federal government agencies, 1979* Washington, US Government Printing Office

US Bureau of Economic Analysis (1984) *The detailed input–output structure of the U.S. economy, 1977* vol 2, Washington, US Government Printing Office

US Department of Defense (1986) *Prime contract awards by region and state, fiscal years 1978, 1979, and 1980* Washington, US Department of Defense

US Department of Defense (1986) *Prime contract awards by region and state, fiscal years 1984, 1985, and 1986* Washington, US Government Printing Office

US General Accounting Office (1987) *University funding: patterns of distribution of federal research funds to universities* Washington, US General Accounting Office

Utterback J M (1979) The dynamics of product and process innovation in industry. In Hill C T, Utterback J M (eds) *Technological innovation for a dynamic economy* New York, Pergamon: 40–65

Utterback J M (1987) Innovation and industrial evolution in manufacturing industries. In Guile B R, Brooks H (eds) *Technology and global industry: companies and nations in the world economy* Washington, National Academy Press: 16–48

Utterback J M, Abernathy W J (1975) A dynamic model of process and product innovation. *Omega* 3: 639–56

Utterback J M, Meyer M, Roberts E, Reitberger G (1988) Technology and industrial innovation in Sweden: a study of technology-based firms formed between 1965 and 1980. *Research Policy* **17**: 15–26

Vakil C N, Brahmananda P R (1987) Technical knowledge and managerial capacity as limiting factors on industrial expansion in underdeveloped countries. In Robinson A (ed) *Economic progress* 2nd edn New York, St Martin's Press: 153–72

Valéry N (1989) Thinking ahead: Japanese technology. *The Economist* December 2

van der Meer J B H, Calori R (1989) Strategic management in technology-intensive industries. *International Journal of Technology Management* **4**: 127–39

van Duijn J J (1983) *The long wave in economic life* London, George Allen & Unwin

van Tilburg J J (1985) Technopolis Twente: a basis for international marketing. In Gibb J M (ed) *Science parks and innovation centres: their economic and social impact* Amsterdam, Elsevier: 268–73

Varaiya P, Wiseman M (1981) Investment and employment in manufacturing in U.S. metropolitan areas 1960–1976. *Regional Science and Urban Economics* **11**: 431–69

Vartiainen H J (1988) Entrepreneurship and local government strategies in Finland: a regional perspective. In Weiss J W (ed) *Regional cultures, managerial behavior, and entrepreneurship: an international perspective* New York, Quorum Books: 87–98

Vaughan R J (1977) *The urban impact of federal policies* vol. 2: *Economic development* Santa Monica, CA, Rand Corporation

Vaughan R J, Pollard R (1986) State and federal policies for high-technology development. In Rees J (ed) *Technology, regions, and policy* Totowa, NJ, Rowman and Littlefield: 268–81

Vazquez-Barquero A (1987) Local development and the regional state in Spain. *Papers of the Regional Science Association* **61**: 65–78

Vernon R (1966) International investment and international trade in the product cycle. *Quarterly Journal of Economics* **80**: 190–207

Vernon R (1974) The location of economic activity. In Dunning J H (ed) *Economic analysis and the multinational enterprise* New York, Praeger: 89–114

Vernon R (1979) The product cycle in a new international environment. *Oxford Bulletin of Economics and Statistics* **41**: 255–67

Vernon R (1989) *Technological development: the historical experience* Washington, World Bank

Vesper K H (1984) Three faces of corporate entrepreneurship: a pilot study. In Hornaday J A, Tarpley F A, Timmons J A, Vesper K H (eds) *Frontiers of entrepreneurship research 1984* Wellesley, MA, Babson College, Center for Entrepreneurial Studies: 294–320

Vickery G (1988) A survey of international technology licensing. *Science Technology Industry Review* **4**: 7–49

Vickery G (1989) Recent developments in the consumer electronics industry. *Science Technology Industry Review* 5: 113–28

Vickery G, Blair L (1989) *Government policies and the diffusion of microelectronics* Paris, Organisation for Economic Co-operation and Development

Villamil J J (ed) (1979) *Transnational capitalism and national development* Atlantic Highlands, NJ, Humanities Press

Vining D R (1987) Review of *Regional dynamics: studies in adjustment theory. Economic Geography* 63: 358

Vining D R, Kontuly T (1978) Population dispersal from major metropolitan regions: an international comparison. *International Regional Science Review* 3: 49–73

Volland C S (1987) A comprehensive theory of long wave cycles. *Technological Forecasting and Social Change* 32: 123–45

von Böventer E (1975) Regional growth theory. *Urban Studies* 12: 1–29

von Glinow M A (1988) *The new professionals: managing today's high-tech employees* Cambridge, MA, Ballinger

von Hippel E (1979) A customer-active paradigm for industrial products idea generation. In Baker M J (ed) *Industrial innovation* London, Macmillan: 82–110

von Hippel E (1983) Increasing innovators' returns from innovation. In Rosenbloom R S (ed) *Research on innovation, management and policy* vol 1, Greenwich, CT, JAI Press: 35–53

von Hippel E (1987) Cooperation between rivals: informal know-how trading. *Research Policy* 16: 291–302

von Hippel E (1988) *The sources of innovation* New York, Oxford University Press

von Hippel E (1989) New product ideas from 'lead users'. *Research-Technology Management* 32(3): 24–7

Voss C A (1984) Technology push and need pull: a new perspective. *R&D Management.* 14: 147–51

Voss C A (1985) The need for a field of study of implementation of innovations. *Journal of Product Innovation Management.* 2: 266–71

Wad A (1984) Science, technology and industrialisation in Africa. *Third World Quarterly.* 6: 327–50

Waite C A (1988) Service sector: its importance and prospects for the future. In Candilis W O (ed) *United States service industries handbook* New York, Praeger: 1–22

Walker R A (1985) Technological determination and determinism: industrial growth and location. In Castells M (ed) *High technology, space, and society* Beverly Hills, CA, Sage: 226–64

Walker R (1988) The geographical organization of production-systems. *Environment and Planning D: Society and Space.* 6: 377–408

Walker R, Storper M (1981) Capital and industrial location. *Progress in Human Geography* 5: 473–509

Walker W B (1980) Britain's industrial performance 1850–1950: a failure to adjust. In Pavitt K (ed) *Technical innovation and British economic performance* London, Macmillan: 19–37

Walker W, Graham M, Harbor B (1988) From components to integrated systems: technological diversity and interactions between the military and civilian sectors. In Gummett P, Reppy J (eds) *The relations between defence and civil technologies* Dordrecht, Kluwer: 17–37

Wallerstein I (1979) *The capitalist world-economy* Cambridge, Cambridge University Press

Wall Street Journal (1989) Texas Instruments Inc. November 8: A4

Walsh V (1984) Invention and innovation in the chemical industry: demand-pull or discovery-push? *Research Policy* 13: 211–34

Walsh V (1987) Technology, competitiveness and the special problems of small countries. *Science Technology Industry Review* 2: 81–133

Walton R E, Susman G I (1987) People policies for the new machines. *Harvard Business Review* 65 (2): 98–106

Wan V (1989) Financing high technology: the Australian venture capital market. *Technovation* 9: 337–55

Wang J, Bradbury J H (1986) The changing industrial geography of the Chinese special economic zones. *Economic Geography* 62: 307–20

Warf B (1989) Telecommunications and the globalization of financial services. *Professional Geographer* 41: 257–71

Warndorf P R, Merchant M E (1986) Development and future trends in computer-integrated manufacturing in the USA. *International Journal of Technology Management* 1: 161–78

Warner T N (1987) Information technology as a competitive burden. *Sloan Management Review* 29 (1): 55–61

Watkins A J (1980) *The practice of urban economics* Beverly Hills, CA, Sage

Watts H D (1980) *The large industrial enterprise* London, Croom Helm

Watts H D (1981) *The branch plant economy* London, Longman

Watts H D (1987) *Industrial geography* London, Longman

Watts H D, Stafford H A (1986) Plant closures and the multiplant firm: some conceptual issues. *Progress in Human Geography* 10: 206–29

Weaver J, Jameson K (1981) *Economic development: competing paradigms* Washington, University Press of America

Webber M J (1987a) Rates of profit and interregional flows of capital. *Annals of the Association of American Geographers* 77: 63–75

Webber M J (1987b) Quantitative measurement of some Marxist categories. *Environment and Planning A* 19: 1303–21

Webber M, Foot S P H (1988) Profitability and accumulation. *Economic Geography* 64: 335–51

Webber M, Tonkin S (1987) Technical changes and the rate of profit in the Canadian food industry. *Environment and Planning A* 19: 1579–96

Webber M, Tonkin S (1988a) Technical changes and the rate of profit in the Canadian textile, knitting, and clothing industries. *Environment and Planning A* 20: 1487–505

Webber M, Tonkin S (1988b) Technical changes and the rate of profit in the Canadian wood, furniture, and paper industries. *Environment and Planning A* 20: 1623–43

Wegener M (1986) Transport network equilibrium and regional deconcentration. *Environment and Planning A* **18**: 437–56

Weiner E, Foust D, Yang D J (1988) Why made-in-America is back in style. *Business Week* November 7: 116–20

Weinstein B L, Gross H T (1988) The rise and fall of sun, rust, and frost belts. *Economic Development Quarterly* **2**: 9–18

Weinstein B L, Gross H T, Rees J (1985) *Regional growth and decline in the United States* 2nd edn New York, Praeger

Weiss L (1988) *Creating capitalism: the state and small business since 1945* Oxford, Basil Blackwell

Weitz R (1986) *New roads to development* Westport, CT, Greenwood Press

Wells P (1987) The military scientific infrastructure and regional development. *Environment and Planning A* **19**: 1631–58

Wenk E (1986) *Tradeoffs: imperatives of choice in a high-tech world* Baltimore, Johns Hopkins University Press

Westney D E, Sakakibara K (1986) The role of Japan-based R&D in global technology strategy. In Horwitch M (ed) *Technology in the modern corporation: a strategic perspective* New York, Pergamon: 145–66

Westphal L E (1987) Industrial development in East Asia's 'gang of four'. *Issues in Science and Technology* **3**(3): 78–88

Westphal L E, Kim L, Dahlman C J (1985) Reflections on the Republic of Korea's acquisition of technological capability. In Rosenberg N, Frischtak C (eds) *International technology transfer: concepts, measures, and comparisons* New York, Praeger: 167–221

Westphal L E, Rhee Y W, Pursell G (1984) Sources of technological capability in South Korea. In Fransman M, King K (eds) *Technological capability in the Third World* London, Macmillan: 279–300

Westwood A R C (1984) R&D linkages in a multi-industry corporation. *Research Management* **27**(3): 23–36

Westwood A R C, Sekine Y (1988) Fostering creativity and innovation in an industrial R&D laboratory. *Research-Technology Management* **31**(4): 16–20

Wetzel W E (1983) 'Angels' and informal risk capital. *Sloan Management Review* **24**(4): 23–34

Wetzel W E (1986) Informal risk capital: knowns and unknowns. In Sexton D L, Smilor R W (eds) *The art and science of entrepreneurship* Cambridge, MA, Ballinger: 85–108

Wetzel W E (1987) The informal venture capital market: aspects of scale and market efficiency. *Journal of Business Venturing* **2**: 299–313

Wever E (1986) New firm formation in the Netherlands. In Keeble D, Wever E (eds) *New firms and regional development in Europe* London, Croom Helm: 54–74

Whalley P (1986) *The social production of technical work: the case of British engineers* Albany, State University of New York Press

Whittington D (ed) (1985) *High hopes for high tech: microelectronics policy in North Carolina* Chapel Hill, University of North Carolina Press

Wicker A W, King J C (1988) Young and numerous: retail/service establishments in the greater Los Angeles area. In Churchill N C, Hornaday J A, Kirchhoff B

A, Krasner O J, Vesper K H (eds) *Frontiers of entrepreneurship research 1987* Wellesley, MA, Babson College, Center for Entrepreneurial Studies: 124–37

Wienert H, Slater J (1986) *East-west technology transfer: the trade and economic aspects* Paris, Organisation for Economic Co-operation and Development

Wijesekera R O B (1979) Building national scientific and technological research capability in the context of underdevelopment. In Thomas D B, Wionczek M S (eds) *Integration of science and technology with development* New York, Pergamon: 17–30

Wilber C K (ed) (1984) *The political economy of development and underdevelopment* 3rd edn New York, Random House

Wilkinson F (ed) (1981) *The dynamics of labour market segmentation* London, Academic Press

Williams A M, Shaw G (1988) Tourism and development: introduction. In Williams A M, Shaw G (eds) *Tourism and economic development: Western European experience* London, Belhaven Press: 1–11

Williamson J G (1965) Regional inequality and the process of national development: a description of the patterns. *Economic Development and Cultural Change* 13: 3–45

Williamson J G (1980) Unbalanced growth, inequality, and regional development: some lessons from U.S. history. In Arnold V L (ed) *Alternatives to confrontation* Lexington MA, Lexington Books: 3–61

Willingale M C (1984) Ship-operator port-routeing behaviour and the development process. In Hoyle B S, Hilling D (eds) *Seaport systems and spatial change* Chichester, John Wiley: 43–59

Willinger M, Zuscovitch E (1988) Towards the economics of information-intensive production systems: the case of advanced materials. In Dosi G, Freeman C, Nelson R, Silverberg G, Soete L (eds) *Technical change and economic theory* London, Pinter: 239–55

Willman P (1987) Industrial relations issues in advanced manufacturing technology. In Wall T D, Clegg C W, Kemp N J (eds) *The human side of advanced manufacturing technology* New York, John Wiley: 135–51

Wilson J W (1985) *The new venturers* Reading, MA, Addison-Wesley

Wilson R W, Ashton P K, Egan T P (1980) *Innovation, competition, and government policy in the semiconductor industry* Lexington, MA, Lexington Books

Winiecki J (1987) Soviet-type economies' strategy for catching-up through technology imports – an anatomy of failure. *Technovation* 6: 115–45

Wionczek M S (1979) Science and technology planning in LDCs. In Thomas D B, Wionczek M S (eds) *Integration of science and technology with development* New York, Pergamon: 167–77

Wolch J R, Geiger R K (1986) Urban restructuring and the not-for-profit sector. *Economic Geography* 62: 1–18

Wolff M F (1989) Forging technology alliances. *Research-Technology Management* 32(3): 9–11

Wood S (1988) Between Fordism and flexibility? The US car industry. In Hyman R, Streeck W (eds) *New technology and industrial relations* Oxford, Basil Blackwell: 101–27

Woods S (1987) *Western Europe: technology and the future* Atlantic paper no. 63, London, Croom Helm

Woodward J (1965) *Industrial organization: theory and practice* Oxford, Oxford University Press

World Bank (1988) *World development report 1988* Oxford, Oxford University Press

Wulf H (1986) India: the unfulfilled quest for self-sufficiency. In Brzoska M, Ohlson T (eds) *Arms production in the Third World* London, Taylor and Francis: 125–45

Wysocki B (1987) Gene squad: Japanese now target another field the U.S. leads: biotechnology. *Wall Street Journal* December 17: A1, A25

Yang D J (1989) Taiwan isn't just for cloning anymore. *Business Week* September 25: 208–12

Yang D J, Gross N, Holstein W J, Bennett D, Nakarmi L (1989) Japan builds a new power base. *Business Week* April 10: 42–4

Yap L Y L (1977) The attraction of cities: a review of the migration literature. *Journal of Development Economics* 4: 239–64

Ybarra J-A (1989) Informalization in the Valecian economy: a model for underdevelopment. In Portes A, Castells M, Benton L A (eds) *The informal economy: studies in advanced and less developed countries* Baltimore, Johns Hopkins University Press: 216–27

Yonemoto K (1986) Robotization in Japan – an examination of the socio-economic impacts. *International Journal of Technology Management* 1: 179–96

Yoshikawa A (1988) Technology transfer and national science policy: biotechnology policy in Japan. *International Journal of Technology Management* 3: 735–43

Young R C, Francis J D (1989) Who helps small manufacturing firms get started? *Rural Development Perspectives* 6(1): 21–5

Young S, Hood N, Dunlop S (1988) Global strategies, multinational subsidiary roles and economic impact in Scotland. *Regional Studies* 22: 487–97

Zeitlin J, Totterdill P (1989) Markets, technology and local intervention: the case of clothing. In Hirst P, Zeitlin J (eds) *Reversing industrial decline? Industrial structure and policy in Britain and her competitors* Oxford, Berg: 155–90

Zumeta B W (1966) How many jobs can one job make? *Business Review, Federal Reserve Bank of Philadelphia* (June): 9–15

Zweig D (1987) From village to city: reforming urban-rural relations in China. *International Regional Science Review* 11: 43–58

Zymelman M (1982) Labor, education, and development: whither Latin America? In Syrquin M, Teitel S (eds) *Trade, stability, technology, and equity in Latin America* New York, Academic Press: 435–58

Index